Contents

Introduction to Environmental Impact Assessment

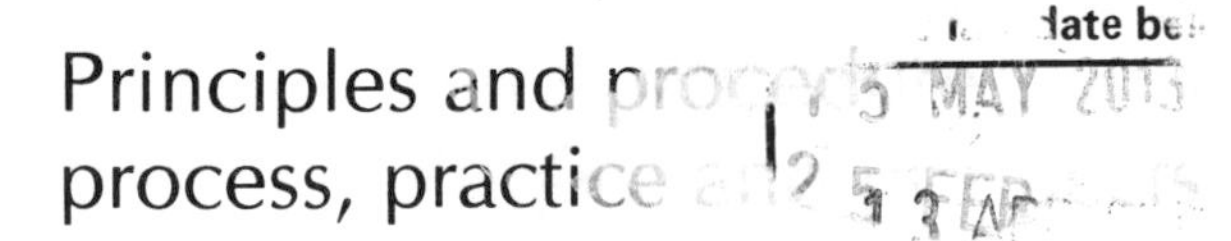

Principles and
process, practic

John Glasson
Riki Therivel
Andrew Chadwick

UCL PRESS

First published in 1994 by UCL Press.
Fourth impression 1996.
Fifth impression 1997.

UCL Press Limited
1 Gunpowder Square
London EC4A 3DE

ISBNs: 1-85728-118-7 PB

British Library Cataloguing in Publication Data.
A catalogue record for this book is available from the British Library.

Typeset in Times New Roman and Omega.
Printed and bound by
Page Bros (Norwich) Ltd, England.

Preface

There has been a remarkable and refreshing interest in environmental issues over the past few years. A major impetus was provided by the 1987 Report of the World Commission on the Environment and Development (the Brundtland Report); the Rio Summit in 1992 sought to accelerate the impetus. Much of the discussion on environmental issues and on sustainable development is about the better management of current activity in harmony with the environment. However, there will always be pressure for new development. How much better it would be to avoid or mitigate the potential harmful effects of future development on the environment at the planning stage. Environmental impact assessment (EIA) seeks to assess the impacts of planned activity on the environment in advance thereby allowing avoidance measures to be taken: prevention is better than cure.

Environmental impact assessment was first formally established in the USA in 1969. It has spread worldwide and received a significant boost in Europe with the introduction of an EC Directive on EIA in 1985. This was implemented in the UK in 1988. Subsequently there has been a rapid growth in EIA activity, and over 300 environmental impact statements (EISs) are now produced in the UK each year. EIA is an approach in good currency. It is also an area where many of the practitioners have limited experience. This text seeks to provide a comprehensive introduction to the various dimensions of EIA. It has been written with the requirements of both undergraduate and postgraduate students in mind. It should also be of considerable value to those in practice – planners, developers and to various interest groups. EIA is on a rapid "learning curve"; this text is offered as a point on the curve.

The book is structured into four parts. The first provides an introduction to the principles of EIA, and an overview of its development and agency and legislative context. Part Two provides a step by step discussion and critique of the EIA process. Part Three examines current practice, broadly in the UK and in several other countries, and in more detail through selected UK case studies. Part Four considers possible future developments. It is likely that much more of the EIA iceberg will become visible in the 1990s and beyond. An outline of important and associated developments in environmental auditing and in strategic environmental assessment concludes the text.

Although the book has a clear UK orientation, it does draw extensively on EIA experience worldwide and it should be of interest to readers from many countries. The book seeks to highlight best practice and to offer enough insight to methods, and to supporting references, to provide valuable guidance to the practitioner. For information on detailed methods for assessment of impacts in particular topic areas (e.g. landscape, air quality, traffic impacts), the reader is referred to the complementary volume, *Methods of environmental impact assessment* (Morris & Therivel 1995).

JOHN GLASSON   RIKI THERIVEL   ANDREW CHADWICK
Oxford Brookes University

Dedicated to our families

Acknowledgements

Our grateful thanks are due to many people without whose help this book would not have been produced. We are particularly grateful to Carol Glasson who typed and retyped several drafts to tight deadlines and to high quality, and who provided invaluable assistance in bringing together the disparate contributions of the three authors. Our thanks also go to Rob Woodward for his production of many of the illustrations. We are very grateful to our consultancy clients and research sponsors, who have underpinned the work of the Impact Assessment Unit in the School of Planning at Oxford Brookes University (formerly Oxford Polytechnic). Michael Gammon provided the initiative and constant support; Phil Saunders and Andrew Hammond maintained the positive link with the electricity supply industry. Other valuable support has been provided by the Royal Society for the Protection of Birds, Economic and Social Research Council and PCFC.

Our students at Oxford Brookes University on both undergraduate and postgraduate programmes have critically tested many of our ideas. In this respect we should like to acknowledge, in particular, the students on the MSc course in Environmental Assessment and Management. We are very grateful for the advice received on the book contents and presentation from Roger Jones, Kate Williams, Peter Roberts and especially from Richard Williams. We have benefited from the support of colleagues in the Schools of Planning and Biological and Molecular Sciences, and from the wider community of EIA academics, researchers and consultants who help to keep us on our toes.

We are also grateful for permission to use material from the following sources in the book:

- Environmental Data Service (Figs 3.2 , 3.3).
- British Association of Nature Conservationists (cartoons: Parts 2 and 3).
- Rendel Planning (Fig. 4.2).
- UNEP Industry and Environment Office (Fig. 4.4 and Table 4.2).
- University of Manchester, Department of Planning and Landscape, EIA Centre (Tables 5.5, 8.1. 8.2, 8.3).
- John Wiley & Sons (Tables 6.1, 6.2).
- Baseline Environmental Consulting, West Berkeley, California (Fig. 7.2).
- David Tyldesley and Associates (Fig. 9.1 and Table 9.6).
- UK Department of Transport (Table 10.1).
- *Planning* newspaper (cartoon: Part 4).
- Project Appraisal; Graham Pinfield, Environmental Policy Unit, Lancashire County Council (Fig. 13.2).
- Kent County Council Planning Department (Figs 13.3, 13.4).

Abbreviations

AEE	assessment of environmental effects
AONB	Area of Outstanding Natural Beauty
BATNEEC	best available technique not entailing excessive costs
BPEO	best practicable environmental option
CBA	cost–benefit analysis
CC	county council
CEC	Commission of the European Communities
CEGB	Central Electricity Generating Board
CEQ	Council on Environmental Quality (US)
CEQA	California Environmental Quality Act
CHP	combined heat and power
CIE	community impact evaluation
CPRE	Council for the Protection of Rural England
CVM	contingent valuation method
DC	district council
DoE	Department of the Environment
DoEn	Department of Energy
DoT	Department of Transport
DTI	Department of Trade and Industry
EA	environmental assessment
EC	European Community
EES	environmental evaluation system
EIA	environmental impact assessment
EIR	environmental impact report
EIS	environmental impact statement
EPA	Environmental Protection Act
ES	environmental statement
ESI	electricity supply industry
ESRC	Economic and Social Research Council
FGD	flue gas desulphurization
FoE	Friends of the Earth
FONSI	finding of no significant impact
GAM	goals achievement matrix
GIS	geographical information system
GNP	gross national product
ha	hectares
HMIP	Her Majesty's Inspectorate of Pollution
HMSO	Her Majesty's Stationery Office
IAU	Impact Assessment Unit
IEA	Institute of Environmental Assessment

IPC	integrated pollution control
km	kilometre
LCP	large combustion plant
LPA	local planning authority
LULU	locally unacceptable land use
MAFF	Ministry of Agriculture, Fisheries and Food
MAUT	multi-attribute utility theory
MEA	Manual of environmental appraisal
mw	megawatts
NEPA	National Environmental Policy Act (US)
NEPP	National Environmental Policy Plan (Netherlands)
NGO	non-government organization
NIMBY	not in my back yard
NRA	National Rivers Authority
PADC	project appraisal for development control
PBS	planning balance sheet
PPG	Planning Policy Guidance note
PWR	pressurised water reactor
RA	risk assessment
RSPB	Royal Society for the Protection of Birds
RTPI	Royal Town Planning Institute
SACTRA	Standing Advisory Committee on Trunk Road Assessment
SDD	Scottish Development Department
SEA	strategic environmental assessment
SIA	social impact assessment
SoS	Secretary of State
SSSI	Site of Special Scientific Interest
T&CP	town and country planning
UK	United Kingdom
UNCED	United Nations Conference on Environment and Development
UNECE	United Nations Economic Commission for Europe
US	United States
WRAM	Water Resources Assessment Method

Part 1
Principles and procedures

CHAPTER 1
Introduction and principles

1.1 Introduction

In recent years there has been a remarkable growth of interest in environmental issues – in sustainability and the better management of development in harmony with the environment. Associated with this growth of interest has been the introduction of new legislation, emanating from national and international sources, such as the European Commission, that seeks to influence the relationship between development and the environment. Environmental impact assessment (EIA) is an important example. EIA legislation was introduced in the USA over 20 years ago. A European Community Directive in 1985 has accelerated its application in EC Member States and, since its introduction in the UK in 1988, it has been a major growth area for planning practice. The originally anticipated 20 environmental impact statements (EIS) per year in the UK has already escalated to over 300, and this is only the tip of the iceberg. The scope of EIA will widen greatly in the 1990s.

It is therefore perhaps surprising that the introduction of EIA met with strong resistance from many quarters, particularly in the UK. Planners argued, with partial justification, that they were already making such assessments. Many developers saw it as yet another costly and time-consuming constraint on development, and central government was also unenthusiastic. Interestingly, current UK legislation refers to environmental assessment (EA), leaving out the apparently politically sensitive, negative sounding, reference to "impacts". Much of the terminology is still at the formative stage. This first chapter therefore seeks to provide some introduction – to EIA as a process, to the purposes of this process, to types of development, environment and impacts, and to current issues in EIA.

1.2 The nature of environmental impact assessment

Definitions

Definitions of environmental impact assessment abound. They range from the oft-quoted and broad definition of Munn (1979), which refers to the need "to identify and predict the impact on the environment and on man's health and well-being of legislative proposals, policies, programmes, projects and operational procedures, and to interpret and communicate information about the impacts" to the narrow UK DOE (1989) operational definition: "The term 'environmental assessment' describes a technique and a process by which information about the environmental effects of a project is collected, both by the developer and from other sources, and taken into account by the planning authority in forming their judgements on whether the development should go ahead". The UN Economic Commission for Europe (1991) has an altogether more succinct and pithy definition: "an assessment of the impact of a planned activity on the environment".

Environmental impact assessment: a process

In essence, EIA is a *process*, a systematic process that examines the environmental consequences of development actions, in advance. The emphasis, compared with many other mechanisms for environmental protection, is on prevention. Of course planners have traditionally assessed the impacts of developments on the environment, but invariably not in the systematic, holistic and multidisciplinary way required by EIA. The process involves a number of steps, as outlined in Figure 1.1. These are briefly described below, pending a much fuller discussion in Chapters 4 to 7. It should be clearly noted at this stage that, although the steps are outlined in linear fashion, EIA should be a cyclical activity, with feedback and interaction between the various steps. It should also be noted that practice can and does vary considerably from the process illustrated in Figure 1.1. For example, current UK EIA legislation does not require some of the steps, including the consideration of alternatives, and post-decision monitoring (DOE 1989). The order of the steps in the process may also vary.

- *Project screening* narrows the application of EIA to those projects that may have significant environmental impacts. Screening may be partly determined by the EIA regulations operating in a country at the time of assessment.
- *Scoping* seeks to identify at an early stage, from all of a project's possible impacts and from all the alternatives that could be addressed, those that are the key, significant issues.
- *Consideration of alternatives* seeks to ensure that the proponent has considered other feasible approaches, including alternative project locations, scales, processes, layouts, operating conditions, and the "no action" option.
- *Description of the project/development action* includes a clarification of the purpose and rationale of the project, and an understanding of its various char-

acteristics – including stages of development, location and processes.

- *Description of the environmental baseline* includes the establishment of both the present and future state of the environment, in the absence of the project,

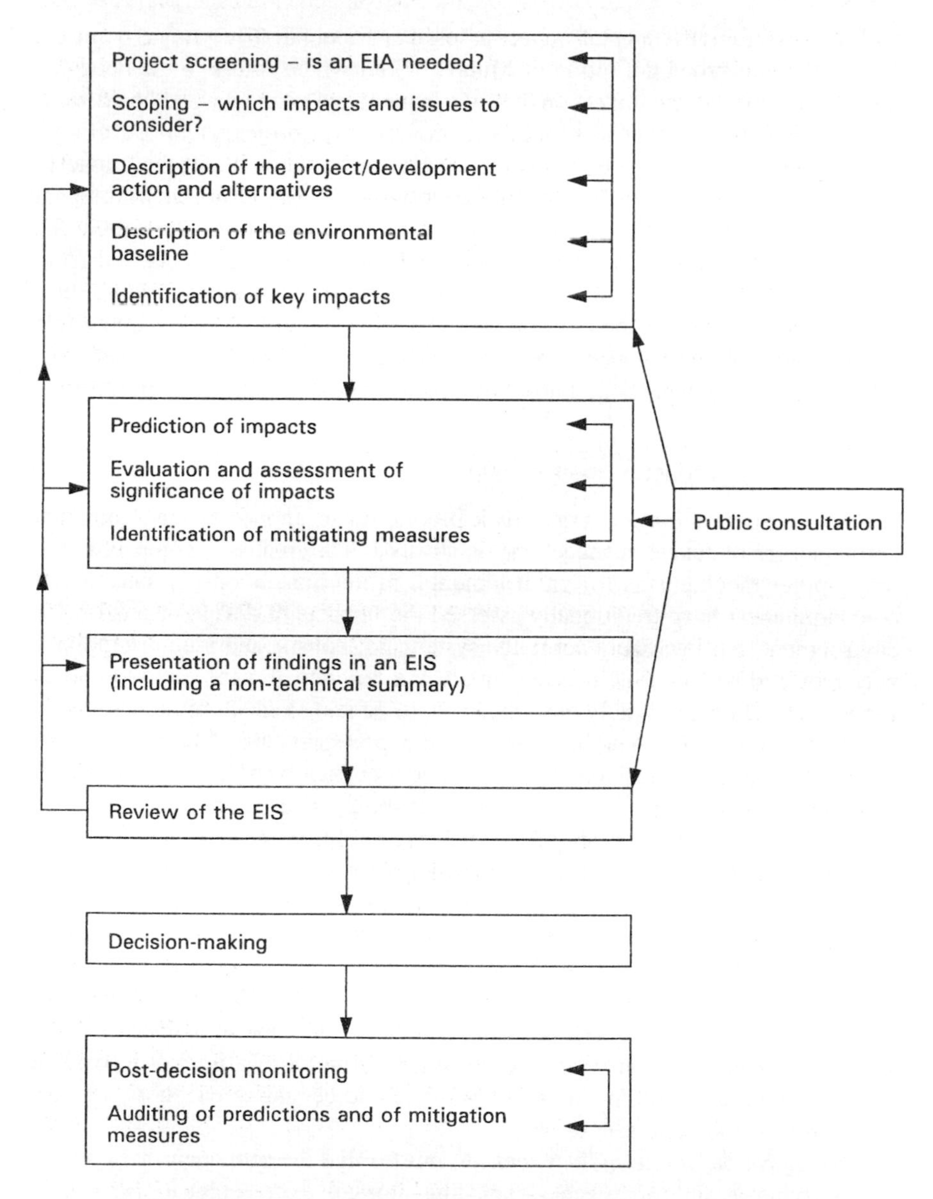

Figure 1.1 Important steps in the EIA process. Note: EIA should be a cyclical process with considerable interaction between the various steps. For example, public participation can be useful at most stages of the process; monitoring systems should relate to parameters established in the initial project and baseline descriptions.

taking into account changes resulting from natural events and from other human activities.

- *Identification of key impacts* brings together the previous steps with the aims of ensuring that all potentially significant environmental impacts (adverse and beneficial) are identified and taken into account in the process.
- *The prediction of impacts* aims to identify the magnitude and other dimensions of identified change in the environment with a project/action, by comparison with the situation without that project/action.
- *Evaluation and assessment of significance* seeks to assess the relative significance of the predicted impacts to allow a focus on key adverse impacts.
- *Mitigation* involves the introduction of measures to avoid, reduce, remedy or compensate for any significant adverse impacts.
- *Public consultation and participation* aims to assure the quality, comprehensiveness and effectiveness of the EIA, as well as to ensure that the public's views are adequately taken into consideration in the decision-making process.
- *EIS presentation* is a vital step in the process. If done badly, much good work in the EIA may be negated.
- *Review* involves a systematic appraisal of the quality of the EIS, as a contribution to the decision-making process.
- *Decision-making* on the project involves a consideration by the relevant authority of the EIS (including consultation responses) together with other material considerations.
- *Post-decision monitoring* involves the recording of outcomes associated with development impacts, after a decision to proceed. It can contribute to effective project management.
- *Auditing* follows from monitoring. It can involve comparing actual outcomes with predicted outcomes, and can be used to assess the quality of predictions and the effectiveness of mitigation. It provides a vital step in the EIA learning process.

Environmental impact statements: the documentation

The environmental impact statement provides *documentation* of the information and estimates of impacts derived from the various steps in the process. Prevention is better than cure; an EIS revealing many significant unavoidable adverse impacts would provide valuable information that could contribute to the abandonment or substantial modification of a proposed development action. Where adverse impacts can be successfully reduced through mitigation measures, there may be a different decision. Table 1.1 provides an example of the content of an EIS for a project.

The *non-technical summary* is an important element in the documentation; EIA can be complex and the summary can help to improve communication with the various parties involved. Reflecting the potential complexity of the process, a *methods statement*, at the beginning, provides an opportunity to clarify some basic

Table 1.1 An EIS for a project – example of contents.

Non-technical summary
Part 1: Methods and key issues
1. Methods statement
2. Summary of key issues; monitoring programme statement
Part 2: Background to the proposed development
3. Preliminary studies: need, planning, alternatives, site selection
4. Site description/baseline conditions
5. Description of proposed development
6. Construction activities and programme
Part 3: Environmental impact assessment – topic areas
7. Land use, landscape and visual quality
8. Geology, topography and soils
9. Hydrology and water quality
10. Air quality and climate
11. Ecology: terrestrial and aquatic
12. Noise
13. Transport
14. Socio-economic
15. Interrelationships between effects

information (e.g. who is the developer, who has produced the EIS, who has been consulted and how, what methods have been used, what difficulties have been encountered and what are the limitations of the EIA?). A *summary statement of key issues*, up-front, can also help to improve communications. More enlightened EISs would also include a *monitoring programme*, either here or at the end of the document. The *background to the proposed development* covers the early steps in the EIA process, including clear descriptions of the project, and baseline conditions (including relevant planning policies and plans). Within each of the *topic areas* of the EIS there would normally be discussion of existing conditions, predicted impacts, scope for mitigation and residual impacts.

EIA and EIS practice vary from study to study, from country to country, and best practice is constantly evolving. A recent UN study of EIA practice in several countries advocated changes in the process and documentation (United Nations Economic Commission for Europe 1991). These included giving a greater emphasis to the socio-economic dimension, to public participation, and to "after the decision" activity, such as monitoring.

Other relevant definitions

Development actions may have impacts not only on the physical environment but also on the social and economic environment. Typically, employment opportunities, services (e.g. health, education) and community structures, life-styles and values may be affected. *Socio-economic impact assessment* or *social impact assessment* (SIA), is regarded here as an integral part of EIA. However, in some countries it is regarded as a separate process, sometimes parallel to EIA, and the reader should be aware of its existence (Carley & Bustelo 1984, Finsterbusch 1985).

Strategic environmental assessment (SEA) expands EIA from projects to policies, plans and programmes. Development actions may be for a project (e.g. a nuclear power station), for a programme (e.g. a number of pressurized water reactor (PWR) nuclear power stations), for a plan (e.g. in the town and country planning system in England and Wales, for local plans and structure plans), or for a policy (e.g. the development of renewable energy). EIA to date has generally been used for individual projects, and that rôle is the focus of this book. But EIA for programmes, plans and policies, otherwise known as strategic environmental assessment, is currently generating much interest in the EC and beyond (Therivel et al. 1992). SEA informs a higher, earlier, more strategic tier of decision-making. In theory, EIA should be carried out first for policies, then plans, programmes, and finally for projects.

Risk assessment (RA) is another term sometimes found associated with EIA. Partly in response to events such as the chemicals factory explosion at Flixborough (UK), and nuclear power station accidents at Three Mile Island (USA) and Chernobyl (Ukraine), risk assessment has developed as an approach to the analysis of risks associated with various types of development. The major study of the array of petrochemicals and other industrial developments at Canvey Island in the UK provides an example of this approach (Health and Safety Commission 1978).

1.3 The purposes of environmental impact assessment

An aid to decision-making

Environmental impact assessment is a process with several important purposes. It is an aid to decision-making. For the decision-maker, for example the local authority, it provides a systematic examination of the environmental implications of a proposed action, and sometimes alternatives, before a decision is taken. The EIS can be considered by the decision-maker along with other documentation related to the planned activity. EIA is normally wider in scope and less quantitative than other techniques, such as cost–benefit analysis. It is not a substitute for decision-making, but it does help to clarify some of the trade-offs associated with a proposed development action, which should lead to more rational and structured decision-making. The EIA process has the potential, not always taken up, to be a basis for negotiation between the developer, public interest groups and the planning regulator. This can lead to an outcome that balances the interests of the development action and the environment.

An aid to the formulation of development actions

Many developers no doubt see EIA as another set of hurdles for them to jump in order to proceed with their various activities; the process can be seen as yet another costly and time-consuming activity in the permission process. However, EIA can be of great benefit to them since it can provide a framework for considering location and design issues and environmental issues in parallel. It can be an aid to the formulation of development actions, indicating areas where the project can be modified to minimize or eliminate altogether the adverse impacts on the environment. The consideration of environmental impacts early in the planning life of a development can lead to environmentally sensitive development; to improved relations between the developer, the planning authority and the local communities; to a smoother planning permission process; and sometimes, as argued by developers such as British Gas, to a worthwhile financial return on the extra expenditure incurred (Breakell & Glasson 1981). O'Riordan (1990) links such concepts of negotiation and redesign to the current dominant environmental themes of "Green consumerism" and "Green capitalism". The emergence of a growing demand by consumers for goods that do no environmental damage, plus a growing market for clean technologies, is generating a developer response. EIA can be the signal to the developer of potential conflict; wise developers may use the process to negotiate "Green gain" solutions that may eliminate or offset negative environmental impacts, reduce local opposition and avoid costly public inquiries.

An instrument for sustainable development

Underlying such purposes is of course the central rôle of EIA as one of the instruments to be used to achieve sustainable development: development that does not cost the Earth! Existing environmentally harmful developments have to be managed as best they can. In extreme cases, they may be closed down, but they can still leave residual environmental problems for decades to come. How much better it would be to mitigate the harmful effects in advance, at the planning stage, or in some cases avoid the particular development altogether. Prevention is better than cure.

Economic and social development must be placed in their environmental context. Boulding (1966) vividly portrays the dichotomy between the "throughput economy" and the "spaceship economy" (Fig. 1.2). The economic goal of increased GNP, using more inputs to produce more goods and services, contains the seeds of its own destruction. Increased output brings with it not only goods and services but also more waste products. Increased inputs demand more resources. The natural environment is the "sink" for the wastes and the "source" for the resources. Environmental pollution and resource depletion are invariably the ancillaries to economic development.

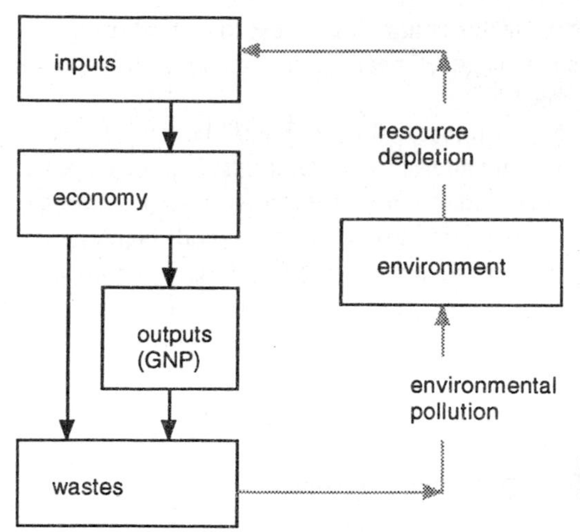

Figure 1.2 The economic development process in its environmental context. (Adapted from: Boulding 1966)

The interaction of economic and social development and the natural environment, and the reciprocal impacts between human actions and the biophysical world, have been recognized by governments from local to international levels. Attempts have been made to manage the interaction better, but a recent European Community report, *Towards sustainability* (CEC 1992), reveals disquieting trends that could have devastating consequences for the quality of the environment. Such EC trends include: a 25% increase in energy consumption by 2010 if there is no change in current energy demand growth rates; a 25% increase in car ownership and a 17% increase in mileage by 2000; a 13% increase in municipal waste between 1987 and 1992 despite increased recycling; a 35% increase in the EC's average water withdrawal rate between 1970 and 1985; and a 60% projected increase in Mediterranean tourism between 1990 and 2000. These trends are likely to be even more pronounced in developing countries where, because of greater rates of population growth and lower current living standards, there will be more pressure on environmental resources. The revelation of the state of the environment in many central and eastern European countries, and worldwide, adds weight to the assertion in the same EC report that "the great environmental struggles will be won or lost during this decade; by the next century it could be too late."

The 1987 Report of the World Commission on Environment and Development (usually referred to as the Brundtland Report) defined sustainable development as "development which meets the needs of the present generation without compromising the ability of future generations to meet their own needs" (UN World Commission on Environment and Development 1987). Sustainable development means handing down to future generations not only "man-made capital" such as roads, schools and historic buildings and "human capital" such as knowledge and skills, but also "natural/environmental capital" such as clean air, fresh water, rain forests, the ozone layer and biological diversity. The Brundtland Report identified the following key characteristics of sustainable development: it maintains the overall quality of life, it maintains continuing access to natural resources, and it avoids lasting environmental damage. It means living on the Earth's income rather than eroding its capital (DoE 1990).

There is, however, a danger of "sustainable development" becoming a weak catch-all phrase; there are already many alternative definitions. Turner & Pearce (1992) and Pearce (1992) have drawn attention to alternative interpretations of maintaining the capital stock. A policy of conserving the overall capital stock (man-made, human and natural) is consistent with running down any part of it, as long as there is substitutability between capital degradation in one area and investment in another. This can be interpreted as a "weak sustainability" position. In contrast, a "strong sustainability" position would argue that it is not acceptable to run down environmental assets, for several reasons: uncertainty (we do not know the full consequences for human beings), irreversibility (lost species cannot be replaced), life-support (some ecological assets serve life support functions) and loss aversion (people are highly averse to environmental losses). The "strong sustainability" position has much to commend it, but institutional responses have varied.

Institutional responses to meet the goal of sustainable development are required at several levels. Issues of global concern, such as ozone-layer depletion, climate change, deforestation and biodiversity loss, require a global political commitment to action. The United Nations Conference on Environment and Development (UNCED) held in Rio de Janeiro in 1992 was an example of international concern, but also of the problems of securing concerted action to deal with such issues. "Agenda 21", an 800-page action plan for the international community into the 21st century, sets out what nations should do to achieve sustainable development. It includes topics such as biodiversity, desertification, deforestation, toxic wastes, sewage, ocean and atmosphere. Unfortunately it is not legally binding. The Rio Conference called for a Sustainable Development Commission to be established to progress the implementation of Agenda 21 (Lovejoy 1992).

Within the EC four Community Action Programmes on the Environment were implemented between 1972 and 1992. These gave rise to specific legislation on a wide range of topics, including waste management, pollution of the atmosphere, protection of nature and environmental impact assessment. The Fifth Programme, "Towards sustainability", is set in the context of the completion of the Single European Market. The latter, with its emphasis on major changes in economic development resulting from the removal of all remaining fiscal, material and technological barriers between Member States, could pose additional threats to the environment. The Fifth Programme recognizes the need for the clear integration of performance targets – in relation to environmental protection – for several sectors, including manufacturing, energy, transport and tourism. EC policy on the environment will be based on the "precautionary principle", that preventive action should be taken, that environmental damage should be rectified at source, and that the polluter should pay. Whereas previous EC programmes relied almost exclusively on legislative instruments, the Fifth Programme advocates a broader mix including "market-based instruments", such as the internalization of environmental costs through the application of fiscal measures, and "horizontal, supporting instruments", such as improved baseline and statistical data and improved spatial and sectoral planning. Figure 1.3 illustrates the interdependence between resources, sectors and policy areas. EIA has a clear rôle to play.

In the UK, the publication of *This common inheritance: Britain's environmental strategy* (DoE 1990) provided the country's first comprehensive White Paper on the Environment. The report includes discussion of the greenhouse effect, town and country, pollution control, and awareness and organization with regard to environmental issues. Throughout there is an emphasis on responsibility for our environment being shared between the government, business and the public. The range of policy instruments advocated includes legislation, standards, planning and economic measures. The last, building on work by Pearce (1989), include charges,

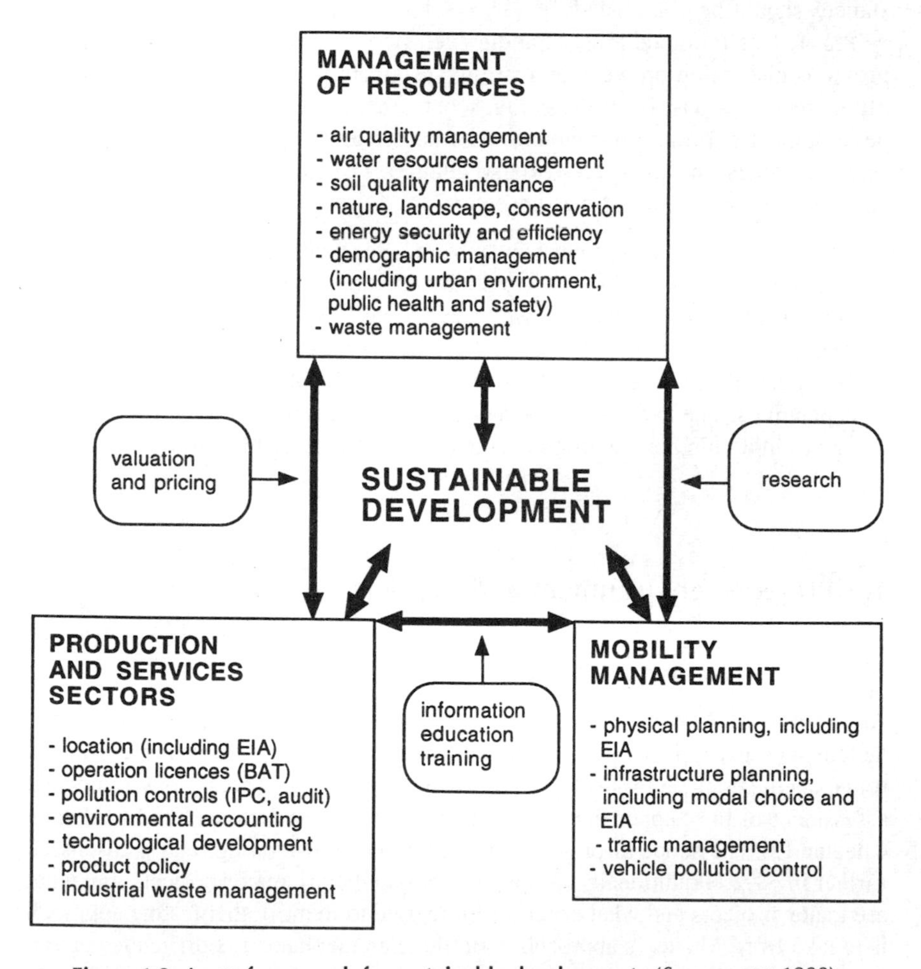

Figure 1.3 An EC framework for sustainable development. (*Source:* CEC 1992)

subsidies, market creation and enforcement incentives. The report also notes, cautiously, the recent addition of EIA to the "toolbox" of instruments.

Changing perspectives on EIA rôles

The arguments for EIA vary in time, space and according to the perspective of those involved. From a minimalist defensive perspective, developers, and possibly also some parts of government, might see EIA as a necessary evil, an administrative exercise, something to be gone through that might result in some minor, often cosmetic, changes to a development that would probably have happened anyway. For the "deep ecologists" or "deep Greens", EIA cannot provide the 100% certainty about the environmental consequences of development proposals; they feel that any projects carried out under uncertain or risky circumstances should be abandoned.

EIA and its methods must straddle such perspectives, partly reflecting the previous discussion on weak and strong sustainability. EIA can be, and is now often, seen as a positive process that seeks a harmonious relationship between development and the environment. The nature and use of EIA will change as relative values and perspectives also change. O'Riordan (1990) provides an appropriate conclusion to this subsection:

> One can see that EIA is moving away from being a defensive tool of the kind that dominated the 1970s to a potentially exciting environmental and social betterment technique that may well come to take over the 1990s. . . . If one sees EIA not so much as a technique, rather as a process that is constantly changing in the face of shifting environmental politics and managerial capabilities, one can visualize it as a sensitive barometer of environmental values in a complex environmental society. Long may EIA thrive.

1.4 Projects, environment and impacts

The nature of major projects

As noted in §1.2, EIA is relevant to a broad spectrum of development actions, including policies, plans, programmes and projects. The focus here is on projects, reflecting the dominant rôle of project EIA in practice. The strategic environmental assessment of the "upper tiers" of development actions is considered further in Chapter 13. The scope of projects covered by EIA is widening, and is discussed further in §4.2. Traditionally, project EIA has applied to major projects, but what are major projects and what criteria can be used to identify them? One approach is to take Lord Morley's approach to defining an elephant: it's difficult, but you easily recognize one when you see it. In a similar vein, the acronym LULU (locally unacceptable land-uses) has been applied in the USA to many major projects, such as energy, transport and various manufacturing projects, clearly reflecting the public perception of the negative impacts associated with such developments. There is no easy definition, but it is possible to highlight some key characteristics (Table 1.2).

Table 1.2 Characteristics of major projects.

- Substantial capital investment
- Cover large areas; employ large numbers (construction and/or operation)
- Complex array of organizational links
- Wide-ranging impacts (geographical and by type)
- Significant environmental impacts
- Require special procedures
- Extractive and primary (including agriculture); services; infrastructure and utilities
- Band, point

Most major projects involve considerable investment. In the UK context, "mega-projects" such as the Sizewell B PWR nuclear power station (budgeted to cost about £2 billion), the Channel Tunnel (about £6 billion) and the proposed Severn Barrage (about £8 billion) constitute one end of the spectrum. At the other end may be industrial estate developments, small stretches of road, various waste-disposal facilities, with considerably smaller, but still substantial, price tags. Such projects often cover large areas and employ many workers, usually in construction, and also in operation for some projects. They also invariably generate a complex array of inter- and intra-organizational activity during the various stages of the project's life. The developments may have wide-ranging, long-term and often very significant impacts on the environment. The definition of significance with regard to environmental effects is a key issue in EIA. It may relate, *inter alia*, to scale of development, to sensitivity of location and to the nature of adverse effects; it will be discussed further in later chapters. Like a large stone thrown into a pond, a major project can create major ripples with impacts spreading far and wide. In many respects major projects tend to be regarded as exceptional, requiring special procedures. In the UK, the latter have included public inquiries, hybrid bills that have to be passed through Parliament (for example, for the Channel Tunnel) and EIA procedures.

Major projects can also be defined according to type of activity. They include manufacturing and extractive projects, such as petrochemicals plants, steelworks, mines and quarries; services projects, such as leisure developments, out-of-town shopping centres, new settlements, and education and health facilities; and utilities and infrastructure, such as power stations, roads, reservoirs, pipelines and barrages. An EC study adopted a further definition of infrastructure into "band" and "point" infrastructure. Point infrastructure would include, for example, power stations, bridges and harbours; "band" or linear infrastructure would include electricity transmission lines, roads and canals (CEC 1982).

Major projects also have a planning and development life cycle, including a variety of stages. It is important to recognize such stages because impacts can vary considerably between stages. The major stages in a project's life cycle are outlined in Figure 1.4. There may be variations in timing between each stage, and internal variations within each stage, but there is a broadly common sequence of

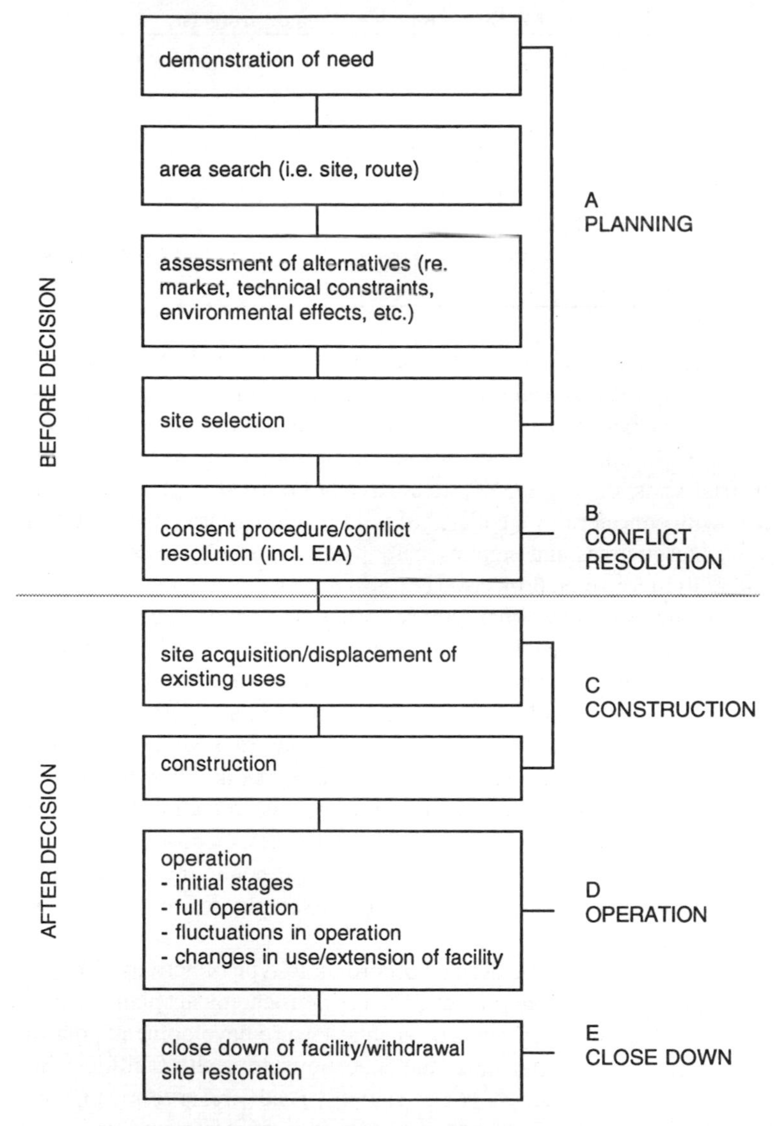

Figure 1.4 Generalized planning and development life-cycle for major projects (with particular reference to impact assessment on host area). (Adapted from Breese et al. 1965)

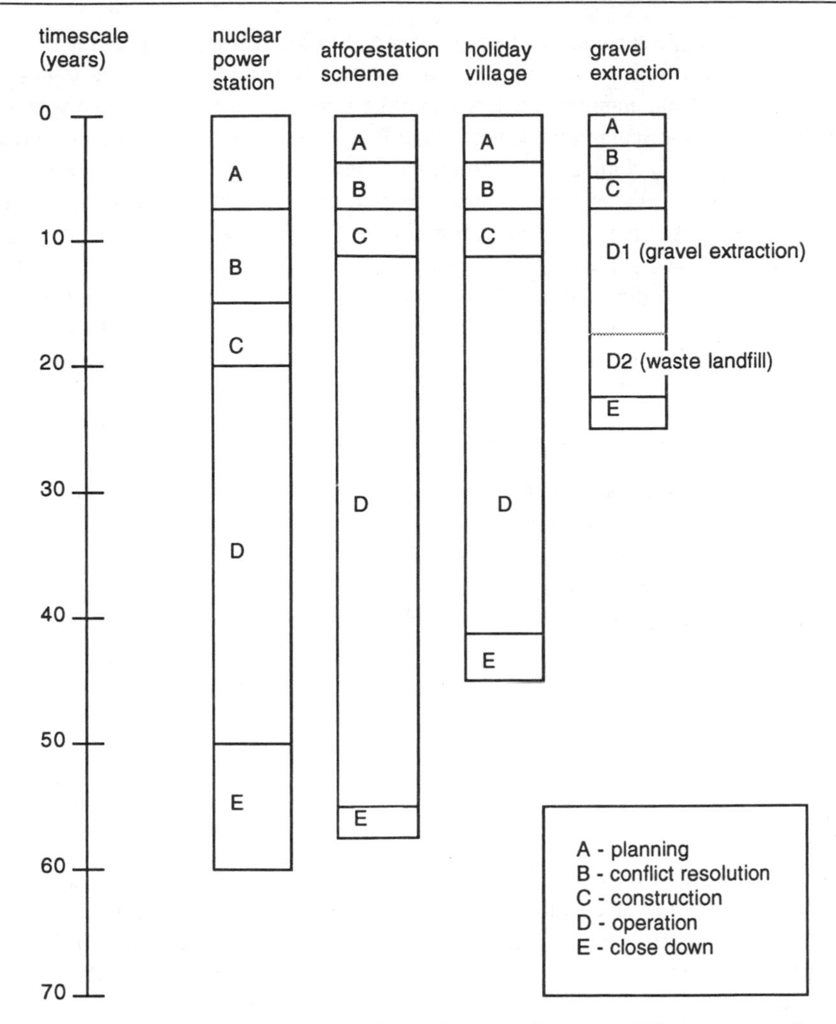

Figure 1.5 Broad variations in life-cycle stages between different types of projects.

events. With regard to EIA, an important distinction is that of "before the decision" (stages A and B), and "after the decision" (stages C, D and E). As noted in §1.2, the monitoring and auditing of the implementation of a project following approval are often absent from the EIA process.

Projects are initiated in several ways. Many are responses to market opportunities (e.g. a holiday village, a subregional shopping centre, a gas-fired power station); others may be seen as necessities (for example, the Thames Barrier); others may have an explicit prestige rôle (for example the programme of Grands Travaux in Paris including the Bastille Opera, Musée d'Orsay and Great Arch). Many major projects are public-sector initiatives, but with the move towards privatization in many countries, there has been a shift in funding towards the private sector, exemplified by such projects as the Mersey Barrage and the Channel Tunnel. The initial planning stage A may take several years, leading to a specific proposal for a particular site. It is at stage B that the various control and regulatory procedures, including EIA, normally come into play. The construction stage can be particularly disruptive, and may last up to ten years for some projects. Major projects invariably have long operational lives, although extractive projects can be short by comparison with infrastructure projects. The environmental impact of the eventual close-down of a facility should not be forgotten; for nuclear power facilities it is a major undertaking. Figure 1.5 illustrates possible broad variations in stages in life-cycles between project types.

Dimensions of the environment

The environment can be structured in several ways – including components, scale/space and time. A narrow definition of environmental components would focus primarily on the physical environment. For example, the UK Department of the Environment takes the term to include all media susceptible to pollution, including air, water, soil; flora and fauna, and human beings; landscape, urban and rural conservation and the built heritage (DoE 1991). The DoE checklist of environmental receptors is outlined in Table 1.3. However, as already noted in §1.2, there are important economic and socio-cultural dimensions to the environment. These include economic structure, labour markets, demography, housing, services (education, health, police, fire, etc.), life-styles and values, and these are added to the checklist in Table 1.3.

The environment can also be analyzed at various scales (Fig. 1.6). Many of the spatial impacts of projects affect the local environment, although the nature of "local" may vary according to the aspect of environment under consideration and to the stage in a project's life. However, some impacts are more than local. Traffic noise, for example, may be a local issue, but changes in traffic flows caused by a project may have a regional impact, and the associated CO_2 pollution contributes to the global greenhouse problem. The environment also has a time dimension. Baseline data on the state of the environment are needed at the time a project is being considered. This in itself may be a daunting request. In the UK, local development plans and national statistical sources, such as the Digest of Environmental Protection and Water Standards, may provide some relevant data. However, tailor-made state of the environment reports/audits are still in limited supply (see Ch. 12 for further information). Even more limited are time-series data highlighting trends in environmental quality. The environmental baseline is constantly changing, irrespective of the development under consideration, and it requires a dynamic rather than a static analysis.

Table 1.3 Environmental components.

Physical environment (adapted from DoE 1991)	
Air and atmosphere	air quality
Water resources and water bodies	water quality and quantity
Soil and geology	classification, risks (e.g. erosion, contamination)
Flora and fauna	birds, mammals, fish, etc.; aquatic and terrestrial vegetation
Human beings	physical and mental health and wellbeing
Landscape	characteristics and quality of landscape
Cultural heritage	conservation areas; built heritage; historic and archaeological sites
Climate	temperature, rainfall, wind, etc.
Energy	light, noise, vibration, etc.
Socio-economic environment	
Economic base – direct	direct employment; labour market characteristics; local/non-local trends
Economic base – indirect	non-basic/services employment; labour supply and demand
Demography	population structure and trends
Housing	supply and demand
Local services	supply and demand of services: health, education, police, etc.
Socio-cultural	lifestyles/quality of life; social problems (e.g. crime); community stress and conflict

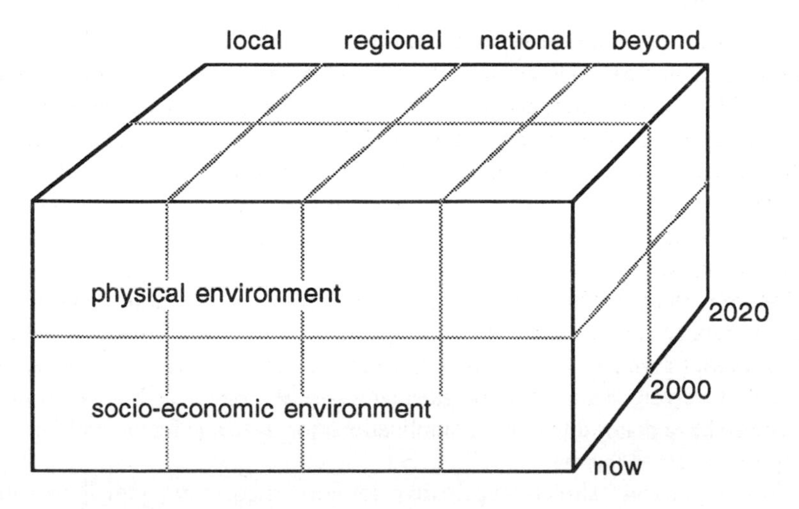

Figure 1.6 Environment: components, scale and time dimensions.

The nature of impacts

The environmental impacts of a project are those resultant changes in environmental parameters, in space and time, compared with what would have happened had the project not been undertaken. The parameters may be any of the type of environmental receptors noted previously: air quality, water quality, noise, levels of local unemployment and crime, for example. Figure 1.7 provides a simple illustration of the concept.

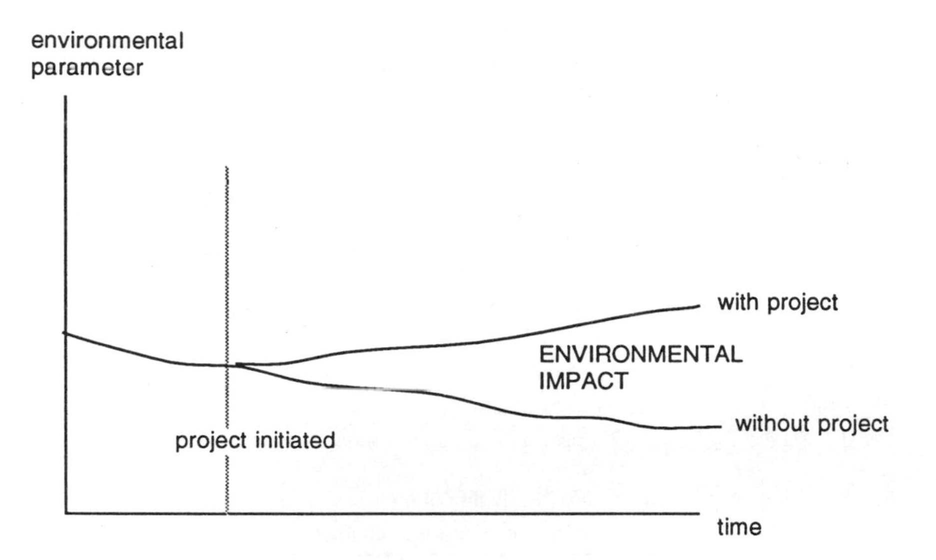

Figure 1.7 The nature of an environmental impact.

Table 1.4 provides a summary of some of the types of impacts that may be encountered in EIA. The physical and socio-economic effects have already been noted. These are often seen as synonymous with adverse and beneficial. Thus, new developments may produce harmful waste streams, but also produce much-needed employment in areas of high unemployment. However, the correlation does not always apply. A project may bring physical benefits when, for example, previously polluted and derelict land is brought back into productive use; similarly the socio-economic impacts of a major project on a community could include pressure on local health services and on the local housing market, and increases in community conflict and crime. Projects may also have immediate and direct impacts that give rise to secondary and indirect impacts over time. A reservoir based on a river system not only takes land for the immediate body of water but may also have major downstream implications for flora and fauna, and for human activities such as fishing and sailing.

The direct and indirect impacts may sometimes correlate with short-run and long-run impacts. The short-run/long-run distinction may also relate to the

Table 1.4 Types of impacts.

- Physical and socio-economic
- Direct and indirect
- Short-run and long-run
- Local and strategic (including regional, national and beyond)
- Adverse and beneficial
- Reversible and irreversible
- Quantitative and qualitative
- Distribution by group and/or area
- Actual and perceived
- Relative to other developments

project construction/operational stage distinction for some impacts; however, other construction stage impacts, such as change in land-use, are much more permanent. Impacts also have a spatial dimension. One distinction is between local and strategic, with the latter covering impacts affecting areas beyond the immediate locality. These are often regional, but may sometimes be of national or international significance.

Environmental resources cannot always be replaced; once destroyed, some may be lost forever. The distinction between reversible and irreversible impacts is a very important one and the identification of irreversible impacts, not susceptible to mitigation, can be a significant issue in an EIA. It may be possible to substitute, compensate for or reconstruct a lost resource in some cases, but substitutions are rarely ideal. The loss of a resource may become more serious over time, and valuations need to allow for this. Some impacts can be quantified, others are less tangible. The latter should not be ignored. Nor should the distributional impacts of a proposed development be ignored. Impacts do not fall evenly on affected parties and areas. Although a particular project may be assessed as bringing an overall benefit, some groups and/or geographical areas may be receiving the bulk of any adverse effects, with the main benefits going to others elsewhere. There is also the distinction between actual and perceived impacts. Individual "subjective" perceptions of impacts may be significant in influencing the responses and decisions of individuals towards a proposed development. They constitute an important source of information to be considered alongside more "objective" predictions of impacts. Finally, all impacts should be assessed relative to the "do-nothing" situation, with the state of the environment predicted without the project. This can be widened to include comparisons with anticipated impacts from alternative development scenarios for an area.

1.5 Current issues in environmental impact assessment

Although EIA now has almost 25 years of history in the USA, elsewhere the development of concepts and practice is more recent. Development is moving apace in many other countries, including the UK and the other EC Member States. Such progress has not been without its problems, and a number of the current issues in EIA are highlighted here and will be discussed more fully in later chapters.

Coverage of project assessment

Whereas legislators may seek to limit coverage, best practice may lead to its widening. For example, project EIA may only be mandatory for a limited set of major projects. In practice many others have been included. But which projects should have assessments? In the UK, case law is now building up, but the criteria for the inclusion or exclusion of a project for EIA are still developing. In a similar vein, there is a case for widening the dimensions of the environment under consideration to include socio-economic impacts more fully. The trade-off between adverse physical impacts and beneficial socio-economic impacts of a development often constitutes the key dilemma for decision-makers. Coverage can also be widened to include other types of impacts only very partially covered to date. Distributional impacts would fall into this category. Lichfield and others are seeking to counter this problem (see Lichfield 1989).

Nature of methods of assessment

As noted in §1.2, some of the key steps in the EIA process (e.g. consideration of alternatives, monitoring) may be missing from many studies. There may also be problems with the steps that are included. The prediction of impacts raises various conceptual and technical problems. The problem of establishing the environmental baseline position has already been noted. There may also be difficulties in clearly establishing the dimensions and development stages of the project under consideration. Further conceptual problems include establishing what would have happened in the relevant environment without the project clarifying the complexity of interactions of phenomena, and making trade-offs in an integrated way (i.e. assessing the trade-offs between economic apples, social oranges and physical bananas!). Other technical problems relate to the general lack of data, and the tendency to focus on the quantitative and often on single indicators in some areas. There may also be time lags and discontinuities between cause and effect, and non-continuity of projects and policies. The lack of auditing of predictive techniques limits the feedback on the effectiveness of methods. Nevertheless, innovative methods are being developed to predict impacts, ranging from simple checklists and matrices to complex mathematical models. These methods are not neutral, in the sense that the more complex they become, the more difficult it becomes for the general public to participate in the EIA process.

Relative rôles of participants in the process

The various "actors" in the EIA process – the developer, the affected parties and general public, and the regulators at various levels of government – have differential access to, and influence on, the outcome. Many would argue that in countries such as the UK, the process is too developer-orientated. The developer/developer's consultant carries out the EIA and prepares the EIS, and is unlikely to predict that the project will be an environmental disaster. Notwithstanding this, developers themselves are concerned about the potential delays associated with the requirement to submit an EIS. They are also concerned about cost. Details on costs are difficult to obtain. Clark (1984) estimates EIA costs of 0.5–2.0% of project value. Hart (1984) and Wathern (1988) suggest figures of a similar order. More recent estimates by Coles et al. (1992) suggest a much wider range, from 0.000025% to 5%, for EISs in the UK.

Procedures for and practice of public participation in the EIA process vary between, and sometimes within, countries, from the very comprehensive to the very partial and largely cosmetic. A key issue concerns the stages in the EIA process to which the public should have access. Government rôles in the EIA process may be conditioned by caution at extending systems, by limited experience and expertise in this new and rapidly developing area, and by resource considerations. For central government, this may be evident in limited guidance on best practice, and inconsistency in decisions. For local government, this may be revealed in difficulties in handling the scope and complexity of content of EISs when they are submitted for particular proposals.

The quality of assessments

Many EISs fail to meet even minimum standards. For example, a survey by Jones et al. (1991) of the EISs published under UK environmental impact assessment regulations highlighted shortcomings. They found that "one-third of the EISs did not appear to contain the required non-technical summary, that, in a quarter of the cases, they were judged not to contain the data needed to assess the likely environmental effects of the development, and that in the great majority of cases, the more complex, interactive impacts were neglected". Quality may vary between types of projects. It may also vary between countries supposedly operating under the same legislation.

Beyond the decision

Many EISs are for one-off projects, and there is little incentive to audit the quality of the assessment predictions and to monitor impacts as an input to a better assessment for the next project. EIA up to and no further than the decision on a project is a very partial linear process, with little opportunity for a cyclical learning process. In some areas of the world (e.g. California), monitoring of impacts is mandatory, and monitoring procedures must be included in the EIS. The extension of such approaches constitutes another significant current issue in the largely project-based EIA process.

Beyond project assessment

As noted in §1.2, strategic environmental assessment (SEA) of policies, plans and programmes represents a logical extension of project assessment. SEA can cope better with cumulative impacts, alternatives and mitigation measures than project assessment. SEA systems already exist in California and the Netherlands, and to a lesser extent in Canada, Germany and New Zealand. Discussions are in hand to introduce an EC-wide system from 1995 onwards (Therivel et al. 1992). The Fifth Action Programme on the Environment states: "Given the goal of achieving sustainable development, it seems only logical, if not essential, to apply an assessment of the environmental implications of all relevant policies, plans and programmes" (CEC 1992).

1.6 Outline of subsequent parts and chapters

This book is in four parts. The first establishes the context of EIA in the growth of concern about environmental issues and in relevant legislation, with particular reference to the UK. Following from this first chapter, which provides an introduction to EIA and an overview of principles, Chapter 2 focuses on the origins of EIA under the US National Environmental Policy Act (NEPA) of 1969, on interim developments in the UK, and on the subsequent introduction of EC Directive 85/337. The details of the UK legislative framework for EIA, under town and country planning and other legislation, are discussed in Chapter 3.

Part Two provides a rigorous step-by-step approach through the EIA process. This is the core of the text. Chapter 4 covers the early starting-up stages: clarifying the type of developments for EIA, and outlining approaches to scoping, consideration of alternatives, project description, establishing the baseline, and impact identification. Chapter 5 explores the central issues of prediction, assessment of significance, and mitigation of adverse impacts. The approach draws out broad principles affecting prediction exercises, exemplified with reference to particular cases. Chapter 6 provides coverage of the important issue identified above of participation in the EIA process. Communication in the EIA process, EIS presentation and EIS review are also covered in this chapter. Chapter 7 takes the process beyond the decision on a project, and examines the importance of, and approaches to, monitoring and auditing in the EIA process.

Part Three exemplifies the process in practice. Chapter 8 provides an overview of UK practice to date, including quantitative and qualitative analyses of the EISs prepared. Chapters 9 and 10 provide case studies of current practice in particular sectors. Chapter 9 includes analyses of several new settlement proposals, produced under the Town and Country Planning (Assessment of Environmental Effects) Regulations. New settlements include a variety of activities and land-uses and provide some of the most comprehensive projects, akin to development plans, for the new procedures. Chapter 10 includes analyses of major road proposals and power station proposals, which are produced under associated legislation: respectively, the Highways (Assessment of Environmental Effects) Regulations and Electricity and Pipe-line Works (Assessment of Environmental Effects) Regulations. Chapter 11 draws on comparative experience from five different countries – the Netherlands, Canada, Australia, Japan and China – presented to highlight some of the strengths and weaknesses of other systems in practice.

Part Four looks to the future. It focuses on many of the issues noted in §1.5. Chapter 12 focuses on improving the effectiveness of the current system of project assessment. Particular emphasis is given to the development of environmental auditing to provide better baseline data, to various procedural developments and to the achievement of compatibility of EIA systems in Europe. Chapter 13 discusses the extension of assessment to policies, plans and programmes, concluding full circle with a further consideration of EIA, SEA and sustainable development.

A set of appendices provide details of legislation and practice not considered appropriate to the main text. Key references to further reading are included there.

References

Boulding, K. 1966. The economics of the coming Spaceship Earth. In *Environmental quality in a growing economy*, H. Jarrett (ed.), 3–14. Baltimore: Johns Hopkins University Press.

Breakell, M. & J. Glasson (eds) 1981. *Environmental impact assessment: from theory to practice*. School of Planning, Oxford Polytechnic.

Breese, G. et al. 1965. *The impact of large installations on nearby urban areas*. Los Angeles: Sage.

Carley, M. J. & E. S. Bustelo 1984. *Social impact assessment and monitoring: a guide to the literature*. Boulder, Colorado: Westview Press.

CEC (Commission of the European Communities) 1982. *The contribution of infrastructure to regional development*. Brussels: CEC.

CEC 1992. *Towards sustainability: a European Community programme of policy and action in relation to the environment and sustainable development*, vol. II. Brussels: CEC.

Clark, B. D. 1984. Environmental impact assessment (EIA): scope and objectives. In *Perspectives on environmental impact assessment*, B. D. Clark et al. (eds). Dordrecht: Reidel.

Coles, T., K. Fuller, M. Slater 1992. *Practical experience of environmental assessment in the UK*. East Kirkby, Lincolnshire: Institute of Environmental Assessment.

DoE (Department of the Environment) 1989. *Environmental assessment: a guide to the*

procedures. London: HMSO.
DoE et al. 1990. *This common inheritance: Britain's environmental strategy* (Cmnd 1200). London: HMSO.
DoE 1991. *Policy appraisal and the environment*. London: HMSO.
Finsterbusch, K. 1985. State of the art in social impact assessment. *Environment and Behaviour* **17**, 192–221.
Hart, S. L. 1984. The costs of environmental review. In *Improving impact assessment*, S. L. Hart et al. (eds). Boulder, Colorado: Westview Press.
Health and Safety Commission 1978. *Canvey: an investigation of potential hazards from the operations in the Canvey Island/Thurrock area*. London: HMSO.
Jones, C. E., N. Lee, C. M. Wood 1991. *UK environmental statements 1988–1990: an analysis*. Occasional Paper 29, Department of Planning and Landscape, University of Manchester.
Lichfield, N. 1989. Environmental assessment. *Journal of Planning and Environmental Law* (November), 32–50.
Lovejoy, D. 1992. What happened at Rio? *The Planner* **78**(15), 13–14.
Munn, R. E. 1979. *Environmental impact assessment: principles and procedures*, 2nd edn. New York: John Wiley.
O'Riordan, T. 1990. EIA from the environmentalist's perspective. *VIA* **4** (March), 13.
Pearce, D. W. 1992. *Towards sustainable development through environment assessment*. Working Paper PA92-11, Centre for Social and Economic Research in the Global Environment, University College London.
Pearce, D., A. Markandya, E. Barbier 1989. *Blueprint for a green economy*. London: Earthscan.
Therivel, R., E. Wilson, S. Thompson, D. Heaney, D. Pritchard 1992. *Strategic environmental assessment*. London: RSPB/Earthscan.
Turner, R. K. & D. W. Pearce 1992. *Sustainable development: ethics and economics*. Working Paper PA92-09, Centre for Social and Economic Research in the Global Environment, University College London.
United Nations Economic Commission for Europe 1991. *Policies and systems of environmental impact assessment*. Geneva: United Nations.
UN World Commission on Environment and Development 1987. *Our common future*. Oxford: Oxford University Press.
Wathern, P. (ed.) 1988. *Environmental impact assessment: theory and practice*. London: Unwin Hyman.

CHAPTER 2
Origins and development

2.1 Introduction

Environmental impact assessment was first formally established in the USA in 1969 and has since spread, in various forms, to most other countries. In the UK, EIA was initially an ad hoc procedure carried out by local planning authorities and developers, primarily for oil- and gas-related developments. A 1985 European Community Directive on EIA (Directive 85/337) introduced broadly uniform requirements for EIA to all EC Member States, and significantly affected the development of EIA in the UK. At present, several years after the 1988 implementation of the EC Directive, Member States are carrying out EIA in a variety of forms, and discrepancies in implementation are becoming obvious. The nature of EIA systems – e.g. mandatory or discretionary, level of public participation, types of actions requiring EIA – and their implementation in practice vary widely from country to country. However, the rapid spread of the concept of EIA, and its central rôle in many countries' programmes of environmental protection, attest to its universal validity as a proactive planning tool.

This chapter begins with a discussion of the evolution of the US system of EIA. The present status of EIA worldwide is then briefly reviewed; Chapter 11 will consider a number of countries' systems of EIA in greater depth. EIA in the UK and the EC are then discussed. Finally, the various systems of EIA in the EC Member States are reviewed.

2.2 The National Environmental Policy Act and subsequent US systems

The US National Environmental Policy Act of 1969, also known as NEPA, was the first legislation to require EIAs to be carried out. Consequently it has become an important model for other EIA systems, both because it was a radically new form of environmental policy, and in the successes and failures of its subsequent development. Since its enactment, NEPA has resulted in the preparation of well over 15,000 EIAs, which have influenced countless decisions and represent a powerful base of environmental information. On the other hand, NEPA is unique. Other countries have shied away from the form it takes and the procedures it sets out, not least because they are unwilling to face a situation like that in the USA, where there has been extensive litigation over the interpretation and workings of the EIA system.

This section covers NEPA's legislative history, namely the early development before it became law; the interpretation of NEPA by the courts and the Council on Environmental Quality (CEQ); the main EIA procedures arising from NEPA; and likely future developments. The reader is referred to Anderson et al. (1984), Bear (1990), Orloff (1980) and the annual reports of the CEQ for further information.

Legislative history

NEPA is in many ways a fluke, strengthened by what should have been amendments weakening it, and interpreted by the courts to have powers that were not originally intended. The legislative history of NEPA is interesting not only in itself but also because it explains many of the anomalies of its operation and touches on some of the major issues involved in designing an EIA system. Several proposals to establish a national environmental policy were discussed in the US Senate and House of Representatives in the early 1960s. These proposals all included some form of unified environmental policy and the establishment of a high-level committee to foster it. In February 1969, Bill S1075 was introduced in the Senate which proposed a programme of federally funded ecological research and the establishment of a Council on Environmental Quality. A similar Bill, HR6750, introduced in the House of Representatives, proposed the formation of a CEQ and a brief statement on national environmental policy. Subsequent discussions in both chambers of Congress focused on several points:

- the need for a declaration of national environmental policy (now Title I of NEPA);
- a proposed statement that "each person has a fundamental and inalienable right to a healthful environment" (which would put environmental health on a par with, say, free speech). This was later weakened to the statement in §101(c) that "each person should enjoy a healthful environment";
- "action-forcing" provisions similar to those then being proposed for the Water Quality Improvement Act, which would require federal officials to prepare a detailed statement concerning the probable environmental impacts of any major action; this was to evolve into NEPA's §102(2)(C) which requires EIA. The initial wording of the Bill had required a "finding", which would have been subject to review by those responsible for environmental protection, rather

than a "detailed statement" subject to inter-agency review. The Senate had intended to weaken the Bill by requiring only a "detailed statement". Instead, the "detailed assessment" became the subject of external review and challenge; the public availability of the "detailed statements" became a major force shaping the law's implementation in its early years.

NEPA became operational on 1 January 1970. Table 2.1 summarizes its main points.

Table 2.1 Main points of NEPA.

NEPA consists of two titles. Title I establishes a national policy on the protection and restoration of environmental quality. Title II sets up a three-member Council on Environmental Quality (CEQ) to review environmental programmes and progress, and to advise the President on these matters. It also requires the President to submit an annual "Environmental Quality Report" to Congress. The provisions of Title I are the main determinants of EIA in the USA, and they are summarized here.

Section 101 contains requirements of a substantive nature. It states that the Federal Government has a continuing responsibility to "create and maintain conditions under which man and nature can exist in productive harmony, and fulfil the social, economic and other requirements of present and future generations of Americans". As such the government is to use all practicable means, "consistent with other essential considerations of national policy", to minimize adverse environmental impact and to preserve and enhance the environment through federal plans and programmes. Finally, "each person should enjoy a healthful environment", and citizens have a responsibility for environmental preservation.

Section 102 requirements are of a procedural nature. Federal agencies are required to make full analyses of all the environmental effects of implementing their programmes or actions. Section 102(1) directs agencies to interpret and administer policies, regulations, and laws in accordance with the policies of NEPA. Section 102(2) requires federal agencies to:

- use "a systematic and interdisciplinary approach" to ensure that social, natural and environmental sciences are used in planning and decision-making;
- identify and develop procedures and methods so that "presently unquantified environmental amenities and values may be given appropriate consideration in decision-making along with traditional economic and technical considerations"; and
- "include in every recommendation or report on proposals for legislation and other *major Federal actions significantly affecting the quality of the human environment*, a *detailed statement* by the responsible official on:
  - the environmental impact of the proposed action,
  - any adverse environmental effects which cannot be avoided should the proposal be implemented,
  - alternatives to the proposed action,
  - the relationship between local short-term uses of man's environment and the maintenance and enhancement of long-term productivity, and
  - any irreversible and irretrievable commitments of resources which would be involved in the proposed action should it be implemented. (authors' emphases)

Section 103 requires federal agencies to review their regulations and procedures for adherence with NEPA, and to suggest any necessary remedial measures.

Interpretation of NEPA

NEPA is a generally worded law that required substantial early interpretation. The CEQ, which was set up by NEPA, prepared guidelines to assist in the Act's interpretation. However, much of the strength of NEPA came from early court rulings. NEPA was immediately seen by environmental activists as a significant vehicle for preventing environmental harm, and the early 1970s saw a series of influential lawsuits and court decisions based on it. These lawsuits were of three broad types, as described by Orloff (1980):

- Challenging an agency's decision not to prepare an EIA: this generally raised issues such as whether a project was "major", "federal", an "action", or with significant environmental impacts (see NEPA §102(2)(C)). For instance, the issue of whether an action is "federal" came into question in some lawsuits concerning federal funding of local government projects.[1]
- Challenging the adequacy of an agency's EIS: this raised issues such as whether the EIS adequately addressed alternatives, and whether it covered the full range of significant environmental impacts. A famous early court case concerned the Chesapeake Environmental Protection Association's claim that the Atomic Energy Commission did not adequately consider the water quality impacts of its proposed nuclear power plants, particularly in the EIA for the Calvert Cliffs power plant.[2] The Commission argued that NEPA merely required the consideration of water quality standards; opponents argued that it required an assessment beyond mere compliance with standards. The courts sided with the opponents.
- Challenging an agency's substantive decision, namely its decision to allow or not to allow a project to proceed in light of the contents of its EIS. Another influential early court ruling[3] laid guidelines for the judicial review of agency decisions, noting that the court's only function was to ensure that the agency had taken a "hard look" at environmental consequences, not to substitute its judgement for that of the agency.

The early proactive rôle of the courts greatly strengthened the power of environmental movements and caused many projects to be stopped or substantially amended. In many cases the lawsuits delayed construction for long enough to make them economically infeasible or to allow the areas where projects would have been sited to be designated as national parks or wildlife areas (Turner 1988). More recent decisions have been less clearly pro-environment than the earliest decisions. The flood of early lawsuits, and the delays and costs involved, was a lesson to other countries in how *not* to set up an EIA system. As will be shown later, many countries carefully distanced their EIA systems from the possibility of lawsuits.

The CEQ was also instrumental in establishing guidelines to interpret NEPA, producing interim guidelines in 1970, and guidelines in 1971 and 1973. Generally the courts adhered closely to these guidelines when making their rulings. However, the guidelines were problematic: they were not detailed enough, and were interpreted by the federal agencies as being discretionary rather than binding. To combat these limitations, President Carter issued Executive Order 11992 in 1977, giving the CEQ authority to set enforceable regulations for implementing NEPA. These were issued in 1978 (CEQ 1978) and sought to make the NEPA process more useful for decision-makers and the public, reduce paperwork and delay, and emphasize real environmental issues and alternatives.

Summary of NEPA procedures

The process of EIA established by NEPA, and developed further in the CEQ regulations, is summarized in Figure 2.1. The following citations are from the CEQ regulations (CEQ 1978).

> [The EIA process begins] as close as possible to the time the agency is developing or is presented with a proposal . . . The statement shall be prepared early enough so that it can serve practically as an important contribution to the decision-making process and will not be used to rationalize or justify decisions already made. (§1502.5)

A "lead agency" is designated that co-ordinates the EIA process. The lead agency first determines whether the proposal requires the preparation of a full EIS, no EIS at all, or a more concise "environmental assessment", which in turn would allow the agency to determine whether an EIS is needed or whether the preparation of a "finding of no significant impact" (FONSI) is appropriate.

If a FONSI is prepared, then a permit would usually be granted following public discussion. If a full EIS is found to be needed, the lead agency publishes a "Notice of Intent" and the *process of scoping begins*. The aim of the scoping exercise is to determine the issues to be addressed in the EIA: to eliminate insignificant issues, focus on those that are significant, and identify alternatives to be addressed. The lead agency invites the participation of the proponent of the action, affected parties and other interested persons.

> [The alternatives] section is the heart of the environmental impact statement . . . [It] should present the environmental impacts of the proposal and the alternatives in comparative form, thus sharply defining the issues and providing a clear basis for choice . . . (§1502.14)

A draft EIS is then prepared, and is reviewed and commented on by the relevant agencies and the public. These comments are taken into account in the subsequent preparation of a final EIS. An EIS is normally presented in the format shown in Table 2.2. In an attempt to be comprehensive, early EISs tended to be so bulky as to be virtually unreadable. The CEQ guidelines consequently emphasize the need to concentrate only on important issues and to prepare readable documents:

> The text of final environmental impact statements shall normally be less than 150 pages . . . Environmental impact statements shall be written in plain language . . . (§1502. 7–8)

The public is involved in this process, both at the scoping stage and after publication of the draft and final EISs.

> Agencies shall: (a) Make diligent efforts to involve the public in preparing and implementing NEPA procedures . . . (b) Provide public notice of NEPA-related hearings, public meetings and the availability of environmental documents . . . (c) Hold or sponsor public hearings . . . whenever appropriate . . . (d) Solicit appropriate information from the public. (e) Explain in its procedures where interested persons can get information or status reports . . . (f) Make environmental impact statements, the comments received, and any underlying documents available to the public pursuant to the provisions of the Freedom of Information Act . . . (§1506.6)

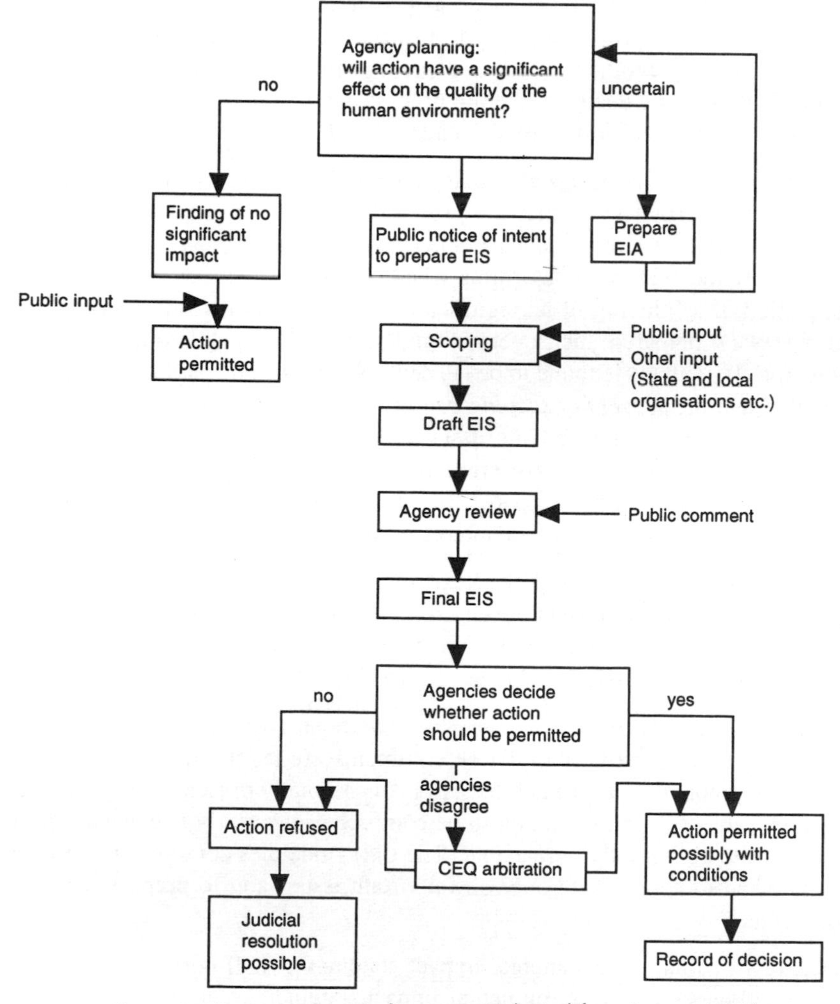

Figure 2.1 Process of EIA under NEPA. (Adapted from Legore 1984)

Finally, a decision is made about whether the proposed action should be permitted:

> Agencies shall adopt procedures to ensure that decisions are made in accordance with the policies and purposes of the Act. Such procedures shall include but not be limited to: (a) Implementing procedures under section

Table 2.2 Typical format for an EIS under NEPA.

(a) Cover sheet
- list of responsible agencies
- title of the proposed action
- contact persons at the agencies
- designation of EIS as draft, final, or supplement
- abstract of the EIS
- the date by which comments must be received

(b) Summary (usually 15 pages or less)
- major conclusions
- areas of controversy
- issues to be resolved

(c) Table of contents

(d) Purpose of and need for action

(e) Alternatives including proposed action

(f) Affected environment

(g) Environmental consequences
- environmental impacts of the alternatives including the proposed action
- any adverse environmental effects which cannot be avoided if the proposal is implemented
- mitigation measures to be used, and residual effects of mitigation
- the relationship between short-term uses of the environment and the maintenance and enhancement of long-term productivity
- any irreversible or irretrievable commitments of resources if the proposal is implemented
- discussion of:
  - direct and indirect effects and their significance
  - possible conflicts between the proposed action and the objectives of relevant land-use plans, policies and controls
  - effects of alternatives including the proposed action
  - energy requirements and conservation potential of various alternatives and mitigation measures
  - natural or depletable resource requirements and conservation of various alternatives and mitigation measures
  - effects on urban quality, historic and cultural resources, and the built environment
  - means to mitigate adverse impacts

(h) List of preparers

(i) List of agencies, etc. to which copies of the EIS are sent

(j) Index

(k) Appendices, including supporting data

102(2) to achieve the requirements of sections 101 and 102(1) . . .
(e) Requiring that . . . the decision-maker consider the alternatives described in the environmental impact statement. (§1505.1)

Where all relevant agencies agree that the action should not go ahead, permission is denied, and a judicial resolution may be attempted. Where agencies agree that the action can proceed, permission is given, possibly subject to specified conditions (e.g. monitoring, mitigation). Where the relevant agencies disagree, the CEQ acts as arbiter (§1504). Until a decision is made, "no action concerning the proposals shall be taken which could: (1) have an adverse environmental impact; or (2) limit the choice of reasonable alternatives . . . " (§1506.1).

Recent trends

During the first 10 years of NEPA's implementation, about 1,000 EISs were prepared annually. Recently, negotiated improvements to the environmental impacts of proposed actions have become increasingly common during the preparation of "environmental assessments". This has led to many "mitigated findings of no significant impact" (no nice acronym exists for this), reducing the number of EISs prepared: whereas 1,273 EISs were prepared in 1979, only 456 were prepared in 1991 (CEQ 1992). This trend can be viewed positively, since it means that environmental impacts are considered earlier in the decision-making process, and since it reduces the costs of preparing EISs. However, the fact that this abbreviated process allows less public participation causes some concern. Of the 456 EISs in 1991, 145 were filed by the Department of Agriculture (primarily for forestry and range management), and 87 were filed by the Department of Transportation (primarily for road construction). Between 1979 and 1991, the number of EISs filed by the Department of Housing and Urban Development fell from 170 to 7! The number of legal cases filed against federal departments and agencies on the basis of NEPA has also fallen, although not in proportion to the fall in EISs, from 139 in 1979 to 85 in 1990. The most common complaints are "no EIS when one should have been prepared" (53 cases in 1990), and "inadequate EIS" (24 cases in 1990).

The twentieth year of operation of NEPA, 1990, was marked by a series of conferences on the Act, and the presentation to Congress of a bill of NEPA amendments. Under the Bill (HR1113), which was not passed, federal actions that take place outside the USA (e.g. projects built in other countries with US federal assistance) would have been subject to EIA, and all EIAs would have been required to consider global climatic change, depletion of the ozone layer, loss of biological diversity and trans-boundary pollution. This latter amendment was controversial: although the need to consider the global impacts of programmes was undisputed, it was felt to be infeasible at the level of project EIA. Finally, the Bill would have required all federal agencies to survey a statistically significant sample of EISs to determine whether mitigation measures promised in the EIS had been implemented and, if so, whether they had been effective.

The context of EIA has also become a matter of concern. EIA is only one part of a broader environmental policy (NEPA), but the procedural provisions set out in NEPA's §102(2)(C) have overshadowed the rest of the Act. It has been argued that mere compliance with these procedures is not enough, and that greater emphasis should be given to the environmental goals and policies stated in §101. EIA must also be seen in the light of other environmental legislation. In the USA, many laws dealing with specific aspects of the environment were enacted or strengthened in the 1970s, including the Clean Water Act and the Clean Air Act. These laws have in many ways superseded NEPA's substantive requirements and have complemented and buttressed its procedural requirements. Compliance with these laws does not necessarily imply compliance with NEPA. However, the permit process associated with these other laws has become a primary method for evaluating project impacts, reducing NEPA's importance except for its occasional rôle as a focus of debate on major projects (Bear 1990).

Little NEPAs and the particular case of California

Many state-level EIA systems have been established in the USA in addition to NEPA. Sixteen of the USA's 50 states[4] have so-called "little NEPAs" that require EIA for state actions, actions that require state funding or permission, and/or projects in sensitive areas. Other states[5] have no specific EIA regulations, but have EIA requirements in addition to those of NEPA.

Of particular interest is the Californian system, established under the California Environmental Quality Act (CEQA) of 1973, and subsequent amendments. This is widely recognized as one of the most advanced EIA systems worldwide. The legislation applies not only to government actions, but also to the activities of private parties that require government agency approval. It is not merely a procedural approach but one that requires state and local agencies to protect the environment by adopting feasible mitigation measures and alternatives in environmental impact reviews (EIRs). The legislation extends beyond projects to higher tiers of actions, and an amendment in 1989 also added mandatory mitigation, monitoring and reporting requirements to CEQA. Annual guidance on the California system is provided in an invaluable publication by the State of California, which sets out the CEQA Statutes and Guidelines in considerable detail (State of California 1992).

2.3 Worldwide spread of EIA

Since the enactment of NEPA, EIA systems have been established in various forms throughout the world, beginning with more developed countries – e.g. Canada in 1973, Australia in 1974, West Germany in 1975, France in 1976 – and later also in the less developed countries. These systems vary greatly. Some are in the form of *mandatory regulations, acts, or statutes*; these are generally enforced by requiring the preparation of an adequate EIS before permission is given for a project to proceed. In other cases, EIA *guidelines* have been established. These are not enforceable but generally impose obligations on the administering agency. Other legislation allows government officials to require EIAs to be prepared at their *discretion*. Still elsewhere, EIAs are prepared in an *ad hoc* manner, often because they are required by funding bodies (e.g. World Bank, USAID) as part of the funding approval process. However, these classifications are not necessarily indicative of how thoroughly EIA is carried out. For instance, the EIA regulations of Brazil and the Philippines are not well carried out or enforced in practice (Abracosa & Ortolano 1987, Moreira 1988), whereas Japan's guidelines are thoroughly implemented and some very good ad hoc EIAs have been prepared in the UK. Figure 2.2 summarizes the present state of EIA systems worldwide to the best of the authors' knowledge.

Another major distinction between types of EIA systems is whether actions that require EIA are given as a *definition* (e.g. the USA's definition of "major federal actions significantly affecting the quality of the human environment") or as a *list of projects* (e.g. roads of more than 10 km in length). Most countries use a list of projects, in part to avoid legal wrangling such as that surrounding NEPA's definition. Yet another distinction is whether EIA is required for *government projects only* (as in NEPA), for *private projects only*, or for both.

Finally, some international development and funding agencies have set up EIA guidelines, including the European Bank for Reconstruction and Development (1992), Overseas Development Administration (1992), Organization for Economic Co-operation and Development (1979), United Nations Environment Programme (1980), United States Agency for International Development (1978), and World Bank (1991).

2.4 Development in the UK

The UK has enacted formal legislation for EIA comparatively recently, in 1988, in the form of several laws that implement European Community Directive 85/337/EEC (Commission on the European Communities 1985). It is quite possible that, without pressure from the European Commission, such legislation would never have been enacted, since the UK government felt that its existing planning system more than adequately controlled environmentally unsuitable developments. However, this does not mean that the UK had no EIA system at all before 1988; many EIAs were prepared voluntarily or at the request of local authorities, and guidelines for EIA preparation were drawn up.

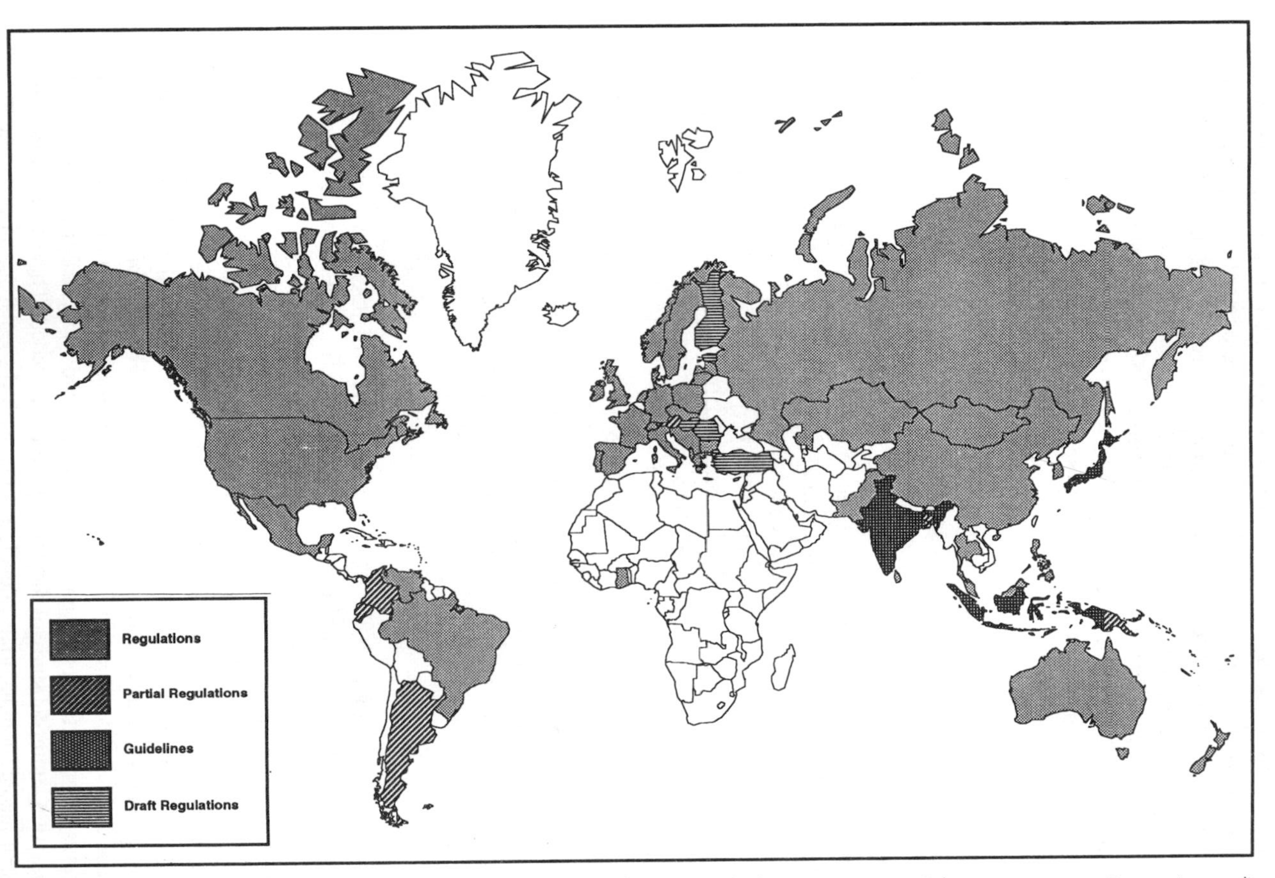

Figure 2.2 **EIA systems worldwide (the authors apologise for any omissions).** (*Sources:* Cabrera 1984, EIA Centre (annual), Wathern 1988, Westman 1985 and others.)

Limitations of the land-use planning system

The UK's statutory land-use planning system has since 1947 required local planning authorities (LPAs) to anticipate likely development pressures, assess their significance, and allocate land, as appropriate, to accommodate them. Environmental factors are a fundamental consideration in this assessment. Most developments require individual planning consent, so environmentally harmful developments can be prevented by the denial of planning consent. This system resulted in the accumulation of considerable planning expertise concerning the likely consequences of development proposals.

After the mid-1960s, however, the planning system began to seem less effective in controlling the impacts of major developments. The increasing scale and complexity of developments, the consequently greater social and physical environmental impacts, and the growing internationalization of developers (e.g. oil and chemicals companies), all outstripped the capability of the development control system to predict and control the impacts of developments. In the late 1960s, public concern about environmental protection also grew considerably, and the relationship between statutory planning controls and the development of major projects came under increasing scrutiny. This became particularly obvious in the case of the proposed Third London Airport. The Roskill Commission was established to select the most suitable site for an airport in southeast England, with the mandate to prepare a cost–benefit analysis of alternative sites. The resulting analysis (HMSO 1971) focused on socio-economic rather than physical environmental impacts; it led to an understanding of the difficulties of expanding cost–benefit analysis to impacts not easily measured in monetary terms, and to the realization that other assessment methods were needed to achieve a balance between socio-economic and physical environmental objectives.

North Sea oil- and gas-related EIA initiatives

The main impetus towards the further development of EIA, however, was the discovery of oil and gas in the North Sea. The extraction of these resources necessitated the construction of large developments in remote areas renowned for their scenic beauty and distinctive way of life (e.g. Shetlands, Orkneys and the Highlands Region). Planning authorities in these areas lacked the experience and resources needed to assess the impacts of such large-scale developments. In response, the Scottish Development Department (SDD) issued a technical advice note to LPAs (Scottish Development Department 1974). *Appraisal of the impacts of oil-related development* noted that these developments and other large-scale and unusual projects need "rigorous appraisal", and suggested that LPAs should commission an impact study of the developments if needed. This was the first government recognition that major developments needed special appraisal. Some EIAs were carried out in the early 1970s, mostly for oil and gas developments. Many of these were sponsored by the SDD and LPAs, and were prepared by environmental consultants, but some (e.g. for the Flotta Oil Terminal and Beatrice Oilfield) were commissioned by the developers. Other early EIAs concerned a coal mine in the Vale of Belvoir, a pumped-storage electricity scheme at Loch Lomond, and various motorway and trunk road proposals (Clark & Turnbull 1984).

In 1973, the Scottish Office and Department of the Environment (DoE) commissioned the University of Aberdeen's Project Appraisal for Development Control (PADC) team to develop a systematic procedure for planning authorities to make a balanced appraisal of environmental, economic and social impacts of major industrial developments. PADC produced an interim report, *The assessment of major industrial applications – a manual* (Clark et al. 1976), which was issued free of charge to all LPAs in the UK and "commended by central government for use by planning authorities, government agencies and developers". The PADC procedure was designed to fit into the existing planning framework, and was used to assess a variety of (primarily private-sector) projects. An extended and updated version of the manual was issued in 1981 (Clark et al. 1981).

In 1974, the Secretaries of State for the Environment, Scotland and Wales commissioned two consultants, J. Catlow and C. G. Thirwall, to investigate the "desirability of introducing a system of impact analysis in Great Britain, the circumstances in which a system should apply, the projects it should cover and the way in which it might be incorporated into the development control system" (Catlow & Thirwall 1976). The resulting report made recommendations on such issues as who should be responsible for preparing and paying for EIAs, and what legislative changes would be needed to institute an EIA system. The report concluded that about 25–50 EIAs per year would be needed, for both public- and private-sector projects. EIA was given additional support by the Dobry Report on the development control system (Dobry 1975), which advocated that LPAs should require developers to submit an impact study for particularly significant development proposals. The report outlined the main topics such a study should address, and information that should be required from developers. Government reactions to the Dobry Report were mixed: the Royal Commission on Environmental Pollution endorsed the report, but the Stevens Committee (1976) on Mineral Workings recommended that a comprehensive standard form for mineral applications be introduced, arguing that such a form would make unnecessary EIAs for mineral workings.

Department of the Environment scepticism

However, overall the DoE remained sceptical about the need, practicality, and cost of EIA. In fact, the government's approach to EIA has been described as being "from the outset grudging and minimalist" (CPRE 1991). In response to the Catlow & Thirwall report, the DoE stated: "Consideration of the report by local authorities should not be allowed to delay normal planning procedures and any new procedures involving additional calls on central or local government finance and manpower are unacceptable during the present period of economic restraint" (DoE

1977). A year later, after much deliberation, the DoE was slightly more positive: "We fully endorse the desirability . . . of ensuring careful evaluation of the possible effects of large developments on the environment . . . The approach suggested by Thirwall/Catlow is already being adopted with many [projects] . . . The sensible use of this approach [should] improve the practice in handling these relatively few large and significant proposals" (DoE 1978). The government's foreword to the PADC manual of 1981 also emphasized the need to minimize the costs of EIA procedures: "It is important that the approach suggested in the report should be used selectively to fit the circumstances of the proposed development and with due economy" (Clark et al. 1981). As will be seen in later chapters, the government still remains sceptical about the value of EIA, and is generally unwilling to extend its remit, as is being suggested by the EC.

By the early 1980s, more than 200 studies on the environmental impacts of projects had been prepared on an ad hoc basis in the UK. These are listed in Petts & Hills (1982). Many of these studies were not full EIAs, but focused on only a few impacts. However, large developers such as British Petroleum, British Gas, the Central Electricity Generating Board and the National Coal Board were preparing a series of increasingly comprehensive statements. In the case of British Gas, these were shown to be a good investment, saving the company £30 million in ten years (House of Lords 1981a).

2.5 EC Directive 85/337

The development and implementation of EC Directive 85/337 greatly influenced the EIA systems of the UK and other EC Member States. In the UK, central government research on a UK system of EIA virtually stopped after the mid-1970s, and attention focused instead on ensuring that any future EC system of EIA would fully incorporate the needs of the UK for flexibility and discretion. Other Member States were eager to ensure that the Directive reflected the requirements of their own more rigorous systems of EIA. Since the Directive's implementation, EIA activity in all the EC Member States has increased dramatically.

Legislative history

The EC had two major reasons for wanting to establish a uniform system of EIA in all its Member States. First, it was concerned about the state of the physical environment and eager to prevent further environmental deterioration. The EC's First Action Programme on the Environment of 1973 (CEC 1973) already advocated the prevention of environmental harm: "the best environmental policy consists of preventing the creation of pollution or nuisances at source, rather than subsequently trying to counteract their effects", and to that end "effects on the environment should be taken into account at the earliest possible stage in all technical planning and decision-making processes". Further Action Programmes of 1977, 1983, 1987 and 1992 have reinforced this emphasis. Land-use planning was seen as a major way of putting these principles into practice, and EIA was viewed as a key technique for incorporating environmental considerations into the planning process.

Secondly, the EC was concerned to ensure that no distortion of competition should arise where one Member State could gain unfair advantage by permitting developments that, for environmental reasons, might be refused by another Member State. In other words, it saw environmental policies as necessary to the maintenance of a level economic playing field. Further motivation for EC action included a desire to encourage best practice across Member States. In addition, pollution problems transcend territorial boundaries (as witnessed by acid rain and river pollution in Europe) and the EC can contribute at least a subcontinental response framework.

The EC began to commission research on EIA in 1975. Five years later and after more than 20 drafts, the Commission presented a draft Directive to the Council of Ministers (CEC 1980), which was circulated throughout the Member States. The 1980 draft attempted to reconcile several conflicting needs. It sought to benefit from the US experience with NEPA, but develop policies appropriate for European need. It also sought to make EIA applicable to all actions likely to have a significant environmental impact, but to ensure that procedures would be practicable. Finally, and perhaps most challenging, it sought to make EIA requirements flexible enough to adapt to the needs and institutional arrangements of the various Member States, but uniform enough to prevent problems arising from widely varying interpretations of the procedures. Harmonization of the types of projects to be subject to EIA, the main obligations of the developers, and the contents of the EIAs were considered particularly important (Lee & Wood 1984, Tomlinson 1986).

As a result, the draft Directive incorporated a number of key features. First, planning permission for projects was to be granted only after an adequate EIA had been completed. Secondly, LPAs and developers were to co-operate in providing information on the environmental impacts of proposed developments. Thirdly, statutory bodies responsible for environmental issues, and other Member States in the case of trans-frontier effects, were to be consulted. Finally, the public was to be informed, and allowed to comment on issues related to project development.

In the UK the draft Directive was examined by the House of Lords Select Committee on the European Commission, where it received widespread support:

> The present draft Directive strikes the right kind of balance: it provides a framework of common administrative practices which will allow Member States with effective planning controls to continue with their system . . . while containing enough detail to ensure that the intention of the draft cannot be evaded . . . The Directive could be implemented in the United Kingdom in a way which would not lead to undue additional delay and costs in

> planning procedures and which need not therefore result in economic and other disadvantages. (House of Lords 1981a)

However, the Parliamentary Under-Secretary of State at the DoE dissented. Although accepting the general need for EIA, he was concerned about the bureaucratic hurdles, delaying objections and litigation that would be associated with the proposed Directive (House of Lords 1981b). The UK Royal Town Planning Institute also commented on several drafts of the Directive. Generally the RTPI favoured the Directive, but was concerned that it might cause the planning system to become overly rigid:

> The Institute welcomes the initiative taken by the European Commission to secure more widespread use of EIA as it believes that the appropriate use of EIA could both speed up and improve the quality of decisions on certain types of development proposals. However, it is seriously concerned that the proposed Directive, as presently drafted, would excessively codify and formalize procedures of which there is limited experience and therefore their benefits are not yet proven. Accordingly the Institute recommends the deletion of Article 4 and annexes of the draft. (House of Lords 1981a)

More generally, slow progress in the implementation of EC legislation was symptomatic of the wide range of interest groups involved, of the lack of public support for increasing the scope of town planning and environmental protection procedures, and of the unwillingness of Member States to adapt their widely varying planning systems and environmental protection legislation to those of other countries (Williams 1988). After considering the many views expressed by the various Member States, the Commission published proposed amendments to the draft Directive in March 1982 (CEC 1982). Approval was expected in November 1983. However, this was delayed by the Danish Government, which was concerned about projects authorized by Acts of Parliament. On 7 March 1985, the Council of Ministers agreed on the proposal; it was formally adopted as a Directive on 27 June 1985 (CEC 1985) and became operational on 3 July 1988.

Since then, the EC's Fifth Action Programme, *Towards sustainability* (CEC 1992), has been published. It stresses the importance of EIA, particularly in helping to achieve sustainable development, and the need to expand the remit of EIA:

> Given the goal of achieving sustainable development it seems only logical, if not essential, to apply an assessment of the environmental implications of all relevant policies, plans and programmes. The integration of environmental assessment within the macro-planning process would not only enhance the protection of the environment and encourage optimization of resource management but would also help to reduce those disparities in the international and inter-regional competition for new development projects which at present arise from disparities in assessment practices in the Member States . . .

The reader is referred to Clark & Turnbull (1984), Lee & Wood (1984), O'Riordan & Sewell (1981), Swaffield (1981), Tomlinson (1986), Williams (1988), and Wood (1981, 1988) for further discussion of the development of EIA in the UK and EC.

Summary of EC Directive 85/337 procedures

Appendix 1 gives the complete wording of EC Directive 85/337. The Directive differs in major respects from NEPA. It requires EIAs to be prepared by both public agencies and private developers, whereas NEPA applies only to federal agencies. It requires EIA for a specified list of projects, whereas NEPA uses the definition "major federal actions . . . ". It specifically lists the impacts that are to be addressed in an EIA, whereas NEPA does not. It does not require a clearly specified scoping stage, which NEPA does. Finally, it includes fewer requirements for public consultation than does NEPA.

Under the provisions of the European Communities Act of 1972, Directive 85/337 is the controlling document, laying down rules for EIA in Member States. Individual Member States enact their own regulations to implement the Directive and have considerable discretion. According to the Directive, EIA is required for two classes of projects, one mandatory (Annex I) and one discretionary (Annex II):

> . . . projects of the classes listed in Annex I shall be made subject to an assessment . . . Projects of the classes listed in Annex II shall be made subject to an assessment . . . where Member States consider that their characteristics so require. To this end Member States may *inter alia* specify certain types of projects as being subject to an assessment or may establish the criteria and/or thresholds necessary to determine which [Annex II projects] are to be subject to an assessment. (Article 4)

Table 2.3 summarizes the projects listed in Annexes I and II.

Similarly, the information required in an EIA is listed in Annex III of the Directive, but must only be provided

> inasmuch as: (a) The Member States consider that the information is relevant to a given stage of the consent procedure and to the specific characteristics of a particular project . . . and of the environmental features likely to be affected, (b) The Member States consider that a developer may reasonably be required to compile this information having regard *inter alia* to current knowledge and methods of assessment. (Article 5.1)

Table 2.4 summarizes the information required by Annex III. The developer is thus required to prepare an EIS that includes the information specified by the relevant Member State's interpretation of Annex III, and to submit it to the "competent authority". This EIS is then circulated to other relevant public authorities and made publicly available.

Table 2.3 Projects requiring EIA under EC Directive 85/337.

Annex I (mandatory)

1. Crude oil refineries, coal/shale gasification and liquefaction
2. Thermal power stations and other combustion installations
3. Radioactive waste storage installations
4. Cast-iron and steel melting works
5. Asbestos extraction, processing, or transformation
6. Integrated chemical installations
7. Construction of motorways, express roads, railways, airports
8. Trading ports and inland waterways
9. Installations for incinerating, treating, or disposing of toxic and dangerous wastes

Annex II (discretionary)

1. Agriculture (e.g. afforestation, poultry-rearing, land reclamation)
2. Extractive industry
3. Energy industry (e.g. storage of natural gas or fossil fuels, hydroelectric energy production)
4. Processing of metals
5. Manufacture of glass
6. Chemical industry
7. Food industry
8. Textile, leather, wood and paper industries
9. Rubber industry
10. Infrastructure projects (e.g. industrial estate developments, ski lifts, yacht marinas)
11. Other projects (e.g. holiday villages, waste-water treatment plants, knackers' yards)
12. Modification or temporary testing of Annex I projects

Table 2.4 Information required in an EIA under EC Directive 85/337.

Annex III

1. Description of the project.
2. Where appropriate, an outline of main alternatives studied and an indication of the main reasons for the final choice.
3. Aspects of the environment likely to be significantly affected by the proposed project, including population, fauna, flora, soil, water, air, climatic factors, material assets, architectural and archaeological heritage, landscape, and the interrelationship between them.
4. Likely significant effects of the proposed project on the environment.
5. Measures to prevent, reduce and where possible offset any significant adverse environmental effects.
6. Non-technical summary.
7. Any difficulties encountered in compiling the required information.

Member States shall take the measures necessary to ensure that the authorities likely to be concerned by the project . . . are given an opportunity to express their opinion . . . (Article 6.1)

Member States shall ensure that

- any request for development consent and any information gathered pursuant to [the Directive's provisions] are made available to the public,
- the public concerned is given the opportunity to express an opinion before the project is initiated.

The detailed arrangements for such information and consultation shall be determined by the Member States. (Article 6.2 and 6.3)

The competent authority must consider the information presented in the EIS, the comments of relevant authorities and the public, and the comments of other Member States (where applicable) in its consent procedure (Article 8). It must then inform the public of the decision and any conditions attached to it (Article 9).

2.6 Overview of EC systems

Given the flexible wording of the EC Directive, differences in its implementation by the Member States are bound to emerge. This section outlines the major differences between the EIA systems established by the Member States in response to the Directive, to the best of the authors' knowledge: legislation is still being rapidly implemented, and the situation will change. Appendix 2 describes the Member States' EIA systems in greater depth (EIA Centre; UNECE 1991).

- The first distinction between Member States' implementation of Directive 85/337 is in their overall implementation of the Directive. They differ on whether the regulations implementing Directive 85/337 come under the broad remit of nature conservation (e.g. France, Greece, the Netherlands, Portugal) or the planning system (e.g. Germany, Ireland, UK), or whether specific EIA regulations were enacted (e.g. Belgium, Italy). In Belgium and to an extent in Germany, the responsibility for EIA has been devolved to the regional level, whereas in most countries the national level retains broad responsibility for EIA. The countries also differ on whether Directive 85/337 has been implemented (relatively) on time (e.g. France, the Netherlands, UK), late (e.g. Belgium, Portugal) or not yet (Luxembourg).
- In all Member States, EIA is mandatory for Annex I projects. However, countries differ in their interpretation of which Annex II projects require EIA: whether only a few Annex II projects require EIA (e.g. Greece, Italy, Portugal), whether the competent authority decides if an EIA is needed on a case-by-case basis (e.g. Ireland, UK), or whether lists are compiled that specify Annex II projects requiring EIA (e.g. France, the Netherlands). This affects the number of EIAs prepared. In France, for instance, thresholds for projects requiring EIA are so

low that over a thousand EIAs are prepared annually. In Denmark, by contrast, only a few dozen EIAs are prepared annually.

- In most Member States, EIAs are carried out and paid for by the developer or consultants commissioned by the developer. However, in Flanders (Belgium) EIAs are carried out by experts approved by the authority responsible for environmental matters, and in Spain the competent authority carries out an EIA based on studies carried out by the developer.
- Scoping is carried out as a discrete and mandatory step in some countries or their regions (e.g. Wallonia in Belgium, the Netherlands), but not in others (e.g. Spain, UK). The consideration of alternatives to a proposed project is mandatory in only some countries or regions of countries (e.g. Wallonia, the Netherlands).
- The Member States vary considerably in the level of public consultation they require in the EIA process. The Directive requires an EIS to be made available after it is handed to the competent authority. However, some Member States or regions of them go well beyond this. In the Netherlands and Wallonia, the public is consulted during the scoping process. In the Netherlands and Flanders, a public hearing must be held after an EIS is handed in. In Spain, the public must be consulted before an EIS is submitted.
- In a few countries or national regions, EIA commissions have been established. In the Netherlands, the commission assists in the scoping process, reviews the adequacy of an EIS, and receives monitoring information from the competent authority. In Flanders, it reviews the qualifications of the people carrying out an EIA, determines an EIA's scope, and reviews an EIS for compliance with legal requirements. Italy also has an EIA commission.
- The decision on whether a project is to proceed is, in the simplest case, the responsibility of the competent authority (e.g. Flanders, Germany, UK). However, in some cases the minister responsible for the environment must (first) decide on whether a project is environmentally compatible (e.g. Denmark, Italy, Portugal).
- Only the Netherlands at present requires systematic monitoring of a project's actual impacts by the competent authority.

As a result of these differences, some countries (e.g. the Netherlands, Belgium) seem to have particularly effective and comprehensive EIA systems, whereas others (e.g. Italy, Luxembourg, Portugal) have considerably weaker systems. However, as mentioned earlier, this situation is changing rapidly, and may be quite different by the year 2000.

2.7 Summary

This chapter has reviewed the development of EIA worldwide, from its unexpectedly successful beginnings in the USA, through to recent developments in the EC.

In practice, EIA ranges from the production of very simple ad hoc reports to extremely bulky and complex documents, from wide-ranging to non-existent consultation with the public, from detailed quantitative predictions to broad statements about likely future trends. All of these systems, however, have the broad aim of improving decision-making by raising decision-makers' awareness of a proposed action's environmental consequences. Over the past twenty years, EIA has become an important tool in project planning, and its applications are likely to expand further (see Ch. 13). The next chapter focuses on EIA in the UK context.

References

Abracosa, R. & L. Ortolano 1987. Environmental impact assessment in the Philippines: 1977–1985. *Environmental Impact Assessment Review* **7**, 293–310.

Anderson, F. R., D. R. Mandelker, A. D. Tarlock 1984. *Environmental protection: law and policy*. Boston: Little, Brown.

Bear, D. 1990. *EIA in the USA after twenty years of NEPA*. EIA newsletter 4, EIA Centre, University of Manchester.

Cabrera, P. R. 1984. *Environmental impact assessment in the Third World*. PhD thesis, University of Reading.

Catlow, J. & C. G. Thirwall 1976. *Environmental impact analysis* (DoE Research Report 11). London: HMSO.

CEC (Commission of the European Communities) 1973. First action programme on the environment. *Official Journal* **C112** (20 December 1973).

CEC 1980. Draft directive concerning the assessment of the environmental effects of certain public and private projects. COM(80), 313 final. *Official Journal* **C169** (9 July 1980).

CEC 1982. Proposal to amend the proposal for a Council directive concerning the environmental effects of certain public and private projects. COM(82), 158 final. *Official Journal* **C110** (1 May 1982).

CEC 1985. On the assessment of the effects of certain public and private projects on the environment. *Official Journal* **L175** (5 July 1985).

CEC 1992. *Towards sustainability*. Brussels: CEC.

CEQ (Council on Environmental Quality) 1978. National Environmental Policy Act. Implementation of procedural provisions: final regulations. *Federal Register* **43**(230), 55977–56007 (29 November 1978).

CEQ 1992. *Environmental quality: 22nd annual report*. Washington DC: US Government Printing Office.

Clark, B. D., K. Chapman, R. Bisset. P. Wathern 1976. *Assessment of major industrial applications: a manual* (DoE Research Report 13). London: HMSO.

Clark, B. D., K. Chapman, R. Bisset, P. Wathern, M. Barrett 1981. *A manual for the assessment of major industrial proposals*. London: HMSO.

Clark, B. D. & R. G. H. Turnbull 1984. Proposals for environmental impact assessment procedures in the UK. In *Planning and ecology*, R. D. Roberts & T. M. Roberts (eds), 135–44. London: Chapman & Hall.

CPRE 1991. *The environmental assessment directive: five years on*. London: Council for the Protection of Rural England.

Dobry, G. 1975. *Review of the development control system: final report*. London: HMSO.

DoE 1977. Press Notice 68. London: Department of the Environment.

DoE 1978. Press Notice 488. London: Department of the Environment.

EIA Centre. *EIA Newsletter* (annual). EIA Centre, University of Manchester.

European Bank for Reconstruction and Development 1992. *Environmental procedures*. London: European Bank for Reconstruction and Development.

HMSO 1971. *Report of the Roskill Commission on the Third London Airport*. London: HMSO.

House of Lords 1981a. *Environmental assessment of projects*. Select Committee on the European Communities, 11th Report, Session 1980–81. London: HMSO.

House of Lords 1981b. *Parliamentary Debates (Hansard) Official Report, Session 1980–81*, 30 April 1981, 1311–47. London: HMSO.

Lee, N. & C. M. Wood 1984. Environmental impact assessment procedures within the European Economic Community. In *Planning and ecology*, R. D. Roberts & T. M. Roberts (eds), 128–34. London: Chapman & Hall.

Legore, S. 1984. Experience with environmental impact assessment in the USA. In *Planning and ecology*, R. D. Roberts & T. M. Roberts (eds), 103–12. London: Chapman & Hall.

Moreira, I. V. 1988. EIA in Latin America. In *Environmental impact assessment: theory and practice*, P. Wathern (ed.), 239–53. London: Unwin Hyman.

National Environmental Policy Act 1970. 42 USC 4321–4347, 1 January, as amended.

Organization for Economic Co-operation and Development 1979. *The assessment of projects with significant impacts on the environment*, C(79)116 (8 May 1979). Paris: OECD.

O'Riordan, T. & W. R. D. Sewell (eds) 1981. *Project appraisal and policy review*. Chichester, England: John Wiley.

Orloff, N. 1980. *The National Environmental Policy Act: cases and materials*. Washington DC: Bureau of National Affairs.

Overseas Development Administration 1992. *Manual of environmental appraisal: revised and updated*. London: ODA.

Petts, J. & P. Hills 1982. *Environmental assessment in the UK*. Institute of Planning Studies, University of Nottingham.

Scottish Development Department 1974. *Appraisal of the impact of oil-related development*, DP/TAN/16. Edinburgh: SDA.

State of California, Governor's Office of Planning and Research 1992. *CEQA: California Environmental Quality Act – statutes and guidelines*. Sacramento: State of California.

Stevens Committee 1976. *Planning control over mineral working*. London: HMSO.

Swaffield, S. 1981. Environmental assessment: by administration or design? *Planning Outlook* **24**(3), 102–9.

Tomlinson, P. 1986. Environmental assessment in the UK: implementation of the EEC Directive. *Town Planning Review* **57**(4), 458–86.

Turner, T. 1988. The legal eagles. *The Amicus Journal* (winter), 25–37.

UNECE (United Nations Economic Commission for Europe) 1991. *Policies and systems of environmental impact assessment*. New York: United Nations.

United Nations Environment Programme 1980. *Guidelines for assessing industrial environmental impact and environmental criteria for the siting of industry* (Industry and Environment Guidelines Series vol. 1). Paris: UNEP.

United States Agency for International Development 1978. Procedures. *Federal Register* **22**(216).

Wathern P (ed.) 1988. *Environmental impact assessment: theory and practice*. London: Unwin Hyman.

Westman, W. E. 1985. *Ecology, impact assessment and environmental planning*. New York: John Wiley.

Williams, R. H. 1988. The environmental assessment directive of the European Communities. In *The rôle of environmental impact assessment in the planning process*, M. Clark & J. Herington (eds), 74–87. London: Mansell.

Wood, C. 1981. The impact of the European Commission's directive on environmental planning in the United Kingdom. *Planning Outlook* **24**(3), 92–8.

Wood, C. 1988. The genesis and implementation of environmental impact assessment in Europe. In *The rôle of environmental impact assessment in the planning process*, M. Clark & J. Herington (eds), 88–102. London: Mansell.

World Bank 1991. *Environmental assessment sourcebook* (Technical Paper 139). Washington DC: World Bank.

Notes

1. E.g. *Ely v. Velds*, 451 F.2d 1130, 4th Cir. 1971; *Carolina Action v. Simon*, 522 F.2d 295, 4th Cir. 1975.
2. *Calvert Cliff's Coordinating Committee, Inc. v. United States Atomic Energy Commission* 449 F.2d 1109, DC Cir. 1971.
3. *Natural Resources Defense Council, Inc. v. Morton*, 458 F.2d 827, DC Cir. 1972.
4. Arkansas, California, Connecticut, Florida, Hawaii, Indiana, Maryland, Massachusetts, Minnesota, Montana, New York, North Carolina, South Dakota, Virginia, Washington and Wisconsin, plus the District of Columbia.
5. Arizona, Delaware, Georgia, Louisiana, Michigan, New Jersey, North Dakota, Oregon, Pennsylvania, Rhode Island and Utah.

CHAPTER 3
UK agency and legislative context

3.1 Introduction

This chapter discusses the legislative framework within which EIA is carried out in the UK. It begins with an outline of the key actors involved in EIA, and in the associated planning and development process. This is followed by an overview of relevant regulations and the types of projects to which they apply. The EIA procedures required by the Town and Country Planning (Assessment of Environmental Effects) Regulations 1988 are then described. These can be considered the "generic" EIA regulations that apply to most projects and provide a model for the other EIA regulations. The other main EIA regulations are then summarized. Readers are referred to Chapter 8 for a discussion of the main effects and limitations of the application of these regulations.

3.2 Key actors

An overview

Any proposed major development has an underlying configuration of interests, strategies and perspectives. But whatever the development, be it a motorway, power station, reservoir or forest, it is possible to divide the "actors" involved in the planning and development process broadly into four major groups. These are:

- the developers;
- those directly or indirectly affected by or having an interest in the development;
- government and regulatory agencies; and
- various intermediaries (consultants, advocates, advisers) with an interest in the

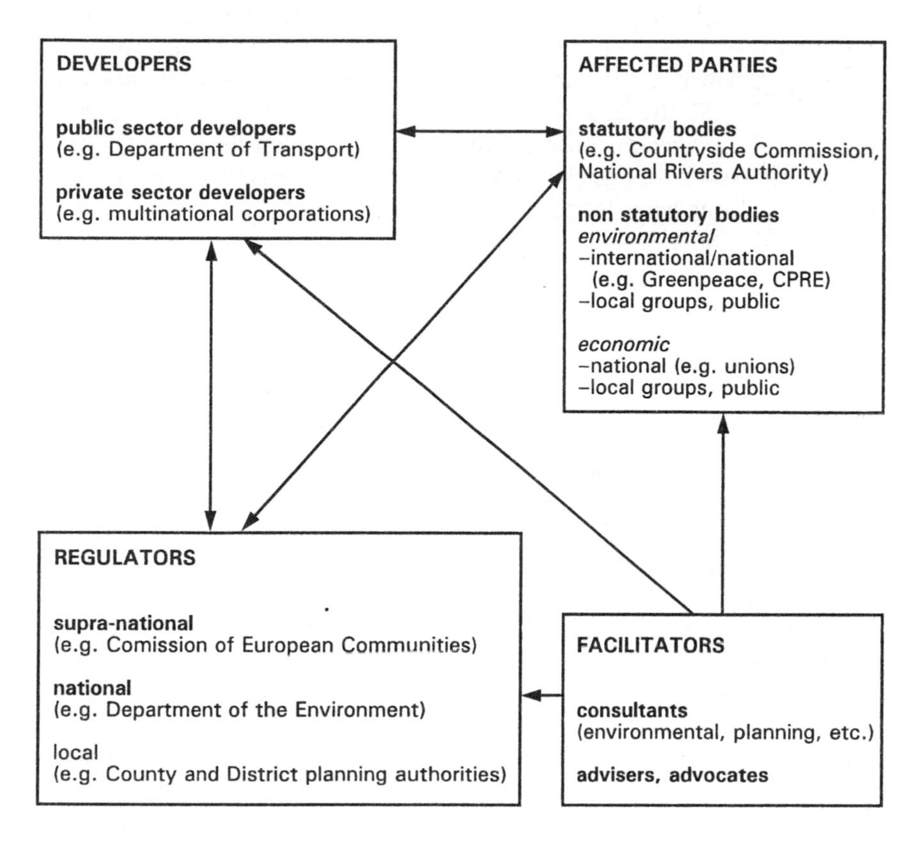

Figure 3.1 Key actors in the EIA and planning and development processes.

interaction between the developer, the affected parties and the regulators (Fig. 3.1).

An introduction to the range of "actors" involved is an important first step in understanding the UK legislative framework for EIA.

Developers

in the UK, EIA applies to both the public and private sectors, although there are notable exemptions, including Ministry of Defence developments and those of the Crown Commission. Public-sector developments are sponsored by central government departments (such as the Department of Transport), by local authorities, and by statutory undertakers, such as the National Rivers Authority (NRA). Some are also sponsored by nationalized industries (such as British Coal, British Rail, and the nuclear industry), but the rapid privatization programme of the 1980s has transferred many former nationalized industries to the private sector. Some, such as the major energy companies (National Power, PowerGen, British Gas) and the regional water authorities, have major and continuing programmes of projects, where it may be possible to develop and refine EIA procedures, learning from experience. Many other private-sector companies, often of multinational form, may also produce a stream of projects. However, for many developers, a major project may be a "one off / once in a lifetime" activity. For them, the EIA process, and the associated planning and development process, may be much less familiar, requiring quick learning and, it is to be hoped, the use of some good advice.

Affected parties

Those parties directly or indirectly affected by such developments are numerous. In Figure 3.1 they have been broadly categorized, according to their rôle or degree of power (e.g. statutory, advisory), level of operation (e.g. international, national, local), or emphasis (e.g. environmental, economic). The growth in environmental groups (such as Greenpeace, Friends of the Earth, the Council for the Protection of Rural England and the Royal Society for the Protection of Birds) is of particular note and is partly associated with the growing public interest in environmental issues. For instance, membership of the RSPB grew from 100,000 in 1970 to over 500,000 in 1990. Such groups, although often limited in resources, may have considerable "moral weight". The accommodation of their interests by a developer is often viewed as an important step in the "legitimization" of a project. As with the developers, some environmental groups, especially at the national level, may have a long-term continuing rôle. Some local amenity groups may also have a continuing rôle and an accumulation of valuable knowledge on the local environment. Others, usually at the local level, may have a short life, being associated with one particular project. In this latter category can be placed local pressure groups, which can spring up quickly to oppose a development. Such groups have sometimes been referred to as NIMBYs ("not in my back yard"), and their aims often include the maintenance of property values and existing life-styles, and the diversion of any necessary development elsewhere.

Statutory consultees are an important group in the EIA process. Such bodies have to be consulted by the planning authority before a decision on a major project requiring an EIA can be made. Statutory consultees in England and Wales include the Countryside Commission, English Nature, Her Majesty's Inspectorate of Pollution (for certain developments), and the principal local council for the area for which the project is proposed. Other consultees often involved include the NRA and the local highway authority. As noted above, non-statutory bodies, such as the RSPB, and the general public, may provide additional valuable information on environmental issues.

Regulators

Government, at various levels, will normally have a significant rôle in regulating and managing the relationship between the groups previously outlined. As discussed in Chapter 2, the Commission of the European Communities has adopted a Directive on EIA procedures (CEC 1985). The UK government has subsequently implemented the EC Directive through an array of regulations and guidance (see §3.3). The principal department involved is the Department of the Environment (DoE) through its London headquarters and regional offices. Notwithstanding government scepticism noted in Chapter 2, William Waldegrave, UK Minister of State for the Environment commented in 1987 that " . . . one of the most important tasks facing Government is to inspire a development process which takes into account not only the nature of any environmental risk but also the perceptions of the risk by the public who must suffer its consequences" (ESRC 1987).

Of particular importance in the EIA process is the local authority, and especially the relevant local planning authority (LPA). In England and Wales, this may involve district and county authorities. In Scotland, there are district and regional authorities. Such authorities act as filters through which schemes proposed by developers usually have to pass. In addition, the local planning authority often opens the door for other agencies to become involved in the development process.

Facilitators

A final group, but one of particular significance in the EIA process, is that which includes the various consultants, advocates and advisers who participate in the EIA and the planning and development processes. Such agents are often employed by developers; occasionally they may be employed by local groups, environmental groups and others to help to mount opposition to a proposal. They may also be employed by regulatory bodies to help them in their examination process.

There has been a massive growth in the number of environmental consultancies in the UK (see Fig. 3.2). The numbers have doubled since the mid-1980s, and it has been estimated that clients in 1992 were spending approximately £400 million on their services. A report by ECOTEC Research and Consultancy has projected an increase in the environmental consulting sector market to more than £1 billion by 1995 (ECOTEC 1990), although the intervening depression may possibly flatten these projections.

Major factors underpinning the consultancy growth have been the advent of the UK Environmental Protection Act (EPA) in 1990, EIA regulations, growing UK business interest in environmental management systems (e.g. BS7750), and proposed EC regulations on eco-auditing. Figure 3.3 provides a summary of the main work areas for environmental consultancies. Although the requirements of the EPA (with its "duty of care" regulations, which came into force in April 1992) and the Water Resources Act of 1991 have concentrated the minds of developers/clients on water pollution and contaminated land in particular, there is no denying the

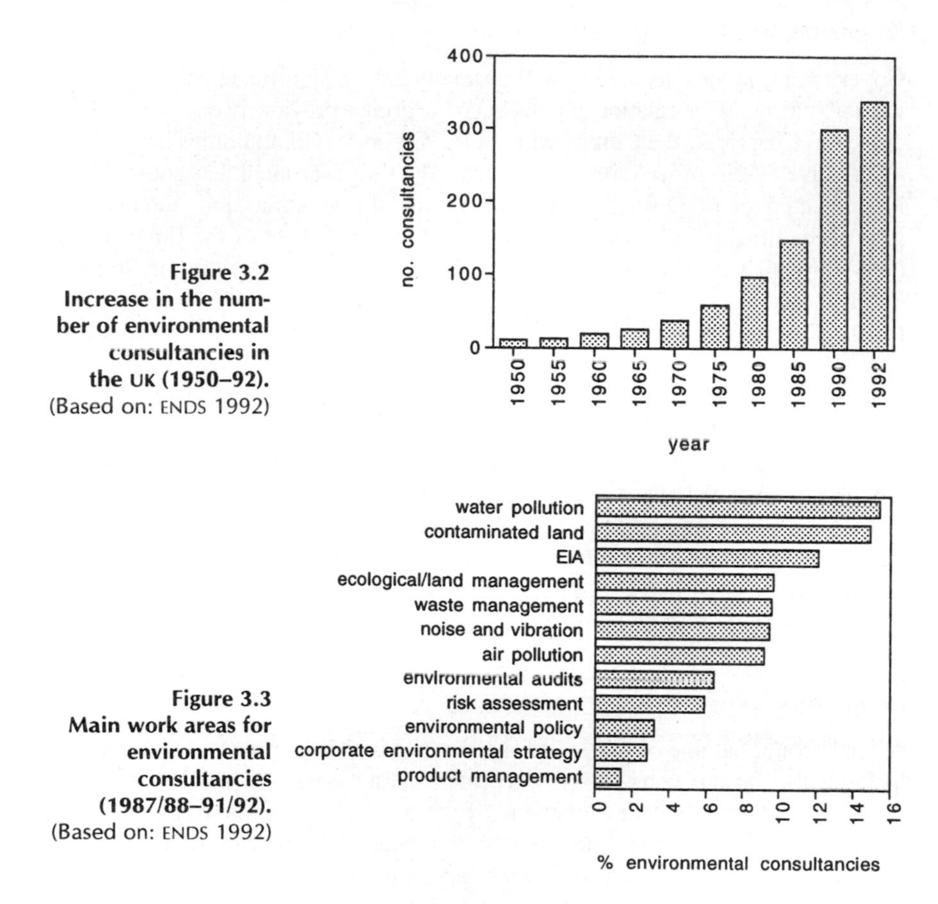

Figure 3.2 Increase in the number of environmental consultancies in the UK (1950–92). (Based on: ENDS 1992)

Figure 3.3 Main work areas for environmental consultancies (1987/88–91/92). (Based on: ENDS 1992)

significance of the EIA boom for consultants. Further characteristics of recent consultancy activity are discussed in Chapter 8.

Agency interaction

The various agencies outlined here represent a complex array of interests and aims, any combination of which may come into play with regard to a particular development. There are several dimensions to this array, and within each there may be a range of (often conflicting) views. For example, there may be conflict between local and national views, between the interests of profit maximization and of environmental conservation, between short-term and long-term perspectives, and between corporate bodies and individuals. The agencies are also linked in various ways. As noted, some links are statutory, others advisory. Some are contractual, others regulatory. The EIA regulations and guidance provide a set of procedures linking the various actors discussed, and these are now outlined.

3.3 EIA regulations: overview

In the UK, EC Directive 85/337 is implemented through twenty different secondary sets of regulations under Section 2(2) of the European Communities Act 1972: these are listed in Table 3.1. The regulations are supplemented by an array of EIA guidance from government and other bodies (see Table 3.2). In addition, the Planning and Compensation Act 1991 allows the government to require EIA for other projects that fall outside the Directive. In contrast to the US system of EIA, that of Directive 85/337 applies to both public- and private-sector developments. The developer carries out the EIA, and the resulting EIS must be handed in with the planning application.

In England and Wales, most of the developments listed in Annexes 1 and 2 of Directive 85/337 fall under the remit of the *planning system*, and are thus covered by the Town and Country Planning (Assessment of Environmental Effects) Regulations 1988 ("the T&CP Regulations"). The Secretary of State (SoS) for the Environment has the power to vary the conditions of the T&CP Regulations to exempt particular developments from EIA, direct that a particular development should come under these regulations even if it is not listed in the regulations, and arbitrate whether or not a proposed development should come under these regulations under the Town and Country Planning General Development Order 1988. Other types of projects listed in the Directive require separate legislation, since they are not governed by the planning system.

Of the various *transport* projects, local highway developments and airports are dealt with under the T&CP Regulations by the local planning (highways) authority, but motorways and trunk roads proposed and regulated by the Department of Transport (DoT) fall under the Highways (AEE) regulations 1988. Applications for harbours made under the harbour revision order are regulated by the DoT under the Harbour Works (AEE) Regulations 1988. Applications for harbour works in England, Wales or Scotland that are below the "low-water mark of medium tides" and are not covered by the previously mentioned regulations are covered by the Harbour Works (AEE) (No. 2) Regulations 1989. New railways are generally promulgated through the private bill procedure.

Energy projects producing less than 50 MW are regulated by the local authority under the T&CP Regulations. Those of 50 MW or over, most electricity power lines, and pipelines (in Scotland as well as England and Wales) were originally controlled by the Department of Energy under the Electricity and Pipeline Works (AEE) Regulations 1989. As a result of the privatization of the electricity industry in 1989, new regulations came into force in March 1990. With the disbanding of the Department of Energy in 1992, energy projects are now under the jurisdiction of the Department of Trade and Industry.

New *land drainage* works, including flood defence and coastal defence works, require planning permission and are thus covered by the T&CP Regulations. Improvements to drainage works carried out by the NRA and other drainage bodies require EIA through the Land Drainage Improvement Works (AEE) regulations 1988, which are regulated by the Ministry of Agriculture, Fisheries and Food (MAFF).

Table 3.1 UK EIA regulations and dates of implementation.

- Town and Country Planning (Assessment of Environmental Effects) Regulations 1988 (ESI 1199) – 15 July 1988
- Town and Country Planning (Assessment of Environmental Effects) (Amendment) Regulations 1990 (ESI 367) – 31 March 1990
- Town and Country Planning General Development (Amendment) Order 1988. Revoked by ESI 1813. The provisions of SI 1272 now form article 14(2) of the 1988 General Development Order – 5 December 1988
- Electricity and Pipe-line Works (Assessment of Environmental Effects) Regulations 1989 (ESI 167). Revoked by Electricity and Pipeline Works (Assessment of Environmental Effects) Regulations 1990 (ESI 442)
- Land Drainage Improvement Works (Assessment of Environmental Effects) Regulations 1988 (ESI 1217) – 16 July 1988
- Highways (Assessment of Environmental Effects) Regulations 1988 (ESI 1241) – 21 July 1988
- Harbour Works (Assessment of Environmental Effects) Regulations 1989 (ESI 1336) – 3 August 1988
- Harbour Works (Assessment of Environmental Effects) (No. 2) Regulations 1989 (ESI 424) – 16 March 1989
- Environmental Assessment (Afforestation) Regulations 1988 (ESI 1207) – 15 July 1988
- Environmental Assessment (Salmon Farming in Marine Waters) Regulations 1988 (ESI 1218) – 15 July 1988
- Environmental Assessment (Scotland) Regulations 1988 (ESI 1221) – 15 July 1988
- Town and Country (General Development) (Scotland) Amendment Order 1988 (ESI 977)
- Town and Country (General Development) (Scotland) Amendment No. 2 Order 1988 (ESI 1249) – August 1988
- The Harbour Works (Assessment of Environmental Effects) Regulations 1992 (ESI 1421)
- The Planning (Assessment of Environmental Effects) Regulations (Northern Ireland) 1989 (SR 20)
- Roads (Assessment of Environmental Effects) Regulations (Northern Ireland) 1988 (SR 344)
- The Environmental Assessment (Afforestation) Regulations (Northern Ireland) 1989 (SR 226)
- The Harbour Works (Assessment of Environmental Effects) Regulations (Northern Ireland) 1990 (SR 181)
- Drainage (Environmental Assessment) Regulations (Northern Ireland) 1991 (SR 376)1988
- Transport and Works Act 1992
- Private Acts of Parliament

Table 3.2 UK Government EIA guidance.

- DOE Circular 15/88 (Welsh Office 23/88) "Environmental Assessment" – 12 July 1988
- SDD Circular 13/88 "Environmental Assessment: Implementation of EC Directive: The Environmental Assessment (Scotland) Regulations 1988" – 12 July 1988
- Scottish Office Circular 26/91 "Environmental Assessment and Private Legislation Procedures"
- Forestry Commission booklet "Environmental Assessment of Afforestation Projects" – 4 August 1988
- Crown Estate Office note "Environmental Assessment of Marine Salmon Farms" – 15 July 1988
- DOE Circular 24/88 (Welsh Office 24/88) "Environmental Assessment of Projects in Simplified Planning Zones and Enterprise Zones" – 25 November 1988
- SDD Circular 26/88 "Environmental Assessment of Projects in Simplified Planning Zones and Enterprise Zones" – 25 November 1988
- DOE memorandum on new towns "Environmental Assessment" – 30 March 1989
- DOT Department Standard HD 18/88 "Environmental Assessment under EC Directive 85/337" – July 1989
- DOE advisory booklet "Environmental Assessment – A Guide to the Procedures" – 6 November 1989
- DOE free leaflet "Environmental Assessment" – October 1989
- Welsh office free leaflet "Environmental Assessment / Asesu'r Amgylchedd"
- Scottish Office free leaflet "Environmental Assessment – a Guide" – June 1990

Forestry projects for which grants are given by the Forestry Authority (previously the Forestry Commission) require EIA under the Environmental Assessment (Afforestation) Regulations 1988. It is possible that in the future EIA will also be required for forestry projects not requiring grants (Anon. 1992: 20).

Salmon farms within 2km of the coast of England, Wales or Scotland do not require planning permission but require a lease from the Crown Estates Commission. For these developments, EIA is required under the Environmental Assessment (Salmon Farming in Marine Waters) Regulations 1988.

Most other developments in *Scotland* are covered by the Environmental Assessment (Scotland) Regulations 1988, including developments related to town and country planning, electricity, roads and bridges, development by planning authorities and land drainage. The British regulations apply to harbours, pipelines and afforestation projects. *Northern Ireland* has separate legislation in parallel with that of England and Wales.

Where a development project is approved as a *private bill*, no formal EIA is necessary, since Directive 85/337 notes that in such a case "the objectives of this Directive, including that of supplying information, are achieved by the legislative process". UK government guidance notes that "where, but for this provision, [EIA] would have been required for a project, the promoter of a [private] Bill should provide an [EIS] . . . The Government will ensure that it will provide such statements in connection with hybrid Bills" (DoE 1989). Some EISs have been prepared for private bills, including those for the Port of Hull, King's Cross Railway, Severn Bridge, Jubilee Line Extension, River Usk Barrage and Manchester Metrolink.

Table 3.3 summarizes the legislation applicable in different areas of the UK to different development types. As will be discussed in Chapter 8, about two-thirds of all the EIAs prepared in the UK fall under the T&CP Regulations, about 10% fall under each of the EA (scotland) Regulations 1988 and the Highways (AEE) Regulations 1988, and almost all of the rest involve land drainage, electricity and pipeline works, afforestation projects in England and Wales, and planning-related developments in Northern Ireland.

The main area of weakness in the UK's implementation of Directive 85/337 is *agriculture*. Although forestry, land drainage and salmon farming in marine waters are all covered by separate EIA regulations, the other categories of agriculture – the cultivation of semi-natural or uncultivated land, land reclamation from the sea, and the restructuring of rural land holdings – have no consent procedure or EIA requirement. The CPRE (1991) argues:

> The latter two classes of project are not generally seen to be relevant to the UK, the re-structuring of rural land holdings being of particular importance to France . . . While there is little land reclamation from the sea at

Table 3.3 EIA regulations and relevant regulators in the UK.

Type of project	England/ Wales	Scotland	Northern Ireland
Planning	ESI 1199 ESI 367 DoE/WO	ESI 1221 SDD	SR 20 DoE(NI)
Trunk roads and motorways	ESI 1241 DoT	ESI 1221 SDD	SR 344 DoE(NI)
Power stations	ESI 442 DTI	ESI 1221 DTI	
Pipelines	ESI 442 DTI	ESI 442	
Afforestation	ESI 1207 FA	ESI 1207 FA	SR 226 DoE(NI)
Land drainage improvements	ESI 1217 MAFF	ESI 1221 SDD	SR 376
Ports and harbours	ESI 1336 ESI 424 MAFF/DoT	ESI 424 SDD	SR 181 DoE(NI)
Marine salmon farming	ESI 1218 CEC	ESI 1218 CEC	
Other	ESI 1272	ESI 977 ESI 1249	

> present in the UK we should, nevertheless, be looking at ways of making provisions for applying [EIA] to such projects should they assume greater importance in future as a result of climate change and rising sea levels. The cultivation for intensive agriculture of uncultivated or semi-natural land is, however, a problem in the UK, particularly with respect to moorlands, heathlands, hay meadows and wetlands . . . In the absence of any notification or consent procedure, the UK government has found it impossible to implement the Directive for such activities.

It is likely that these types of projects will also require EIA in the future (Anon. 1992: 20). The T&CP Regulations were amended in 1993 to include trout farming, water treatment plants, wind generators, motorway and other service areas, coastal protection works, golf courses, and privately financed toll roads (*Journal of Planning and Environment Law*, February 93 Bulletin).

Further information on, and critiques of, the various EIA regulations can be found in Fortlage (1990) and CPRE (1991).

3.4 The Town and Country Planning (Assessment of Environmental Effects) Regulations 1988 (ESI 1199)

The T&CP Regulations implement Directive 85/337 for those projects that require planning permission in England and Wales. They are the central form in which Directive 85/337 is implemented in the UK; the other UK EIA regulations were established to cover projects that are not covered by the T&CP regulations. As a result, the T&CP Regulations are the main focus of discussions on EIA procedures and effectiveness. This section presents the procedures of the T&CP Regulations. Figure 3.4 summarizes these procedures; the letters in the figure correspond to the letters in bold preceding the explanatory paragraphs below. §3.5 considers other main EIA regulations as variances of the T&CP Regulations.

The T&CP Regulations were issued on 15 July 1988, twelve days after Directive 85/337 was to have been implemented. Guidance on the T&CP Regulations, aimed primarily at local planning authorities, is given in DoE Circular 15/88 (Welsh Office Circular 23/88). A guidebook entitled *Environmental assessment: a guide to the procedures*, aimed primarily at developers and their advisers, was released in November 1989. Only the regulations are mandatory: the guidance interprets and advises, but cannot be enforced. However, the reader is strongly advised to read the *Guide*. Further DoE guidance on good practice in carrying out and reviewing EIAs is expected in 1994 and 1995.

Which projects require EIA?

The T&CP Regulations require EIAs to be carried out for two lists of projects, given in Schedules 1 and 2. These broadly[1] correspond to Annexes I and II of

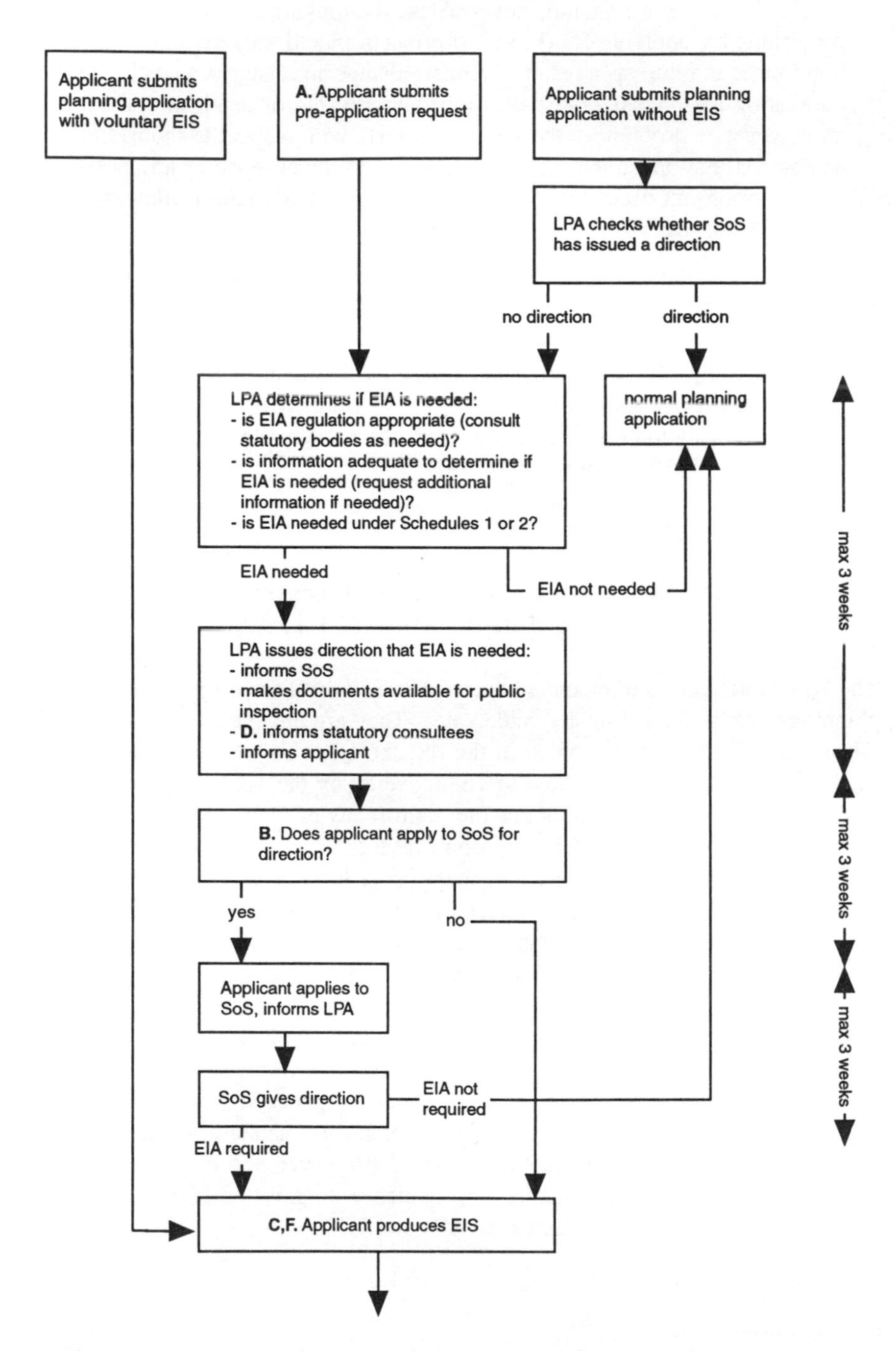

Figure 3.4 Summary of T&CP Regulations EIA procedure. (Based on : DoE 1989)

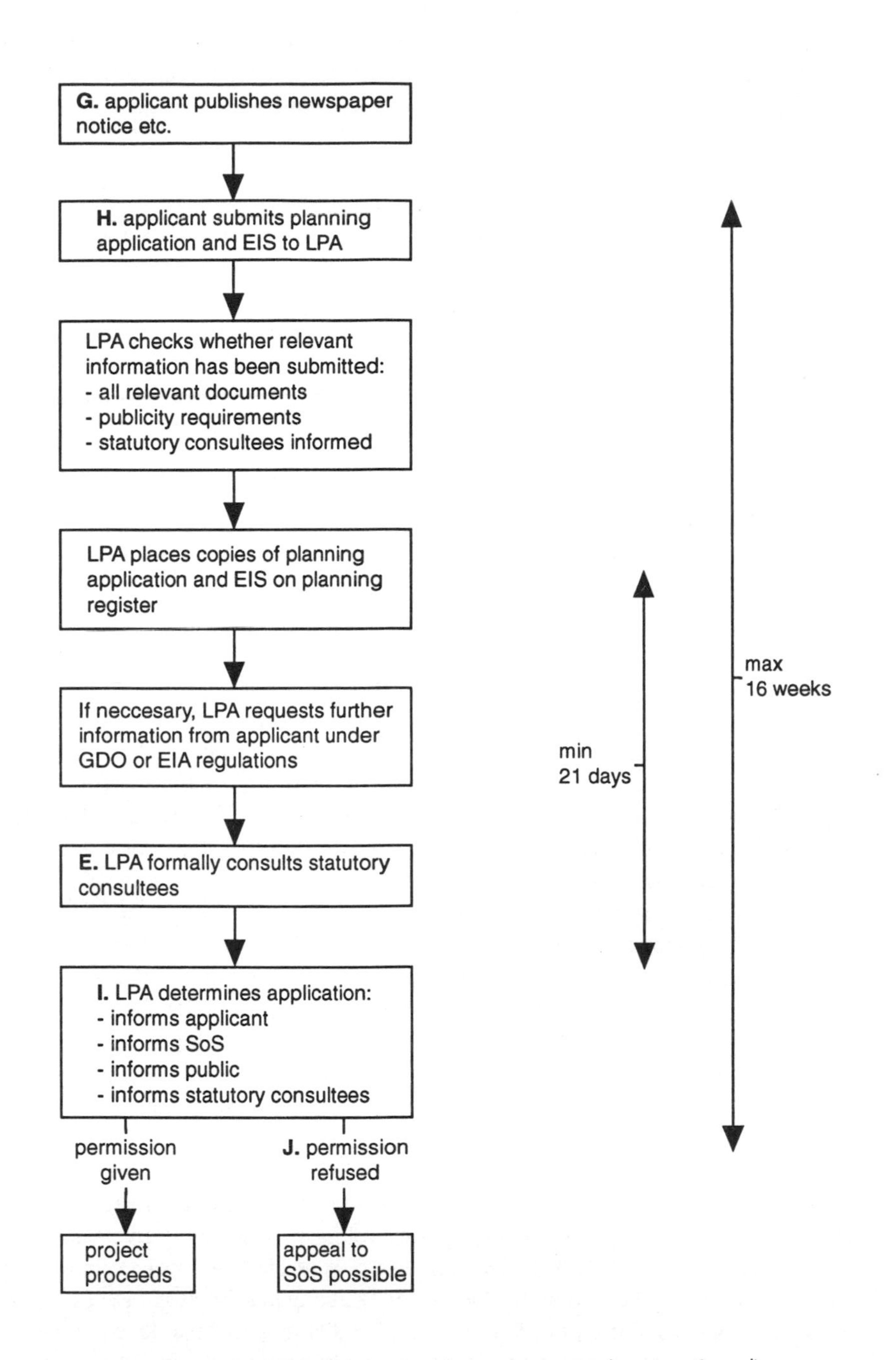

Figure 3.4 Summary of T&CP Regulations EIA procedure (continued).

Directive 85/337 (see Appendix 1), excluding those projects that do not require planning permission. Table 3.4 lists Schedule 1 and Schedule 2 projects. For Schedule 1 projects, EIA is required in every case. For Schedule 2 projects, EIA is required if the project is deemed "likely to give rise to significant environmental effects". The "significance" of a project's environmental effects is determined on the basis of three criteria:

i. whether the project is of more than local importance, principally in terms of physical scale;
ii. whether the project is intended for a particularly sensitive location, for example, a national park or a [Site of Special Scientific Interest] . . .
iii. whether the project is thought likely to give rise to particularly complex or adverse effects, for example, in terms of the discharge of pollutants. (DoE 1989)

The 1989 Guide to the Procedures includes indicative criteria and thresholds for a range of Schedule 2 projects that "are intended to indicate the types of cases in which, in the Secretary of State's view, environmental assessment may be required under the regulations". For instance, pig-rearing installations for more than 400 sows, industrial estate developments of more than 20 ha, and new roads of over 10 km which are not located in a designated area "may" require EIA according to the Circular.

A. A developer may decide that a project requires EIA under the T&CP Regulations, or may want to carry out an EIA even if it is not required. If the developer is uncertain, s/he can ask the LPA to determine if an EIA is needed. To do this they must provide the LPA with a plan showing the development site, a description of the proposed development, and an indication of its possible environmental impacts. The LPA must then give a determination within three weeks. The LPA can ask for more information from the developer, but this does not extend the three-week decision-making period.

If the LPA decides that no EIA is needed, the application is processed as a normal planning application. If instead the LPA decides that an EIA is needed, it must explain why, and make both the developer's information and the decision publicly available. If the LPA receives a planning application without an EIS when it feels that an EIS is needed, the LPA must notify the developer within three weeks, explaining why the EIS is needed. The developer then has three weeks in which to notify the LPA of the intention to either prepare an EIS or appeal to the SoS; if the developer does not do so, the planning application is refused.

B. If the LPA decides that an EIA is needed but the developer disagrees, the developer can refer the matter to the SoS[2] for a ruling. The SoS must give a decision within three weeks. If the SoS decides that an EIA is needed, he must explain why; this explanation is published in the *Journal of Planning and Environment Law*. No explanation is needed if no EIA is required. The SoS may make a decision where the developer has not requested an opinion, and he may rule, usually as a result

Table 3.4 Projects requiring EIA under the T&CP Regulations.

Schedule 1

The following types of development ("Schedule 1 projects") require environmental assessment in every case:

(1) The carrying out of building or other operations, or the change of use of buildings or other land (where a material change) to provide any of the following:

1. A crude-oil refinery (excluding an undertaking manufacturing only lubricants from crude oil) or an installation for the gasification and liquefaction of 500 tonnes or more of coal or bituminous shale per day.
2. A thermal power station or other combustion installation with a heat output of 300 MW or more, other than a nuclear power station or other nuclear reactor.
3. An installation designed solely for the permanent storage or final disposal of radioactive waste.
4. An integrated works for the initial melting of cast-iron and steel.
5. An installation for the extraction of asbestos or for the processing and transformation of asbestos or products containing asbestos:
   a) where the installation produces asbestos–cement products, with an annual production of more than 20,000 tonnes of finished products; or
   b) where the installation produces friction material, with an annual production of more than 50 tonnes of finished products; or
   c) in other cases, where the installation will utilize more than 200 tonnes of asbestos per year.
6. An integrated chemical installation, that is to say, an industrial installation or group of installations where two or more linked chemical or physical processes are employed for the manufacture of olefins from petroleum products, or of sulphuric acid, nitric acid, hydrofluoric acid, chlorine or fluorine.
7. A special road; a line for long-distance railway traffic; or an aerodrome with a basic runway length of 2100 m or more.
8. A trading port, an inland waterway which permits the passage of vessels of over 1,350 tonnes or a port for inland waterway traffic capable of handling such vessels.
9. A waste-disposal installation for the incineration or chemical treatment of special waste.

(2) The carrying out of operations whereby land is filled with special waste, or the change of use of land (where a material change) to use for the deposit of such waste.

Schedule 2

The following types of development ("Schedule 2 projects") require environmental assessment if they are likely to have significant effects of the environment by virtue of factors such as their nature, size or location:

1. Agriculture
   (a) water-management for agriculture
   (b) poultry-rearing
   (c) pig-rearing
   (d) a salmon hatchery
   (e) an installation for the rearing of salmon
   (f) the reclamation of land from the sea
2. Extractive industry
   (a) extracting peat
   (b) deep drilling, including in particular:
   - geothermal drilling

(continued pp. 62–3)

- drilling for the storage of nuclear waste material
- drilling for water supplies

but excluding drilling to investigate the stability of the soil

(c) extracting minerals (other than metalliferous and energy-producing minerals) such as marble, sand, gravel, shale, salt, phosphates and potash
(d) extracting coal or lignite by underground or open-cast mining
(e) extracting petroleum
(f) extracting natural gas
(g) extracting ores
(h) extracting bituminous shale
(i) extracting minerals (other than metalliferous and energy-producing minerals) by open-cast mining
(j) a surface industrial installation for the extraction of coal, petroleum, natural gas or ores or bituminous shale
(k) a coke oven (dry distillation of coal)
(l) an installation for the manufacture of cement

3. Energy industry
(a) a non-nuclear thermal power station, not being an installation falling within Schedule 1, or an installation for the production of electricity, steam and hot water
(b) an industrial installation for carrying gas, steam or hot water; or the transmission of electrical energy by overhead cables
(c) the surface storage of natural gas
(d) the underground storage of combustible gases
(e) the surface storage of fossil fuels
(f) the industrial briquetting of coal or lignite
(g) an installation for the production or enrichment of nuclear fuels
(h) an installation for the reprocessing of irradiated nuclear fuels
(i) an installation for the collection or processing of radioactive waste, not being an installation falling within Schedule 1
(j) an installation for hydroelectric energy production

4. Processing of metals
(a) an ironworks or steelworks including a foundry, forge, drawing plant or rolling mill (not being a works falling within Schedule 1)
(b) an installation for the production (including smelting, refining, drawing and rolling) of non-ferrous metals, other than precious metals
(c) the pressing, drawing or stamping of large castings
(d) the surface treatment and coating of metals
(e) boiler-making or manufacturing reservoirs, tanks and other sheet-metal containers
(f) manufacturing or assembling motor vehicles or manufacturing motor-vehicle engines
(g) a shipyard
(h) an installation for the construction or repair of aircraft
(i) the manufacture of railway equipment
(j) swaging by explosives
(k) an installation for the roasting or sintering of metallic ores

5. Glass making
the manufacture of glass

6. Chemical industry
(a) the treatment of intermediate products and production of chemicals, other than development falling within Schedule 1
(b) the production of pesticides or pharmaceutical products, paints or varnishes, elastomers or peroxides

(c) the storage of petroleum or petrochemical or chemical products

7. Food industry
   (a) the manufacture of vegetable or animal oils or fats
   (b) the packing or canning of animal or vegetable products
   (c) the manufacture of dairy products
   (d) brewing or malting
   (e) confectionery or syrup manufacture
   (f) an installation for the slaughter of animals
   (g) an industrial starch manufacturing installation
   (h) a fish-meal or fish-oil factory
   (i) a sugar factory

8. Textile, leather, wood and paper industries
   (a) a wool scouring, degreasing and bleaching factory
   (b) the manufacture of fibre board, particle board or plywood
   (c) the manufacture of pulp, paper or board
   (d) a fibre-dyeing factory
   (e) a cellulose-processing and production installation
   (f) a tannery or a leather dressing factory

9. Rubber industry
   the manufacture and treatment of elastomer-based products

10. Infrastructure projects
   (a) an industrial estate development project
   (b) an urban development project
   (c) a ski-lift or cable-car
   (d) the construction of a road, or a harbour, including a fishing harbour, or an aerodrome, not being development falling within Schedule 1
   (e) canalization or flood-relief works
   (f) a dam or other installation designed to hold water or store it on a long-term basis
   (g) a tramway, elevated or underground railway, suspended line or similar line, exclusively or mainly for passenger transport
   (h) an oil or gas pipeline installation
   (i) a long-distance aqueduct
   (j) a yacht marina

11. Other projects
   (a) a holiday village or hotel complex
   (b) a permanent racing or test track for cars or motor cycles
   (c) an installation for the disposal of controlled waste or waste from mines and quarries, not being an installation falling within Schedule 1
   (d) a waste water treatment plant
   (e) a site for depositing sludge
   (f) the storage of scrap iron
   (g) a test bench for engines, turbines or reactors
   (h) the manufacture of artificial mineral fibres
   (i) the manufacture, packing, loading or placing in cartridges of gunpowder or other explosives
   (j) a knackers' yard

12. The modification of a development which has been carried out, where that development is within a description mentioned in Schedule 1.

13. Development within a description mentioned in Schedule 1, where it is exclusively or mainly for the development and testing of new methods or products and will not be permitted for longer than one year.

of information made available by other bodies, that an EIA is needed where the LPA has decided that it is not needed.

Contents of the EIA

Schedule 3 of the T&CP Regulations, which is shown at Table 3.5, lists the information that should be included in an EIA. Schedule 3 interprets the requirements of Directive 85/337's Annex III according to the criteria set out in Article 5 of the Directive, namely:

> Member States shall adopt the necessary measures to ensure that the developer supplies in an appropriate form the information specified in Annex III inasmuch as:
> (a) The Member States consider that the information is relevant to a given state of the consent procedure and to the specific characteristics of a particular project or type of project and of the environmental features likely to be affected;
> (b) The Member States consider that a developer may reasonably be required to compile this information having regard *inter alia* to current knowledge and methods of assessment.

In Schedule 3, the information required in Annex III has been interpreted to fall into two categories: "specified information" which must be included in an EIA, and "further information" which *may* be included "by way of explanation or amplification of any specified information". This distinction is important: as will be seen in Chapter 8, the EISs prepared to date have generally not included "further information", despite the fact that this includes such important matters as which alternatives were considered and the expected wastes or emissions from the development. In addition, in Appendix 4 of the 1989 Guide to the Procedures, the DoE has given a longer checklist of matters which may be considered for inclusion in an EIA: this list is for the purposes of guidance only, but is useful for ensuring that all possible significant effects of the development are considered.

C. There is no mandatory requirement in the UK for a formal "scoping" stage at which the LPA, the developer and other interested parties agree on what will be included in the EIA. Indeed, there is no requirement for any kind of consultation between the developer and other bodies before the submission of the formal EIS and planning application. However, the DoE guidance stresses the benefits of early consultation and early agreement on the scope of the EIA. It also notes that the preparation of the EIS is the responsibility of the applicant, although the LPA may put forward its views about what it should include.

Statutory and other consultees

Under the T&CP Regulations, a number of *statutory consultees* are involved in the EIA process, as noted in §3.2. These bodies are involved at two stages of an EIA.

Table 3.5 Content of EIS required by the T&CP Regulations.

The following are the statutory provisions with respect to the content of environmental statements, as set out in Schedule 3 to the Town and Country Planning (Assessment of Environmental Effects) Regulations 1988.

1. An environmental statement comprises a document or series of documents providing for the purpose of assessing the likely impact upon the environment of the development proposed to be carried out, the information specified in paragraph 2 (referred to in this Schedule as "the specified information").
2. The specified information is:
   (a) a description of the development proposed, comprising information about the site and the design and size or scale of the development;
   (b) the data necessary to identify and assess the main effects which that development is likely to have on the environment;
   (c) a description of the likely significant effects, direct and indirect, on the environment of the development, explained by reference to its possible impact on: human beings, soil, fauna, flora, water, air, climate, the landscape, the interaction between any of the foregoing, material assets, and the cultural heritage;
   (d) where significant adverse effects are identified with respect to any of the foregoing, a description of the measures envisaged in order to avoid, reduce or remedy those effects; and
   (e) a summary in non-technical language of the information specified above.
3. An environmental statement may include, by way of explanation or amplification of any specified information, further information on any of the following matters:
   (a) the physical characteristics of the proposed development, and the land-use requirements during the construction and operational phases;
   (b) the main characteristics of the production processes proposed, including the nature and quantity of the materials to be used;
   (c) the estimated type and quantity of expected residues and emissions (including pollutants of water, air or soil, noise, vibration, light, heat and radiation) resulting from the proposed development when in operation;
   (d) (in outline) the main alternatives (if any) studied by the applicant, appellant or authority and an indication of the main reasons for choosing the development proposed, taking into account the environmental effects;
   (e) the likely significant direct and indirect effects on the environment of the development proposed which may result from:
   - the use of natural resources;
   - the emission of pollutants, the creation of nuisances, and the elimination of waste;

   (f) the forecasting methods used to assess any effects on the environment about which information is given under subparagraph (e); and
   (g) any difficulties, such as technical deficiencies or lack of know-how, encountered in compiling any specified information.

In paragraph (e), "effects" includes secondary, cumulative, short-, medium- and long-term, permanent, temporary, positive and negative effects.

4. Where further information is included in an environmental statement pursuant to paragraph 3, a non-technical summary of that information shall also be provided.

D. First, when a LPA determines that an EIA is required, it must inform the statutory consultees of this. The consultees in turn must make available to the developer, if so requested and at a reasonable charge, any relevant environmental information in their possession. This does not include any confidential information or information that the consultees do not already have in their possession.

E. Secondly, once the EIS has been submitted, the LPA or developer must send a free copy to each of the statutory consultees. The consultees may make representations on the EIS to the LPA for at least two weeks after they receive the EIS. The LPA must take account of these representations when deciding whether to grant planning permission. The developer may also contact *other consultees* and the *general public* while preparing the EIS. The DoE guidance explains that these bodies may have particular expertise in the subject or may highlight key environmental issues that could affect the project. The developer is under no obligation to contact any of these groups, but again the DoE guidance stresses the benefits of early and thorough consultation.

Carrying out the EIA, preparing the EIS

F. The DoE gives no formal guidance about what techniques and methodologies should be used in EIA, noting only that these techniques will vary depending on the proposed development, the receiving environment, and information available.

Submission of the EIS and planning application, public consultation

G. When the EIS has been completed, the developer must publish a notice in a local newspaper and post notices at the site. These notices must fulfil the requirements of §26 of the Town and Country Planning Act 1971, and also state that a copy of the EIS is available for public inspection, give a local address where copies of the EIS may be obtained, and state the cost of the EIS if any. The public can make written representations to the LPA for at least 20 days after the publication of the notice, but within 21 days of the LPA's receipt of the planning application.

H. After the EIS has been publicly available for at least 21 days, the developer submits to the LPA the planning application, copies[3] of the EIS, and certification that the requisite public notices have been published and posted. The LPA must then send copies of the EIS to the statutory consultees, inviting written comments within a specified time (at least two weeks from receipt of the EIS), forward another copy to the SoS, and place the EIS on the planning register. It must also determine if any additional information on the project is needed before a decision can be made, and obtain such information from the developer. The clock does not stop in this case: a decision must still be taken within the appropriate time.

Planning decision

I. Before making a decision about the planning application, the LPA must collect written representations from the public within three weeks of receipt of the planning application, and from the statutory consultees at least two weeks from their receipt of the EIS. It must wait at least three weeks after receipt of the planning application to make a decision. In contrast to normal planning applications, which must be decided within eight weeks, those accompanied by an EIS must be decided within 16 weeks. If the LPA has not made a decision after 16 weeks, the applicant can appeal to the SOS for a determination. The LPA cannot consider a planning application to be invalid because the accompanying EIS is felt to be inadequate: it can only ask for further information within the 16-week period.

In making its decision, the LPA must consider the EIS and any comments from the public and statutory consultees, as well as other material considerations. The environmental information is only part of the information that the LPA considers, alongside other material considerations. The decision is essentially still a political one, but with the assurance that the project's environmental implications are understood. The LPA may grant or refuse permission, with or without conditions.

J. If a LPA refuses planning permission, the developer may appeal to the SOS, as for a normal planning application. The SOS may request further information before making a determination.

3.5 Other EIA regulations

This section summarizes the procedures of the other EIA regulations under which a relatively large number of EISs have been prepared to date. The regulations are discussed in descending order of frequency of application to date (see Fig. 8.1):

- Environmental Assessment (Scotland) Regulations 1988
- Highways (AEE) Regulations 1988
- Land Drainage Improvement Works (AEE) Regulations 1988
- Electricity and Pipe-line Works (AEE) Regulations 1990, and
- Environmental Assessment (Afforestation) Regulations 1988.

Environmental Assessment (Scotland) Regulations 1988 (ESI 1221)

The EA (Scotland) Regulations apply to projects covered by the T&CP (Scotland) Act 1972 c. 52, and the T&CP (Development by Planning Authorities) (Scotland) Regulations 1981. They have five major sections, which apply to different types of development. Part II, Planning, resembles the English T&CP regulations, but allows four weeks for decisions by the LPA or SOS instead of three, and allows statutory consultees formally to withdraw from the consultation process. Part III, Electricity Applications, resembles the English electricity regulations (see below). Part IV, New Towns, allows development corporations to act as planning authorities for EIA purposes. Part V, Drainage Works, requires EIA for drainage works that LPAs or statutory bodies consider to be potentially environmentally harmful. Part VI, Trunk Road Projects, requires the SOS to decide whether or not a road proposal is subject to Directive 85/337: if so, the SOS must prepare an EIS, make it available for public consultation, and allow time to receive representations before coming to a decision. The EA (Scotland) Regulations have Schedules very similar to those of the English T&CP Regulations, except that nuclear power stations are included in Schedule 1.

Highways (Assessment of Environmental Effects) Regulations 1988 (ESI 1241)

The Highways (AEE) Regulations apply to motorways and trunk roads proposed by the DoT. The regulations amend the Highways Act 1980 by inserting a new Section 105A that requires the SOS for Transport to publish an EIS for the proposed route when draft orders for certain new highways, or major improvements to existing highways, are published. The SOS determines whether the proposed project comes under Annex I or Annex II of Directive 85/337, and whether an EIA is needed. Motorways and express roads (special roads under the Highways Act 1980) fall within Annex I. DoT Standard HD 18/88 (DoT 1989) gives criteria for other roads that are likely to require EIA under Annex II, including all new trunk roads over 10km; new trunk roads over 1km in length which pass through or within 100m of a national park, Site of Special Scientific Interest (SSSI), conservation area or nature reserve, or through an urban area where 1,500 or more dwellings lie within 100m of the proposed road; and road improvements that are likely to have a significant environmental effect.

The regulations require an EIS to contain:

- i. a description of the published scheme and its site,
- ii. a description of measures proposed to mitigate adverse environmental effects,
- iii. sufficient data to identify and assess the main effects that the scheme is likely to have on the environment, and
- iv. a non-technical summary.

> Where options have been presented for public consultation and have environmental effects significantly different from those of the published scheme, the following additional components shall be included in the [EIS]:
>
> - i. a summary description of the main alternatives presented at the Public Consultation,
> - ii. the reasons for the choice of the Published Scheme.

The regulations further specify the requirements for each of the above points.

They note that data on environmental effects (iii. above) "will be provided by the appraisal framework specified in the Manual of Environmental Appraisal": this framework is discussed further in §10.2. They also note that " . . . individual highway schemes do not have a significant effect on climatic factors and, in most cases, are unlikely to have significant effects on soil or water".

The effectiveness of EIA under the Highways (AEE) regulations has been criticized, e.g. by the Standing Advisory Committee on Trunk Road Assessment (SACTRA 1992) and the CPRE (1991), because of the limited requirements in terms of EIS contents, concern about the adequacy of the Manual of Environmental Appraisal (DoT 1983), the system's inability to consider different modes of transport as viable alternatives in EIA, and the DoT's conflicting rôles as project proponent, EIA producer, and competent authority. These points are discussed further in Chapters 8 and 10.

Land Drainage Improvement Works (Assessment of Environmental Effects) Regulations 1988 (ESI 1217)

The Land Drainage Improvement Works (AEE) Regulations apply to almost all watercourses in England and Wales except public health sewers. If a drainage body (including a local authority acting as a drainage body) determines that its proposed improvement actions are likely to have a significant environmental effect, it must publish a description of the proposed actions in two local newspapers and indicate whether it intends to prepare an EIS. If it does not intend to prepare an EIS, the public can make representations concerning possible environmental impacts of the proposal within 28 days; if no representations are made, the drainage body can proceed without an EIS. If representations are made but the drainage body still wants to proceed without an EIS, MAFF gives a decision at ministerial level on the issue.

The contents required of the EIS under these regulations are virtually identical to those under the T&CP Regulations. When the EIS is complete, the drainage body must publish a notice in two local newspapers, send copies to English Nature, the Countryside Commission and any other relevant bodies, and make copies of the EIS available at a reasonable charge. Representations must be made within 28 days and are considered by the drainage body in making its decision. If all objections are then withdrawn, the works can proceed, but otherwise the minister gives a decision. Overall, these regulations are considerably weaker than the T&CP Regulations, because of their weighting in favour of consent unless objections are raised, and their minimal requirements for consultation with environmental organizations.

Electricity and Pipe-line Works (Assessment of Environmental Effects) Regulations 1990 (SI 442)

The Electricity and Pipe-line Works (AEE) Regulations require an EIA to be carried out for the construction or extension of all nuclear power stations and of other generating stations of 300MW or more in England and Wales under §36 of the Electricity Act 1989. They also require EIA for:

- the construction or extension of a non-nuclear generating station of less than 300MW in England and Wales under §36 of the Electricity Act;
- the installation of an overhead transmission line in England or Wales under §37 of the Electricity Act; and
- the construction or diversion of a pipeline in Great Britain under the Pipe-lines Act 1962

where the SoS for the Department of Trade and Industry determines that the development would be "likely to have significant effects on the environment by virtue of factors such as its nature, size or location".

The regulations allow developers to make a written request to the SoS for a determination as to whether an EIA is needed. The SoS must consult with the LPA before making a decision. When a developer gives notice that s/he will be preparing an EIS, the SoS must notify the LPA or the principal council for the relevant area, the Countryside Commission, English Nature, and Her Majesty's Inspectorate of Pollution (HMIP) in the case of a power station, so that they can provide relevant information to the applicant.

The contents required of the EIA are virtually identical to those listed in the T&CP Regulations. When the EIS and application are handed in, the developer must publish a notice in one or more local papers for two successive weeks, giving details of the proposed project and/or a local address where copies of the EIS can be obtained. The SoS must advise the statutory consultees that the EIS has been completed, determine whether they want copies of the EIS, and inform them that they may make representations. However, the regulations have no clear procedures for consultation with environmental organizations or the public after the EIS is published. Chapter 10 provides further discussion.

Environmental Assessment (Afforestation) Regulations 1988 (ESI 1207)

These regulations, which are further explained in a booklet by the Forestry Authority entitled *Environmental assessment of afforestation projects*, apply to applications for grants or loans for afforestation projects, where the Forestry Authority thinks that the project is likely to have significant environmental impacts. Afforestation projects that do not require a grant, and projects carried out by the Forestry Authority itself, do not require EIA, although this may change in the future (Anon. 1992: 20).

When the Forestry Authority receives an application for a forestry planting grant, it informs the applicant if an EIA is needed. An EIA is likely to be needed for any new planting in a national nature reserve or an SSSI, or if the planting is expected to have a major impact because of its size, nature or location. The applicant may appeal to the Minister of the MAFF in case of disagreement. The contents required of the EIA are almost identical to those required by the T&CP Regulations. The lack of EIA requirements for forestry projects that do not require a grant, the fact that the Forestry Authority reviews EIAs despite its primary rôle as a promoter of forestry and afforestation, and the lack of requirement for the Forestry Authority to carry out EIAs on its own projects have been criticized (e.g. by the CPRE 1991).

3.6 Summary

Directive 85/337 is implemented in the UK through twenty regulations that link the various actors involved - developers, affected parties, regulators and facilitators - in a variety of ways. The Town and Country Planning (Assessment of Environmental Effects) Regulations 1988 are central. Other regulations cover projects that do not fall under the English and Welsh planning system, such as motorways and trunk roads, power stations, electricity transmission lines, pipelines, land drainage works, afforestation projects, and development projects in Scotland and Northern Ireland. The Planning and Compensation Act 1991 allows other projects not listed in Directive 85/337 also to be subject to EIA.

The T&CP Regulations include three Schedules, which broadly correspond to the Annexes of EC Directive 85/337. Schedule 3, which lists the contents required of an EIS, distinguishes between mandatory "specified information" and discretionary "further information". As will be discussed in Chapter 8, this distinction weakens the UK's EIA system considerably. Local planning authorities have discretion in determining which Schedule 2 projects require EIA and in recommending the contents of the EIS, but the developer is ultimately responsible for preparing the EIS. Statutory consultees must be sent copies of the EIS, and the public must be allowed to purchase copies; both groups can make representations to the LPA about the EIS. The LPA must consider the EIS and any representations when deciding whether to grant or refuse planning permission. The developer can appeal to the SOS in cases of disagreement with the LPA.

The other EIA regulations are broadly similar to those of the T&CP Regulations, with some differences, mainly in terms of screening and public consultation. The EA (Scotland) Regulations cover more project types, and have similar procedures. The Highways (AEE) Regulations require less in terms of EIS contents, and rely heavily on the pre-Directive appraisal framework given in the DOT *Manual of environmental appraisal*. The Land Drainage Improvement Works (AEE) regulations do not automatically require EIAs to be carried out; in cases where the drainage body does not propose to carry out an EIA, the EIA process is triggered only where objections are raised to a development proposal. As with the Electricity and

Pipe-line Works (AEE) Regulations, they have few requirements for consultation with environmental organizations and the public. The Environmental Assessment (Afforestation) Regulations cover only afforestation projects requiring Forestry Authority grants.

References

Anon. 1992. Whitehall to act on rural assessment. *Planning* 996 (27 November 1992), 20

CEC (Commission of the European Communities) 1985. On the assessment of the effects of certain public and private projects on the environment. *Official Journal* L175 (5 July 1985).

CPRE (Council for the Protection of Rural England) 1991. *The Environmental Assessment Directive: five years on*. London: CPRE.

DOE (Department of the Environment) 1989. *Environmental assessment: a guide to the procedures*. London: HMSO.

DoT (Department of Transport) 1983. *Manual of environmental appraisal*. London: HMSO.

DoT 1989. *Departmental standard HD 18/88. Environmental assessment under the EC Directive 85/337*. London: Department of Transport.

ECOTEC 1990. *The impact of environmental management on skills and jobs: a report for the Training Agency*. Birmingham: ECOTEC Research and Consultancy Ltd.

ENDS 1992. *Directory of environmental consultants 1992/93*. London: Environmental Data Services

ESRC 1987. *Newsletter: Environmental Issues*. London: Economic and Social Research Council, February 1987.

Fortlage, C. 1990. *Environmental assessment: a practical guide*. Aldershot: Gower.

SACTRA (Standing Advisory Committee on Trunk Road Assessment) 1992. *Assessing the environmental impact of road schemes*. London: HMSO.

Notes

1. There are some discrepancies. For instance, power stations of 300 MW or more are included in Schedule 1 although they actually fall under the Electricity and Pipe-line Works (AEE) Regulations, and all "special roads" are included although the regulations should actually apply to special roads under local authority jurisdiction.
2. The decision is actually made by the relevant regional office of the DoE; there are ten such offices. As will be discussed in Chapter 8, this has led to some discrepancies where two or more different offices have made different decisions on very similar projects.
3. This includes enough copies for all of the statutory consultees to whom the developer has not already sent copies; one copy for the LPA; and several (dependent on the 1993 amendments to the T&CP Regulations) for the Secretary of State.

PART 2
Process

This illustration by Neil Bennett is reproduced from Bowers, J. (1990), *Economics of the environment: the conservationist's response to the Pearce Report*, British Association of Nature Conservationists, 69 Regent Street, Wellington, Telford, Shropshire TF1 1PE.

CHAPTER 4
Starting up / early stages

4.1 Introduction

This is the first of four chapters that discuss how an EIA is carried out. The focus throughout will be on both the procedures required by UK legislation, and the "ideal" of best practice. Although Chapters 4 to 7 seek to provide a logical step-by-step approach through the environmental impact assessment process, there is no one absolute approach. Process is set within an institutional context and the context will vary from country to country (see Ch. 11). As already noted, even in one country, the UK, there may be a variety of regulations for different projects (see Chs 9 and 10). The various steps in the process can be taken in different sequences. Some steps may be completely missing in certain cases. Also, hopefully, the process will be not just linear but will build in cycles, with feedback from later stages to the earlier stages.

Chapter 4 covers the early stages of the EIA process. These include the clarification of whether an EIA is required at all ("screening"), and an outline of the extent of the EIA ("scoping"), which may involve consultation with several of the key actors outlined in Chapter 3. Early stages of EIA should also include an exploration of possible alternative approaches for the project, although this is not a mandatory requirement in UK legislation and is missing from many studies. Baseline studies, setting out the parameters of the development action (including associated policy positions) and the present and future state of the environment involved, are also included in Chapter 4. However, the major section in the chapter is devoted to impact identification. This is important in the early stages of the process, but, reflecting the cyclical, interactive nature of the process, some of the impact identification methods discussed here may also be employed in the later stages. Conversely, some of the prediction, evaluation, communication and mitigation approaches discussed in Chapter 5 can be used in the early stages, as can the participation approaches outlined in Chapter 6.

4.2 Project screening – is an EIA needed?

The number of projects that could be subject to EIA is potentially very large. Yet many projects have no substantial or significant environmental impact. A screening mechanism seeks to focus on those projects with potentially significant adverse environmental impacts or where the impacts are not fully known. Those with little or no impacts are screened out and are allowed to proceed to the normal planning permission / administrative processes without any additional assessment and without additional loss of time and expense.

Screening can be partly determined by the EIA regulations operating in a country at the time of the assessment. Chapter 3 indicates that in the EC, including the UK, there are some projects (Annex/Schedule 1) that will always be screened out for full assessment, by virtue of their scale and potential environmental impacts (for example: a crude-oil refinery, a sizeable thermal power station and a special road). There are many other projects (Annex/Schedule 2) where the screening decision is less clear. Here two examples of a particular project may be screened in different ways (one "in", one "out" for full assessment) by virtue of a combination of criteria, including project scale, sensitivity of proposed location and expectation of adverse environmental impacts. Chapter 9 provides examples of variations in interpretation of need for EISs with regard to new settlements in the UK. In such cases it is important to have working guidelines, indicative criteria and thresholds on conditions considered likely to give rise to significant environmental impacts (see §3.4).

In California, the list of projects that must always have the full review is determined by project type, development and location. For example, "type" includes, *inter alia*, a proposed local general plan; "development" includes, *inter alia*, a residential development of more than 500 units, a hotel/motel of more than 500 rooms, a commercial office building of more than 250,000 square feet of floor space; and "location" includes, *inter alia*, the Lake Tahoe Basin, the California Coastal Zone and an area within a quarter of a mile of a wild and scenic area (State of California 1992).

Some EIA procedures include an initial outline EIA study to check on likely environmental impacts and on their significance. Under the California Environmental Quality Act a "negative declaration" can be produced by the project proponent, thereby claiming that the project has minimal significant effects and does not require a full EIA. The declaration must be substantiated by an initial study, which is usually a simple checklist against which environmental impacts must be ticked as "yes", "maybe" or "no". If the responses are primarily "no", and most of the "yes" and "maybe" responses can be mitigated, then the project may be screened out from a full EIA. In Canada, the screening procedures are also well developed (see §11.3).

4.3 Scoping – which impacts and issues to consider?

The scope of the EIA is the impacts and issues that it addresses. The process of scoping is that of determining, from all a project's possible impacts and from all the alternatives that could be addressed, those that are the key, significant ones. An initial scoping of possible impacts may identify those impacts thought to be potentially significant, those thought to be non-significant, and those where the position is unclear. Further study should examine impacts in the various categories. Those confirmed by such study to be non-significant are eliminated, those in the uncertain category that may be potentially significant are added to the initial category of other potentially significant impacts. This refining of focus onto the most significant impacts continues throughout the EIA process.

Scoping is generally carried out in discussions between the developer, the competent authority, other relevant agencies and ideally the public. It is often the first stage of negotiations and consultation between a developer and other interested parties. It is an important step in EIA because it enables the limited resources of the team preparing an EIA to be allocated to best effect, and prevents misunderstanding between the parties concerned about the information required in an EIS. Scoping can also identify issues that should later be monitored. Although scoping is an important step in the EIA process, it is not a legally mandated step in the UK. The Department of thc Environment recommends that developers should consult with the competent authority and statutory consultees before preparing the EIS, but in practice this happens in only about half of all cases (Fuller 1992). This lack of early discussion is one of the major limitations to effective EIA to date.

Scoping should begin with the identification of individuals, communities, local authorities and statutory consultees likely to be affected by the project; good practice would be to bring them together in a working group and/or meetings with the developer. One or more of the impact identification techniques discussed in §4.7 can be used to structure a discussion and suggest key issues to consider. Other key issues could include particularly valued environmental attributes; those impacts perceived to be of particular concern to the affected parties; and social, economic, political and environmental issues related to the specific locality. Reference should be made to relevant structure plans, local plans, subject plans, and government policies and guidelines, as discussed in §4.6. Various alternatives should be considered, as discussed in §4.4. The end result of this process of information collection and negotiation should be the identification of key issues and impacts, an explanation of why other issues are not considered significant, and, for each key impact, a defined temporal and spatial boundary within which the impact will be measured.

As will be discussed in Chapter 11, other countries (e.g. Canada and the Netherlands) have a formal scoping stage, in which the developer agrees with the competent authority or an independent EIA commission, sometimes after public consultation, which subjects the EIA will cover. As part of its five-year review of Directive 85/337, the EC is considering including a mandatory scoping stage.

4.4 Consideration of alternatives

If a project is not screened out, and is believed to have potentially significant impacts on the environment, then an EIA is undertaken for the project *and* ideally for feasible alternatives. During the course of project planning, many decisions are made concerning the type and scale of project proposed, its location, and the processes involved. Most of the possible alternatives that arise will be rejected by the developer on economic, technical or regulatory grounds. The rôle of EIA is to ensure that environmental criteria are also considered at these early stages. The US Council on Environmental Quality (CEQ 1978) calls the discussion of alternatives "the heart of the environmental impact statement": how an EIA addresses alternatives will determine its relation to the subsequent decision-making process. A discussion of alternatives ensures that the developer has considered both other approaches to the project and the means of preventing environmental damage. A consideration of alternatives also encourages analysts to focus on *differences* between real choices. It can allow people who were not directly involved in the decision-making process to evaluate various aspects of the proposed project and how they were arrived at. It also provides a framework for the competent authority's decision, rather than merely a justification for a particular action. Finally, if unforeseen difficulties arise during construction or operation of a project, a re-examination of these alternatives may help to provide rapid and cost-effective solutions.

UK regulatory requirements

EC Directive 85/337 states that alternative proposals should be considered in an EIA, subject to the requirements of Article 5 (if the information is relevant and if the developer may reasonably be required to compile this information). Annex III requires "where appropriate, an outline of the main alternatives studied by the developer and an indication of the main reasons for this choice, taking into account the environmental effects". In the UK, this requirement has been interpreted to be discretionary. The Town and Country Planning (Assessment of Environmental Effects) Regulations 1988 note that:

> An environmental statement may include, by way of explanation or amplification of any specified information, further information on . . . (in outline) the main alternatives (if any) studied by the applicant, appellant or authority and an indication of the main reasons for choosing the development proposed, taking into account the environmental effects.

With minor changes of wording, this clause is repeated in the other UK EIA regulations. To date in the UK, about two-thirds of EISs have not considered alternatives at all.[1] The one-third of EISs that have considered alternatives have mostly been for linear developments (e.g. roads, rail, transmission lines) that consider different routes between two given points. A few others mention alternative sites that have been considered for development projects. The Department of Transport's *Manual of environmental appraisal* (DoT 1983) requires alternatives to be considered: without this requirement, which was in existence before Directive 85/337 became operational, very few EISs would consider alternatives at all.

Types of alternatives

A thorough consideration of alternatives would begin early in the planning process, before the type and scale of development and its location have been agreed on. A number of broad types of alternatives can be considered: the "no action" option, alternative locations, alternative scales of the project, alternative processes or equipment, alternative site layouts, alternative operating conditions, or alternative ways of dealing with environmental impacts. The last of these will be discussed in §5.4.

The *"no action" option* refers to environmental conditions if the project did not go ahead. Consideration of this option is required in some countries, e.g. the USA.[2] In essence, consideration of the "no action" option is equivalent to a discussion of the need for the project: do the benefits of the project outweigh its costs? This option is rarely discussed in UK EISs.

The consideration of alternative *locations* is an essential component of the project planning process. In some cases, the project location is constrained in varying degrees: for instance gravel extraction can take place only in areas with sufficient gravel deposits, and wind-farms require locations with sufficient wind speed. In other cases, the best location can be chosen to maximize criteria such as economic, planning and environmental considerations. For industrial projects, for instance, economic criteria such as land values, the availability of infrastructure, the distance from sources and markets, and the labour supply, are likely to be important (Fortlage 1990). For road projects, engineering criteria strongly influence the alignment. In all these cases, however, siting in "environmentally robust" areas, or away from designated or environmentally sensitive areas, should be considered.

The consideration of different *scales* of development is also integral to project planning. In some cases, the project scale will be flexible. For instance, the scale of a waste-disposal site can be changed, depending, for example, on the demand for landfill space, the availability of other sites, and the presence of nearby residences or environmentally sensitive sites. A wind-farm could include a wide range of number of turbines. In other cases, the developer will need to decide whether an entire unit should be built or not. For instance, the reactor building of a PWR nuclear power station is a large discrete structure that cannot easily be scaled down. Pipelines or bridges, to be functional, cannot be broken down into smaller sections.

Alternative *processes and equipment* involve the possibility of achieving the same objective through a different method. For instance, 1500 MW of electricity can be generated by one combined-cycle gas turbine power station, by a tidal barrage, by several waste-burning power stations, or in the extreme by thousands of wind turbines. Gravel can be directly extracted or recycled; waste may be incinerated or put in a landfill.

Once the location, scale and processes of a development have been decided upon, different *site layouts* can still have different impacts. For instance, noisy plant can be sited near or away from nearby residences. Power-station cooling towers can be few and tall (using less land) or many and short (causing less visual impact). Buildings can be sited either prominently or to minimize their visual impact. Similarly, *operating conditions* can be changed to minimize impacts. For instance, the same level of noise at night is usually more annoying than during the day, so night-time work could be avoided. Establishing designated routes for project-related traffic can help to minimize disturbance to local residents. Construction can take place at times of year that minimize environmental impacts, for example to migratory or nesting birds.

Presentation and comparison of alternatives

Alternatives have different costs for different groups of people and for different environmental components. Discussions with local residents, statutory consultees and special interest groups may rapidly eliminate some alternatives from consideration and suggest others. However, it is unlikely that one alternative will emerge as being most acceptable to all parties concerned. The EIS should distil information about a (reasonable) number of realistic alternatives into a format that will facilitate public discussion and finally decision-making. Methods for comparing and presenting alternatives span the range from simple non-quantitative descriptions, through increasing levels of quantification, to a complete translation of all impacts into their monetary values.

To date in the UK, those EISs for non-road proposals that *have* discussed alternatives, have merely described them, their main impacts, and reasons for their rejection. Many of the (impact identification) methods discussed later in this chapter are relevant to this stage of decision-making. Overlay maps compare the impacts of various locations in a non-quantitative manner. Checklists or less complex matrices can also be applied to various alternatives and compared; this may be the most effective way of visually presenting the impacts of alternatives. Some of the other techniques used for impact identification – the threshold of concern checklist, weighted matrix, and EES – allow alternatives to be implicitly compared. They broadly do this by assigning quantitative importance weightings to environmental components, rating each alternative (quantitatively) according to its impact on each environmental component, multiplying the ratings by their weightings to obtain a weighted impact, and aggregating these weighted impacts to obtain a total score for each alternative. These scores do not correspond to real-life monetary value, but can be compared against each other to identify preferable alternatives. With the exception of the threshold-of-concern checklist, they do not lend themselves to the clear presentation of the alternatives in question, and none of them clearly states who will be affected by the different alternatives.

The UK Department of Transport tries to tread an uneasy middle path between the various techniques. Its *Manual of environmental appraisal* (DoT 1983) uses a framework to appraise the impacts of road proposals against a "do nothing" option for various affected groups (see §10.2, for further details).

4.5 Understanding the project/development action

Understanding the dimensions of the project

Schedule 3 of the Town and Country Planning (Assessment of Environmental Effects) Regulations 1988, requires "a description of the development proposed, comprising information about the site and the design and scale or size of the development" and "the data necessary to identify and assess the main effects which that development is likely to have on the environment". *Environmental assessment: a guide to procedures* (DoE 1989) provides a brief listing of information that may be used to describe the project. At first glance, this description of the proposed development would appear to be one of the more straightforward steps in the EIA process. However, projects have many dimensions, and relevant information may be limited. As a consequence, this first step may pose some challenges. Key dimensions to be clarified include the purpose of the project, its life-cycle, physical presence, process(es), policy context and associated policies.

An outline of *the purpose and rationale* of a project provides a useful introduction to the project description. This may, for example, set the particular project in a wider context – the missing section of a major motorway, a power station in a programme of developments, a new settlement in an area of major population growth. A discussion of purpose may include the rationale for the particular type of project, for the choice of location and for the timing of the development. It may also provide background information on planning and design activities to date.

As noted in §1.4, all projects have a *life-cycle of activities*, and the project description should clarify the various stages in the life-cycle, and their relative duration, for the project under consideration. A minimum description would usually involve the identification of construction and operational stages and associated activities. Further refinement might include planning and design, project commissioning, expansion, close-down and site rehabilitation stages. The size of the development at various stages in the life-cycle should also be specified. This can include reference to inputs, outputs, physical size and the number of people to be employed.

The *location and physical presence* of the project should also be clarified at an early stage. This should include the general location of the project on a base map in relation to other activities and to administrative areas. A more detailed site layout of the proposed development, again on a (large-scale) base map, should illustrate the land area and the main disposition of the elements of the project (e.g. storage areas, main processing plant, waste collection areas, transport connections to the site). Where the site layout may differ substantially between different stages in the life-cycle, it is valuable to have a sequence of anticipated layouts. Location and physical presence should also identify elements of a project that, although integral, may be detached from the main site (e.g. construction of a barrage in one area may involve opening up a major quarry development in another area). A description of the physical presence of the project is invariably improved with a three-dimensional visual image, which may include a photo-montage of what the site layout may look like at, for example, full operation. A clear presentation of location and physical presence is important for an assessment of change in land-uses, physical disruption to other infrastructures, severance of activities (e.g. agricultural holdings, villages) and visual intrusion and landscape changes.

Understanding the project also involves an understanding of the *processes* integral to the project. The nature of processes varies between industrial, service and infrastructure projects, but many can be described in terms of a flow of inputs through a process and their transformation into outputs. The nature, origins and destinations of the inputs and outputs, and the timescale over which they are expected, should be identified. This systematic identification should be undertaken for both physical and socio-economic characteristics, although the interaction should be clearly recognized, with many of the socio-economic following from the physical.

Physical characteristics may include:

- land take and physical transformation of a site (e.g. clearing, grading) that may vary between different stages of the project life-cycle;
- overall operation of the process involved (usually illustrated with a process-flow diagram);
- types and quantities of resources used (e.g. water abstraction, minerals, energy);
- transportation requirements (of inputs and outputs);
- generation of wastes, including estimates of types, quantity, and strength of aqueous wastes, gaseous and particulate emissions, solid wastes, noise and vibration, heat and light, radiation, etc;
- the potential for accidents, hazards and emergencies;
- processes for the containment, treatment and disposal of wastes and for the containment and handling of accidents; monitoring and surveillance systems.

Socio-economic characteristics may include:

- the labour requirements of the project – including size, duration, sources, particular skills categories and training;
- provision or otherwise by the developer of housing, transport, health and other services for the workforce;
- direct services required from local businesses or other commercial organizations;
- flow of expenditure from the project into the wider community (from the

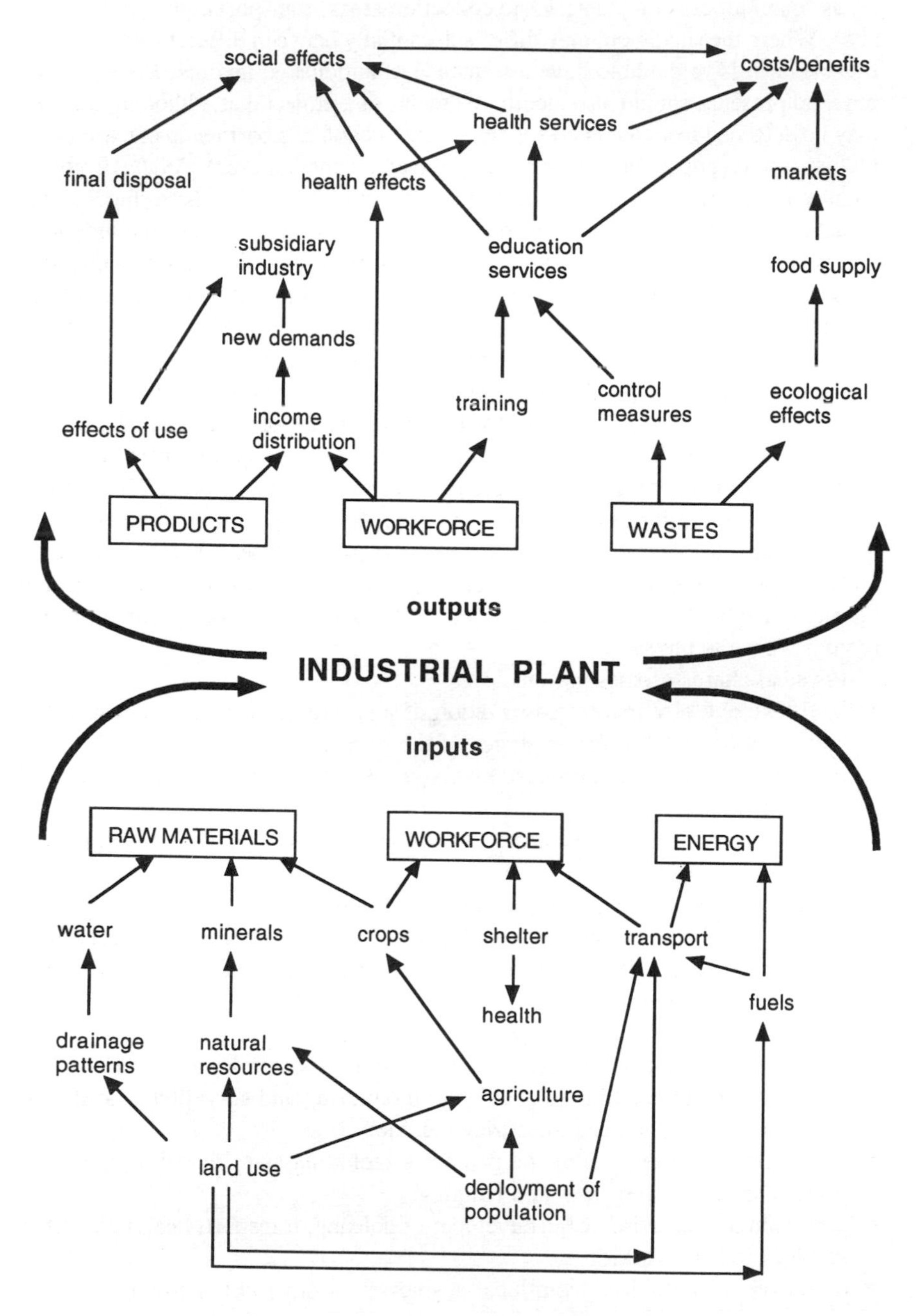

Figure 4.1 Interaction between an industrial plant and its socio-ecological environment. *(Source:* Marstrand 1976)

employees and subcontracting);
- flow of social activities (service demands, community participation, community conflict).

Figure 4.1 (after Martstrand 1976) shows the interaction between the physical (ecological in this case) and socio-economic processes that may be associated with an industrial plant.

Projects should also be seen in their *planning policy context*. In the UK, the main local policy context is outlined and detailed in structure and local plans. The description of location must pay regard to land-use designations and development constraints that may be implicit to some of the designations. Of particular importance is the location in relation to various environmental zones (e.g. Areas of Outstanding Natural Beauty (AONB), Sites of Special Scientific Interest (SSSI), green belts, and local and national nature reserves). Attention should also be given to national planning guidance, provided in the case of the UK by an important set of Department of Environment Planning Policy Guidance Notes (PPGs).

The projects themselves may also have *associated policies*, not obvious from site layouts and process–flow diagrams, that are nevertheless significant for subsequent impacts. For example, shift working will have implications for transport and noise that may be very significant for nearby residents. The use of a construction site hostel/camp/village can be significant in the internalization of impacts on the local housing market and on the local community. The provision of on- or off-site training can significantly affect the mix of local and non-local employment, and the balance of socio-economic effects.

Types and sources of data

Various *types of data* are used. The life-cycle of a project can be illustrated on a linear bar chart. Particular stages may be identified in more detail where the impacts are considered of particular significance; this is often the case for the construction stage of major projects. Location and physical presence are best illustrated on a map base, with varying scales to move from the broad location to the specific site layout. This may be supplemented by aerial photographs, photo-montages and visual mock-ups according to the resources and issues involved (see Figs 4.2–5).

A process diagram for the different activities associated with the project should accompany the location and site-layout maps. This may be presented in the form of a simplified pictorial diagram or in a block flow chart. The latter can be presented simply to show the main interconnections between the element of a project (see Fig. 4.3 for socio-economic processes) or in sufficient detail to provide a comprehensive picture. Figure 4.4 shows a material-flow chart for a petroleum refinery, which outlines all raw materials, additives, end-products, byproducts and atmospheric, liquid and solid wastes. A comprehensive flow chart of a production process should include the types, quantities and locations of resource inputs, intermediate and final product outputs and wastes generated by the total process.

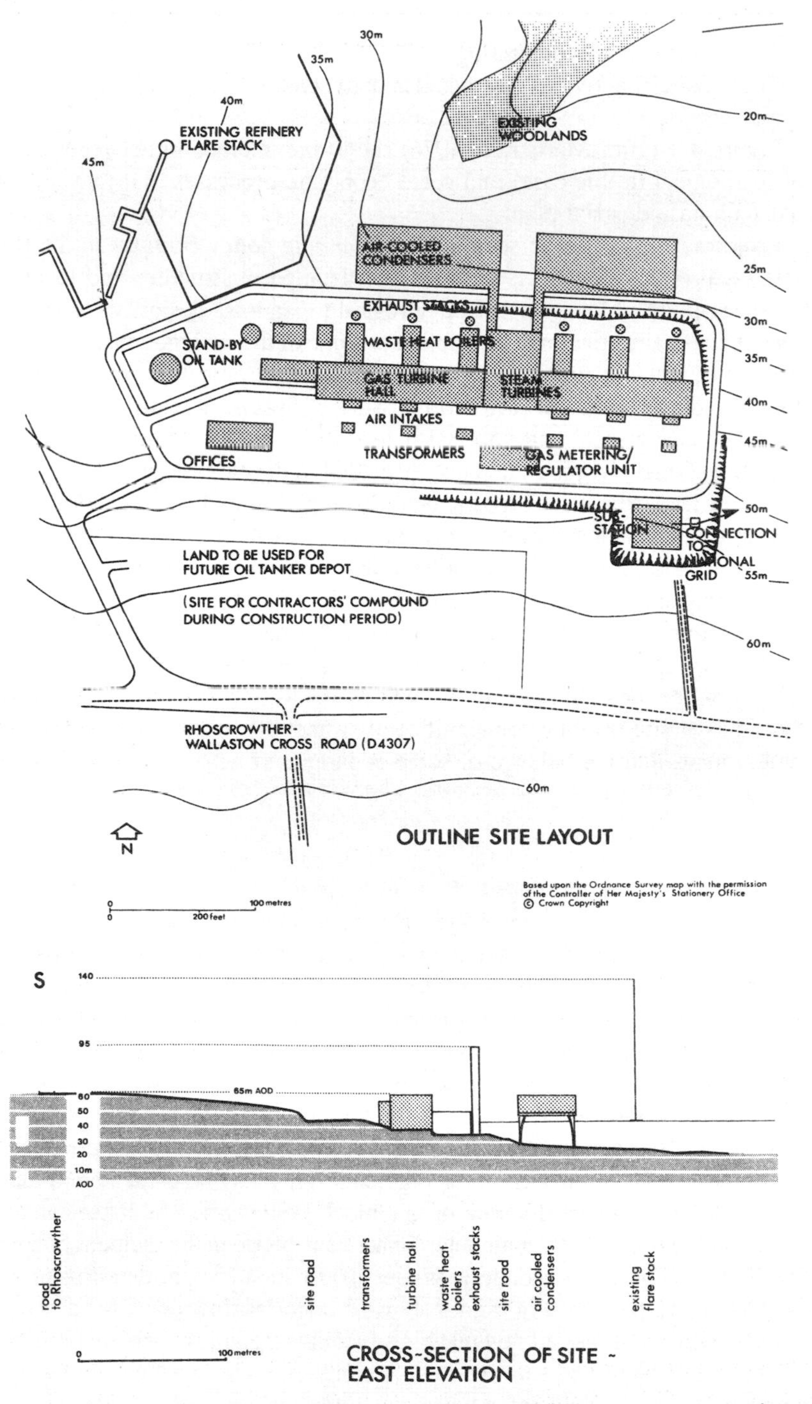

Figure 4.2 Example of a project site layout. *(Source:* Rendel Planning 1990, *Angle Bay Energy Project environmental statement)*

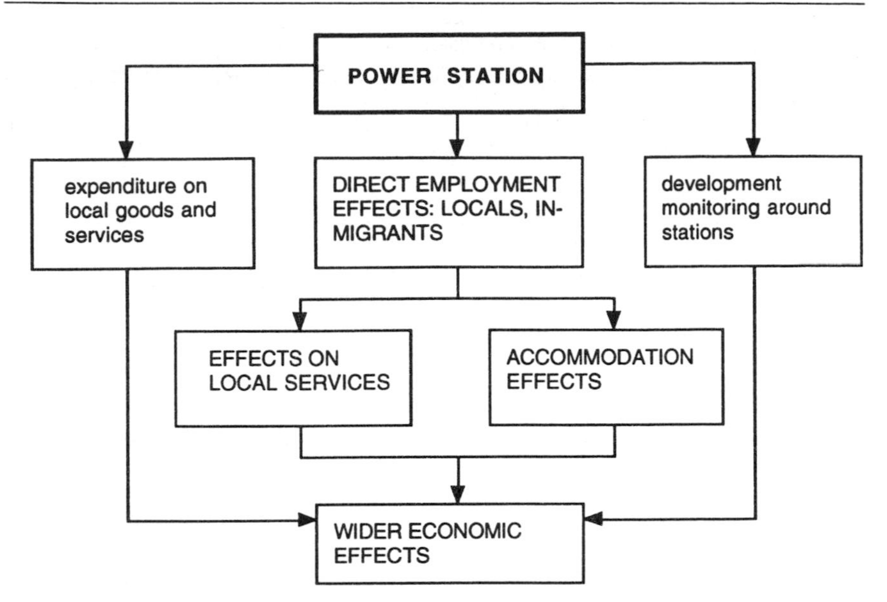

Figure 4.3 Socio-economic process diagram for a major project.

The various information and illustrations should clearly identify major variations between project stages, as already noted. Figure 4.5 provides an illustration of a labour-requirements diagram that identifies the widely differing requirements, in absolute numbers and in skill categories, between the construction and operational stages. In addition, more sophisticated flow diagrams could indicate the type, frequency (normal, batch, intermittent or emergency) and duration (minutes or hours/day or week) of each operation. Seasonal and material variations, including time periods of peak pollution loads, can also be documented.

The form and *sources of data* vary according to the degree of detail required and the stage in the assessment process. Site-layout diagrams and process–flow charts may be only in outline, provisional form at the initial design stage. Subsequent investigations, and identification of sources of potential significant impacts, may lead to changes in layout and process.

The initial brief from the developer provides the starting point. Ideally, the developer may have detailed knowledge of the proposed project's characteristics, likely layout and production processes, drawing on previous experience. With the rapid development of EIA and the production of EISs, the analyst can also supplement such information with reference to comparative studies as sources for project profiles, although the availability of such statements in the UK is still far from satisfactory and their predictions are untested (see Chs 7 and 8). Use can also be made of other published data (e.g. published emission and effluent factors for components of a project, published data on accident rates; TNO 1983, VROM 1985). Site visits can be made to comparable projects, and advice can be gained from consultants with experience of the type of project under consideration. As the project design and assessment process develops, so the developer will be required to provide more detailed information on characteristics specific to the project.

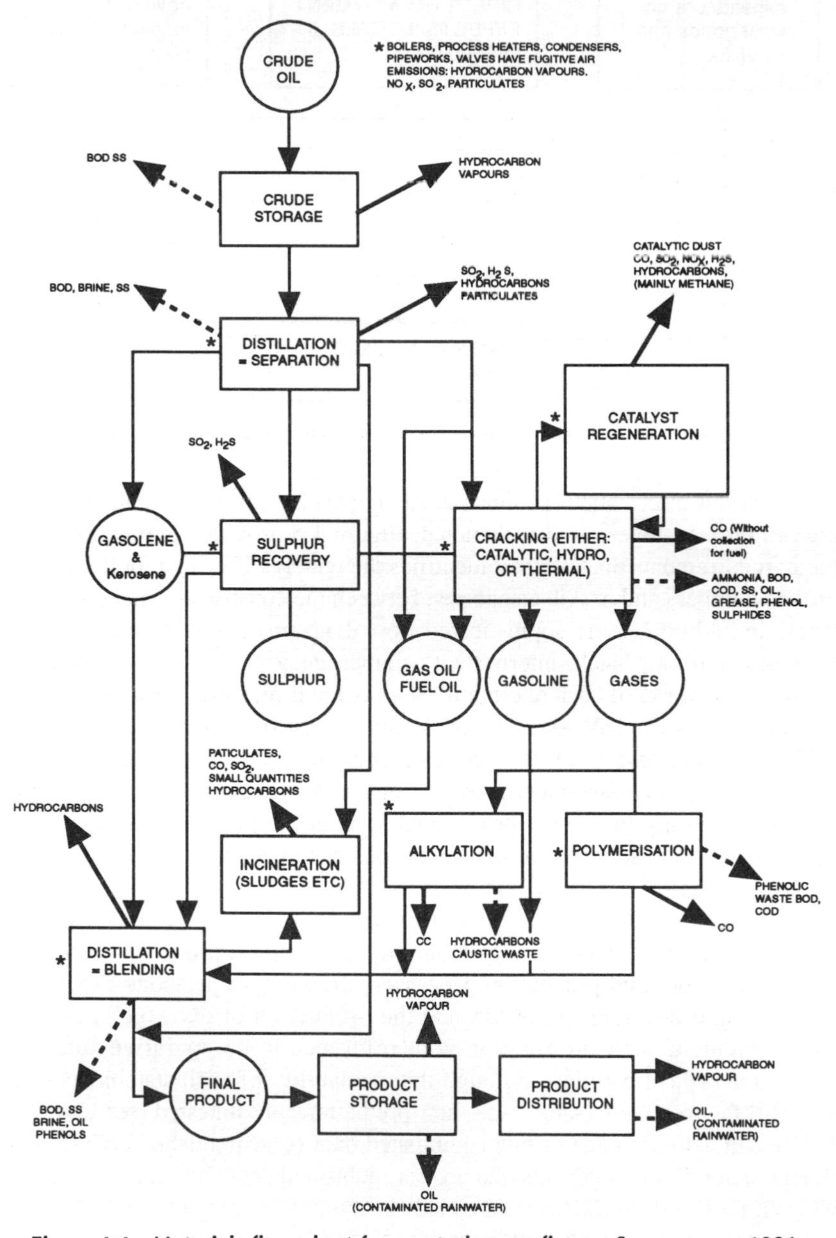

Figure 4.4 Materials flow-chart for a petroleum refinery. *Source:* UNEP 1981.

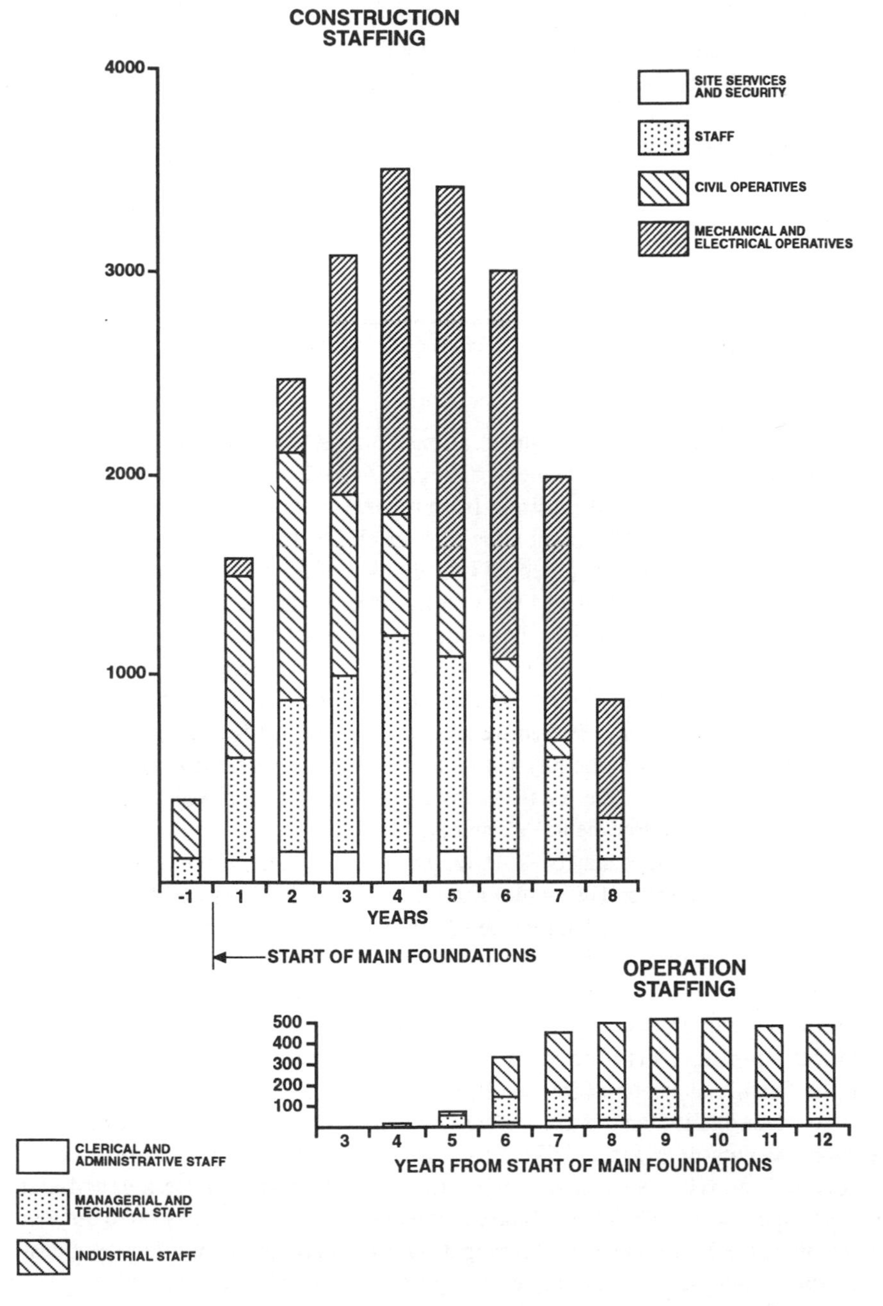

Figure 4.5 Labour requirements for a project over several stages of project life.

Even in the ideal situation, there will be considerable interaction between the analyst and developer to refine the project characteristics. Unfortunately, the situation may often be far from ideal; Mills (1992) provides an interesting example of major changes between the project description in the EIS and the actual implemented action for a power station development. The initial brief may leave a lot to be desired, and the analyst will have to draw on the other sources noted above to clarify the details of the project. The analyst may also draw on EIA literature (books and journals), guidelines, manuals and statistical sources. Further useful sources on understanding the project include United Nations Environment Programme (1981), Lee (1987), and Wood & Lee (1987).

4.6 Establishing the environmental baseline

General considerations

The establishment of the environmental baseline includes both the present and likely future state of the environment, assuming that the project is not undertaken, taking into account changes resulting from natural events and from other human activities. For example, the population of a species of fish in a lake may already be declining even before the potential introduction of an industrial project on the lake shore. Figure 1.6 illustrated the various time, component and scale dimensions of the environment, and all these dimensions need to be considered in establishing the environmental baseline. The timespan for the prediction of the future state of the environment should be comparable with the life of the proposed development, which may necessitate prediction forward for several decades. Components include both the physical and socio-economic environment. Spatial coverage may focus on the local, but with reference to the wider region and beyond for some environmental elements.

Initial baseline studies may be wide-ranging, but comprehensive overviews can be wasteful of resources. The studies should focus as quickly as possible on those aspects of the environment that may be significantly affected by the project, either directly or indirectly. The rationale for the choice of focus should be explained with reference to the nature of the project and to initial scoping and consultation exercises. Although the studies would normally take the various environmental elements separately, it is also important to understand the interaction between them and the functional relationships involved; for instance, flora will be affected by air and water quality, and fauna will be affected by flora. This will facilitate prediction. As with most aspects of the EIA process, establishing the baseline is not a "one-off" activity. Studies will move from broad-brush to more detailed and focused approaches. The identification of new potential impacts may open up new elements of the environment for investigation; the identification of effective measures for mitigating impacts may curtail certain areas of investigation.

Table 4.1 Information describing the site and its environment.

Physical features

1. Population – proximity and numbers.
2. Flora and fauna (including both habitats and species) – in particular, protected species and their habitats.
3. Soil; agricultural quality, geology and geomorphology.
4. Water; aquifers, water courses, shoreline, including the type, quantity, composition and strength of any existing discharges.
5. Air; climatic factors, air quality, etc.
6. Architectural and historic heritage, archaeological sites and features, and other material assets.
7. Landscape and topography.
8. Recreational uses.
9. Any other relevant environmental features.

The policy framework

10. Where applicable, the information considered under this section should include all relevant statutory designations such as national nature reserves, sites of special scientific interest, national parks, areas of outstanding natural beauty, heritage coasts, regional parks, country parks, national forest parks and designated areas, local nature reserves, areas affected by tree preservation orders, water protection zones, nitrate sensitive areas, conservation areas, listed buildings, scheduled ancient monuments, and designated areas of archaeological importance. It should also include references to structure, unitary and local plan policies applying to the site and surrounding area which are relevant to the proposed development.
11. Reference should also be made to international designations, e.g. those under the EC "Wild Birds" Directive, the World Heritage Convention, the UNEP Man and Biosphere Programme and the Ramsar Convention.

Source: DOE 1989.

Baseline studies can be presented in the EIS in a variety of ways. These often involve either a brief overview of both the physical and socio-economic environment for the area of study, following after the project description, with the detailed focused studies in subsequent impact chapters (e.g. air quality, geology, employment), or a more comprehensive set of detailed studies at an early stage providing a point of reference for future and often briefer impact chapters.

Environmental components or elements can be described simply in broad categories, as outlined in Table 1.3 (adapted from DoE 1991). *Environmental assessment: a guide to the procedures* (DoE 1989) also provides a relatively short list (see Table 4.1), including an important distinction between physical features and policy framework. In contrast, Leopold has 88 components in his interactive matrix (see Fig. 4.11) and each of these could be subdivided further. Several UN publications provide a more balanced listing of both the physical and socio-economic elements (see United Nation Environment Programme 1981). Table 4.2 provides an example of a framework for analyzing each baseline sub-element.

Table 4.2 Framework for analysing baseline subelements: example of use.

Subelement	Objectives	Required information/ specialist(s)	Methodology	Findings/ measurements
Water quality	Protection of human health and aquatic life	Existing water quality; possible sources of pollution: run-off, leakage from waste treatment system, surface seepage of pollutants, intrusion of saline or polluted water; capacity of treatment system. Water quality analyst; aquatic biologist; water pollution control engineer; sanitary and civil engineers.	Laboratory analyses or field measurement of water quality; pollution indices.	Potential for degradation of water quality; safety of potable water.
Surface waters	Protection of: plant and animal life; water supply for domestic and industrial needs; natural water purification systems; groundwater recharge and discharge; and recreation and aesthetic values	Location of surface waters – streams, rivers, ponds, lakes, etc.; surface water volume, flow rates, frequency and duration of seasonal variations; 7-day, 10-year low flow; water uses; ecological characteristics; recreation and aesthetic "uses". Hydrologist/ ecologist	Measure proximity of site to surface waters; field measurement of volume, rate and direction of water movement; categories of water usage; ecological assessment – see ECOLOGY	Potential modification of volume, rate and direction of water movement; impact on ecological character; degree and type of water usage

Source: UNEP Industry and Environment Office (1981)

Data sources and issues

Data on environmental conditions vary in availability and in quality. Important data sources for most locations in the UK are the statutory development plans (local plans, structure plans). These plans usually provide a range of very useful data on the physical, social and economic environment, which are reasonably up to date. In a few locations, such data are supplemented by environmental audits or state-of-the-environment reports. The focus of these new studies is normally on the physical environment. The Environmental Audit for Oxfordshire, for example, includes detailed studies on land-use, landscape, open space, forestry, wildlife, agriculture, noise, air quality, water pollution, waste management, transport, energy and environmental management (Aspinwall & Company Ltd 1991). Chapter 12 provides further discussion of environmental audits. Local data can be supplemented in the UK with published data from a wide range of national government sources – including the Census of population, *Regional trends*, *Digest of environmental protection and water standards*, *Transport statistics* – and increasingly from EC sources. Some countries have guidelines and manuals for EIA that list major sources of information for different environmental elements and these can be very useful.

However, much useful information is unpublished or semi-published and internal to various organizations. In the UK, under the EIA regulations, statutory consultees (e.g. the Countryside Commission, English Nature, Her Majesty's Inspectorate of Pollution and the National Rivers Authority) are obliged "to provide the developer (on request) with any information in their possession which is likely to be relevant to the preparation of the environmental statement . . . The obligation on statutory consultees relates only to information already in their possession: they are not required to undertake research on behalf of the developer" (DoE 1989). There are of course many other useful non-statutory consultees, at local and other levels, who may be able to provide valuable information. Local history, conservation and naturalist societies may have a wealth of local information on, for example, flora and fauna, rights of way, and archaeological sites. National bodies, such as the Royal Society for the Protection of Birds (for bird populations and habitats) and the Forestry Authority (for tree surveys), may have particular knowledge and expertise to offer. Consultation with local amenity groups at an early stage in the EIA process can help not only with data but also with the identification of those key environmental issues for which data should be collected.

Every use should be made of data from existing sources, but there will invariably be gaps in the required environmental baseline data for the project under consideration. Environmental monitoring and surveys may be necessary, although the UK DoE notes: "While a careful study of the proposed location will generally be needed (including environmental survey information), original scientific research will not normally be necessary" (DoE 1989). Surveys and monitoring raise a number of issues. They are inevitably constrained by budgets and time, and must be selective. However, such selectivity must ensure that the length of time over which monitoring and survey are undertaken is appropriate to the task in hand. For example, for certain environmental features (e.g. many types of flora and fauna) a survey period of 12 months or more may be needed to take account of seasonal variations. Sampling procedures will often be used for surveys; the extent and implications of the sampling error involved should be clearly established.

The quality and reliability of environmental data vary a great deal, and this can influence the use of such data in the assessment of impacts. Fortlage (1990) clarifies this in the following useful classification:

- "hard" data from reliable sources which can be verified and which are not subject to short-term change, such as geological records and physical surveys of topography and infrastructure;
- "intermediate" data which are reliable but not capable of absolute proof such as water quality, land values, vegetation condition, and traffic counts, which have variable values;
- "soft" data which are a matter of opinion or social values, such as opinion surveys, visual enjoyment of landscape, and numbers of people using amenities, where the responses depend on human attitudes and the climate of public feeling.

A valuable innovation in environmental data provision is the development of computerized data banks, and the increasing use of geographical information systems, for particular sets of data and for particular locations. Such data may, however, have limited accessibility. The analyst should also be wary of the seductive attraction of quantitative data at the expense of qualitative data; both types have a valuable rôle in establishing baseline conditions. Finally, it should be remembered that all data sources have some uncertainty attached to them, and this needs to be explicitly recognized in the prediction of environmental effects (see Ch. 5).

4.7 Impact identification

Aims and methods

Impact identification brings together project characteristics and baseline environmental characteristics with the aim of ensuring that all potentially significant environmental impacts (adverse or favourable) are identified and taken into account in the EIA process. A wide range of methods have been developed. Sorensen & Moss (1973) note that the present diversity "should be considered as a healthy condition in a newly formed and growing discipline".

In choosing a method, the analyst needs to consider more specific aims, some of which conflict:

1. to ensure compliance with regulations
2. to provide a comprehensive coverage of a full range of impacts, including social, economic and physical

3. to distinguish between positive and negative, large and small, long-term and short-term, reversible and irreversible impacts
4. to identify secondary, indirect and cumulative impacts as well as direct impacts
5. to distinguish between significant and insignificant impacts
6. to allow comparison of alternative development proposals
7. to consider impacts within the constraints of an area's carrying capacity
8. to incorporate qualitative as well as quantitative information
9. to be easy and economical to use
10. to be unbiased and to give consistent results, and
11. to be of use in summarizing and presenting impacts in the EIS.

Many of the methods were developed in response to the NEPA and have since been expanded and refined. The simplest involve the use of lists of impacts to ensure that none has been forgotten. The most complex include the use of interactive computer programmes, networks showing energy flows, and schemes to allocate significance weightings to various impacts. Many of the more complex methods were developed for (usually US) government agencies that deal with large numbers of relatively similar project types (e.g. the US Bureau of Land Reclamation and the US Forest Service).

In the UK, the use of impact identification techniques is less well developed. Simple checklists or, at best, simple matrices are used to identify and summarize impacts. This may be attributable to the fact that few EISs have been prepared to date by most developers, to the high degree of flexibility and discretion in the UK's implementation of Directive 85/337, to a general unwillingness in the UK to make the EIA process over-complex, or to disillusionment with the more complex approaches that are available.

The aim of this section is to present a range of these methods, from the simplest checklists needed for compliance with regulations, to complex approaches that developers, consultants and academics who aim to further "best practice" may wish to investigate further. The methods are divided into the following categories:

- checklists
- matrices
- quantitative methods
- networks, and
- overlay maps.

The discussion of methods here relates primarily to impact identification, but most of the approaches are also of considerable (and sometimes more) use in other stages of the EIA process – in impact prediction, evaluation, communication, mitigation, presentation, monitoring and auditing. As such, there is considerable interaction between Chapters 4, 5, 6 and 7, paralleling the interaction in practice between these various stages.

The reader is referred to Bregman & Mackenthun (1992), Bisset (1983, 1989), Wathern (1984), Sorensen & Moss (1973), Munn (1979), Rau & Wooten (1980) and Jain et al. (1977) for further general information on the range of methods available.

Checklists

Most checklists are based on a list of special biophysical, environmental, social and economic factors that may be affected by a development. The *simple checklist* can only help to identify impacts and ensure that impacts are not overlooked. Checklists do not usually include direct cause–effect links to project activities. Nevertheless, they have the advantage of being easy to use. Appendix 3 of the Town and Country Planning (Assessment of Environmental Effects) Regulations 1988 (see Table 5.1) is an example of a simple checklist.

Descriptive checklists (e.g. Schaenman 1976) give guidance on how to assess impacts. They can include data requirements, information sources, and predictive techniques. An example of part of a descriptive checklist is shown in Table 4.3.

Questionnaire checklists are based on a set of questions to be answered. Some of the questions may concern indirect impacts and possible mitigation measures. They may also provide a scale for classifying estimated impacts from highly adverse to highly beneficial. Figure 4.6 shows part of a questionnaire checklist.

Threshold of concern checklists consist of a list of environmental components and, for each component, a threshold at which those assessing a proposal should become concerned with an impact. The implications of alternative proposals can be seen by examining the number of times that an alternative exceeds the threshold of concern. For example, Figure 4.7 shows part of a checklist developed by the US Forest Service that compares three alternative development proposals on the basis of various components. For the component of economic efficiency, a benefit:cost ratio of 1:1 is the threshold of concern; for spotted owls, 35 pairs are the threshold. In the example, alternative X causes two thresholds of concern to be exceeded, alternative Y one, and alternative Z four; this would indicate that alternative Y is the least detrimental. Impacts are also rated according to their duration: A for 1 year or less, B for 1–10 years, C for 10–50 years, and

Table 4.3 Part of a descriptive checklist.

Data required	Information sources, predictive techniques
Nuisance	
Change in occurrence of odour, smoke, haze, etc., and number of people affected.	Expected industrial processes and traffic volumes, citizen surveys.
Water quality	
For each body of water, changes in water uses, and number of people affected.	Current water quality, current and expected effluent.
Noise	
Change in noise levels, frequency of occurrence, and number of people bothered.	Current noise levels, changes in traffic or other noise sources, changes in noise mitigation measures, noise propagation model, citizen surveys.

(Adapted from Schaenman 1976)

Disease vectors

a) Are there known disease problems in the project area transmitted through vector species such as mosquitoes, flies, snails, etc. ?	yes	no	not known
b) Are these vector species associated with:			
– aquatic habitats?	yes	no	not known
– forest habitats?	yes	no	not known
– agricultural habitats?	yes	no	not known
. . .			
f) Will the project provide opportunities for vector control through improved standards of living?	yes	no	not known

Estimated impact on disease vectors?

high adverse ◄------------ insignificant ------------► high benefit

Figure 4.6 Part of a questionnaire checklist. Adapted from US Agency for International Development 1981.

D for irreversible impacts. Of the impacts listed, a reduction in the number of spotted owls would be irreversible, and the other impacts would last 10–50 years (Sassaman 1981).

Environmental component	Criterion	TOC	Alt X Imp	Imp > TOC?	Alt Y Imp	Imp > TOC?	Alt Z Imp	Imp > TOC?
Air quality	emission standards	1	2C	yes	1C	no	2C	yes
Economics	benefit: cost ratio	1:1	3:1	no	4:1	no	2:1	no
Endangered species	no. pairs of spotted owls	35	50D	no	35D	no	20D	yes
Water quality	water quality standards	1	1C	no	2C	yes	2C	yes
Recreation	no. camping sites	5000	2800C	yes	5000C	no	3500C	yes

Figure 4.7 Part of a threshold of concern (TOC) checklist. (Adapted from Sassaman 1981.)

Matrices

Matrices are the most commonly used method of impact identification in EIA. Simple matrices are merely two-dimensional charts showing environmental components on one axis, and development actions on the other. They are, essentially, expansions of checklists that acknowledge the fact that different components of a development project (e.g. construction, operation, decommissioning; buildings, access road) have different impacts. Actions likely to have an impact on an environmental component are identified by placing a cross in the appropriate cell. The main advantage is the incorporation of cause–effect relationships. Figure 4.8 shows an example of a *simple matrix*. Three-dimensional matrices have also been developed, in which the third dimension refers to economic and social institutions: such an approach identifies the institutions from which data are needed for the EIA process, and highlights areas in which knowledge is lacking.

The *time-dependent matrix* (e.g. Parker & Howard 1977) includes a number sequence to represent the timescale (e.g. one figure per year) of the impacts. Figure 4.9 shows an example, where magnitude is represented by numbers from 0 (none) to 4 (high), over the course of seven years.

Magnitude matrices go beyond the mere identification of impacts by describing the impacts according to their magnitude, importance, and/or time frame (e.g. short-, medium, or long-term). Figure 4.10 is an example of a magnitude matrix.

The best known type of quantified matrix is the *Leopold matrix*, which was developed for the US Geological Survey by Leopold et al. (1971). This matrix is based on a horizontal list of 100 project actions, and a vertical list of 88

	Project action				
	Construction		Operation		
Environmental component	Utilities	Residential and commercial buildings	Residential buildings	Commercial buildings	Parks and open spaces
Soil and geology	X	X			
Flora	X	X			X
Fauna	X	X			X
Air quality				X	
Water quality	X	X	X		
Population density			X	X	
Employment		X		X	
Traffic	X	X	X	X	
Housing			X		
Community structure		X	X		X

Figure 4.8 Part of a simple matrix.

Environmental component	Project action				
	Construction (3 years)		Operation (25 years, evens out after 4 years)		
	Utilities	Residential and commercial buildings	Residential buildings	Commercial buildings	Parks and open spaces
Soil and geology	211	321	0000	0000	0001
Flora	221	422	1223	1111	1123
Fauna	221	311	1100	1100	1122
Air quality	000	000	0123	0034	0011
Water quality	010	022	1223	0111	0000
Population density	011	112	2344	0222	0011
Employment	120	342	1111	1334	1111
Traffic	220	332	2333	2333	1111
Housing	010	121	2344	0000	0000
Community structure	010	232	2344	1111	1233

Figure 4.9 Part of a time-dependent matrix.

Environmental component	Project action				
	Construction		Operation		
	Utilities	Residential and commercial buildings	Residential buildings	Commercial buildings	Parks and open spaces
Soil and geology	•	•			
Flora	•	●			O
Fauna	•	•			o
Air quality				•	
Water quality	O	•	•		
Population density			o	o	
Employment		O		O	
Traffic	•	•	•	●	
Housing			O		
Community structure		•	O		o

• = small negative impact
● = large negative impact
o = small positive impact
O = large positive impact

Figure 4.10 Part of a magnitude matrix.

(a)

1. Identify all actions (located across the top of the matrix) that are part of the proposed project
2. Under each of the proposed actions, place a slash at the intersection with each item on the side of the matrix if an impact is possible
3. Having completed the matrix, in the upper left hand corner of each box with a slash, place a number from 1 to 10 which indicates the MAGNITUDE of the possible impact; 10 represents the greatest magnitude of impact and 1, the least (no zeroes). Before each number place + (if the impact would be beneficial). In the lower right hand corner of the box place a number from 1 to 10 which indicates the IMPORTANCE of the possible impact (e.g. regional vs. local); 10 represents the greatest importance and 1 the least (no zeroes)
4. The text which accompanies the matrix should be a discussion of the significant impacts, those columns and rows with large numbers of boxes marked and individual boxes with large numbers

	a	b	c	d	e
a		2/1			8/3
b		7/2	8/8	3/1	9/7

Sample matrix

		A. Modification of regime													B. Land transformation and construction																			C. Resource extraction						
CHEMICAL CHARACTERISTICS	Proposed actions	a. Exotic flora or fauna introduction	b. Biological controls	c. Modification of habitat	d. Alteration of ground cover	e. Alteration of ground water hydrology	f. Alteration of drainage	g. River control and flow modification	h. Canalization	i. Irrigation	j. Weather modification	k. Burning	l. Surface or paving	m. Noise and vibration	a. Urbanization	b. Industrial sites and buildings	c. Airports	d. Highways and bridges	e. Roads and trails	f. Railroads	g. Cables and lifts	h. Transmission lines, pipelines, corridors	i. Barriers including fencing	j. Channel dredging and straightening	k. Channel revetments	l. Canals	m. Dams and impoundments	n. Piers, seawall, marinas and sea terminals	o. Offshore structures	p. Recreational structures	q. Blasting and drilling	r. Cut and fill	s. Tunnels and underground structures	a. Blasting and drilling	b. Surface excavation	c. Subsurface excavation and retorting	d. Well drilling and fluid removal	e. Dredging	f. Clear cutting and other lumbering	g. Commercial fishing and hunting
1. Earth	a. Mineral resources																																							
	b. Construction material																																							
	c. Soils																																							
	d. Land form																																							
	e. Force fields and background radiation																																							
	f. Unique features																																							
2. Water	a. Surface																																							
	b. Ocean																																							
	c. Underground																																							
	d. Quality																																							
	e. Temperature																																							
	f. Recharge																																							
	g. Snow, ice and permafrost																																							

(b)

Part 1. Project actions

A. Modification of regime
a) exotic flora or fauna introduction
b) Biological controls
c) Modification of habitat
d) Alteration of ground cover
e) Alteration of groundwater hydrology
f) Alteration of drainage
g) River control and flow modification
h) Canalization
i) Irrigation
j) Weather modification
k) Burning
l) Surface or paving
m) Noise and vibration

B. Land transformation and construction
a) Urbanization
b) Industrial sites and buildings
c) Airports
d) Highways and bridges
e) Roads and trails
f) Railroads
g) Cables and lifts
h) Transmission lines, pipelines and corridors
i) Barriers, including fencing
j) Channel dredging and straightening
k) Channel revetments
l) Canals
m) Dams and impoundments
n) Piers, seawalls, marinas, and sea terminals
o) Offshore structures
p) Recreational structures
q) Blasting and drilling
r) Cut and fill
s) Tunnels and underground structures

C. Resource extraction
a) Blasting and drilling
b) Surface excavation
c) Subsurface excavation and retorting
d) Well drilling and fluid removal
e) Dredging
f) Clear cutting and other lumbering
g) Commercial fishing and hunting

D. Processing
a) Farming
b) Ranching and grazing
c) Feed lots
d) Dairying
e) Energy generation
f) Mineral processing
g) Metallurgical industry
h) Chemical industry
i) Textile industry
j) Automobile and aircraft
k) Oil refining
l) Food
m) Lumbering
n) Pulp and paper
o) Product storage

E. Land alteration
a) Erosion control and terracing
b) Mine sealing and waste control
c) Strip-mining rehabilitation
d) Landscaping
e) Harbour dredging
f) Marsh fill and drainage

F. Resource renewal
a) Reforestation
b) Wildlife stocking and management
c) Groundwater recharge
d) Fertilization application
e) Waste recycling

G. Changes in traffic
a) Railway
b) Automobile
c) Trucking
d) Shipping
e) Aircraft
f) River and canal traffic
g) Pleasure boating
h) Trails
i) Cables and lifts
j) Communication
k) Pipeline

H. Waste emplacement and treatment
a) Ocean dumping
b) Landfill
c) Emplacement of tailings, spoil and overburden
d) Underground storage
e) Junk disposal
f) Oil well flooding
g) Deep well emplacement
h) Cooling water discharge
i) Municipal waste discharge, including spray irrigation
j) Liquid effluent discharge
k) Stabilization and oxidation ponds
l) Septic tanks, commercial and domestic
m) Stack and exhaust emission
n) Spent lubricants

I. Chemical treatment
a) Fertilization
b) Chemical de-icing of highways, etc.
c) Chemical stabilization of soil
d) Weed control
e) Insect control (pesticides)

J. Accidents
a) Explosions
b) Spills and leaks
c) Operational failure

Others

Part 2. Natural and human environmental elements

A. Physical and chemical characteristics
1. Earth
a) Mineral resources
b) Construction material
c) Soils
d) Landform
e) Force fields and background radiation
f) Unique physical features

2. Water
a) Surface
b) Ocean
c) Underground
d) Quality
e) Temperature
f) Recharge
g) Snow, ice and permafrost

3. Atmosphere
a) Quality (gases, particulates)
b) Climate (micro, macro)
c) Temperature

4. Processes
a) Floods
b) Erosion
c) Deposition (sedimentation, precipitation)
d) Solution
e) Sorption (ion exchange, complexing)
f) Compaction and settling
g) Stability (slides, slumps)
h) Stress-strain (earthquakes)
i) Air movements

B. Biological conditions
1. Flora
a) Trees
b) Shrubs
c) Grass
d) Crops
e) Microflora
f) Aquatic plants
g) Endangered species
h) Barriers
i) Corridors

2. Fauna
a) Birds
b) Land animals, including reptiles
c) Fish and shellfish
d) Benthic organisms
e) Insects
f) Microfauna
g) Endangered species
h) Barriers
i) Corridors

C. Cultural factors
1. Land-use
a) Wilderness and open spaces
b) Wetlands
c) Forestry
d) Grazing
e) Agriculture
f) Residential
g) Commercial
h) Industrial
i) Mining and quarrying

2. Recreation
a) Hunting
b) Fishing
c) Boating
d) Swimming
e) Camping and hiking
f) Picnicking
g) Resorts

3. Aesthetics and human interest
a) Scenic views and vistas
b) Wilderness qualities
c) Open space qualities
d) Landscape design
e) Unique physical features
f) Parks and reserves
g) Monuments
h) Rare and unique species or ecosystems
i) Historical or archaeological sites and objects
j) Presence of misfits

4. Cultural status
a) Cultural patterns, lifestyle
b) Health and safety
c) Employment
d) Population density

5. Man-made facilities and activities
a) Structures
b) Transportation network (movement access)
c) Utility networks
d) Waste disposal
e) Barriers
f) Corridors

D. Ecological relationships, such as
a) Salinization of water resources
b) Eutrophication
c) Disease – insect vectors
d) Food chains
e) Salinization of surficial material
f) Brush encroachment
g) Other

Others

Figure 4.11 (a) Part of a Leopold Matrix; (b) Leopold Matrix elements.

environmental components. Figure 4.11 shows a section of this matrix and lists all the matrix elements. Of the 8,800 possible interactions between project action and environmental component, Leopold et al. estimate that an individual project is likely to result in 25–50. In each appropriate cell, two numbers are recorded. The number in the top left-hand corner represents the impact's magnitude, from +10 (very positive) to -10 (very negative). That in the bottom right-hand corner represents the impact's significance, from 10 (very significant) to 1 (insignificant); there is no negative significance. This distinction between magnitude and significance is important: an impact could be large but insignificant, or small but significant. For instance, in ecological terms, paving over a large field of intensively used farmland may be quite insignificant compared with the destruction of even a small area of a SSSI.

The Leopold matrix is easily understood, can be applied to a wide range of developments, and is reasonably comprehensive for first-order, direct impacts. However, it has disadvantages. The fact that it was designed for use on many different types of projects makes it unwieldy for use on any one project. It cannot reveal indirect effects of developments: like checklists and most other matrices, it does not relate environmental components to one another, so that the complex interactions between ecosystem components that lead to indirect impacts are not assessed. The inclusion of magnitude/significance scores has additional drawbacks: it gives no indication of whether the data on which these values are based are qualitative or quantitative; it does not specify the probability of an impact occurring; it excludes details of the techniques used to predict impacts; and the scoring system is inherently subjective and open to bias. People may also attempt to add the numerical values to produce a composite value for the development's impacts and compare this with that for other developments; this should not be done because the matrix does not assign weightings to different impacts to reflect their relative importance (Clark et al. 1979).

Weighted matrices were developed in an attempt to respond to some of the above problems. Importance weightings are assigned to environmental components, and sometimes to project components. The impact of the project (component) on the environmental component is then assessed and multiplied by the appropriate weighting(s), to obtain an overall total for the project. Figure 4.12 shows a small weighted matrix that compares three alternative project sites. Each environmental component is assigned an importance weighting (a) relative to other environmental components: in the example, air quality is weighted 21% of the total environmental components. The magnitude (c) of the impact of each project on each environmental component is then assessed on a scale of 0–10, and multiplied by (a) to obtain a weighted impact (a×c): for instance, site A has an impact of 3 out of 10 on air quality, which is multiplied by 21 to give the weighted impact. For each site, the weighted impacts can then be added up to give a project total. The site with the lowest total, in this case site B, is the least environmentally harmful.

Figure 4.13 shows a similar abbreviated weighted matrix for a sewage treatment facility, broken down into its components. An importance weighting (b),

Environmental component	(a)	Alternative sites					
		Site A		Site B		Site C	
		(c)	(axc)	(c)	(axc)	(c)	(axc)
Air quality	21	3	63	5	105	3	63
Water quality	42	6	252	2	84	5	210
Noise	9	5	45	7	63	9	81
Ecosystem	28	5	140	4	112	3	84
Total	100		500		364		438

(a) = relative weighting of environmental component (total 100)
(c) = impact of project at particular site on environmental component (0–10)

Figure 4.12 A weighted matrix: alternative project sites.

out of 100, has been determined for each project component, and the magnitude (c) of the impact of that project component on each environmental component is then assessed, out of 10. The factors (b) and (c) are then multiplied with the importance weighting (a) of the relevant environmental component to give the weighted impact of each project component on each environmental component. All of these can be added to represent the total impact of the project. This can then be compared with those of other projects. In the example, the treatment plant is the only project component to affect water quality (b = 100), and it has a large impact (c = 9) on water quality; the weighted impact of the treatment plant on water quality is thus 900. This is multiplied by the importance weighting of water quality, 42, to get the weighted impact on water quality, 37,800. In the case

	Importance weighting (a)	Treatment plant	Pumping station	Interceptor	Outfall	Total
Air quality	21	10(b) 8(c)	0 –	50 7	40 8	15,750
Water quality	42	100 9	0 –	0 –	0 –	37,800
Noise	9	0 –	100 3	0 –	0 –	2700
Ecosystem	28	10 5	20 4	40 8	30 8	19,320
Total	100					75,570

(a) = relative weighting of environmental component (total 100)
(b) = relative weighting of project component (total 100)
(c) = impact of project on environmental component (0–10)

Figure 4.13 A weighted matrix: weighted project components. Based on Wenger & Rhyner (1972).

of air quality, where more than one project component affects air, the weighted impacts of the various components are first added up (e.g. 80 for the treatment plant plus 350 for the interceptor plus 320 for the outfall), and then multiplied by the importance weighting of air quality, 21, to get the weighted impact on air quality, 15,750. The project's total weighted impact could then be compared to that of other project alternatives.

This method has the advantage of allowing various alternatives to be compared numerically. However, the evaluation procedure depends heavily on the weightings and impact scales assigned. The major problems implicit in such weighting approaches are considered further in Chapter 5. The method also does not consider indirect impacts.

Quantitative methods

Quantitative methods attempt to compare the relative importance of all impacts by weighting, standardizing and aggregating impacts to produce a composite index. The best known of these methods is the *environmental evaluation system* (EES) devised by the Battelle Columbus Laboratories for the US Bureau of Land Reclamation to assess water resource developments, highways, nuclear power plants and other projects (Dee et al. 1973). It consists of a checklist of 74 environmental, social and economic parameters that may be affected by a proposal; these are shown in Figure 4.14. It assumes that these parameters can be expressed numerically and that they represent an aspect of environmental quality. For instance, the concentration of dissolved oxygen is a parameter that represents an aspect of the quality of an aquatic environment.

For each parameter, functions were designed by experts to express environmental quality on a scale of 0–1 (degraded – high quality). Two examples are shown in Figure 4.15. For instance, a stream with more than 10 mg/l of dissolved oxygen is felt to have a high level of environmental quality (1.0), whereas one with only 4 mg/l of dissolved oxygen is felt to have an environmental quality of only about 0.35. Impacts are measured in terms of the likely change in environmental quality for each parameter. Two environmental quality scores are determined for each parameter, one for the current state of the environment and one for the state predicted once the project is in operation. If the post-development score is less than the pre-development score, the impact is negative, and vice-versa.

To enable impacts to be compared directly, each parameter is given an importance weighting, which is then multiplied by the appropriate environmental quality score. The importance weightings (shown in parentheses in Fig. 4.14) were determined by having a panel of experts distribute 1,000 points among the parameters. For instance, dissolved oxygen was considered quite important, at 31 points out of 1,000. A composite score for beneficial/adverse effects of a single project, or for the net impact of alternative projects, can be obtained by adding up the weighted impact scores.

As an example of the full use of the EES, assume that the existing deer : range-

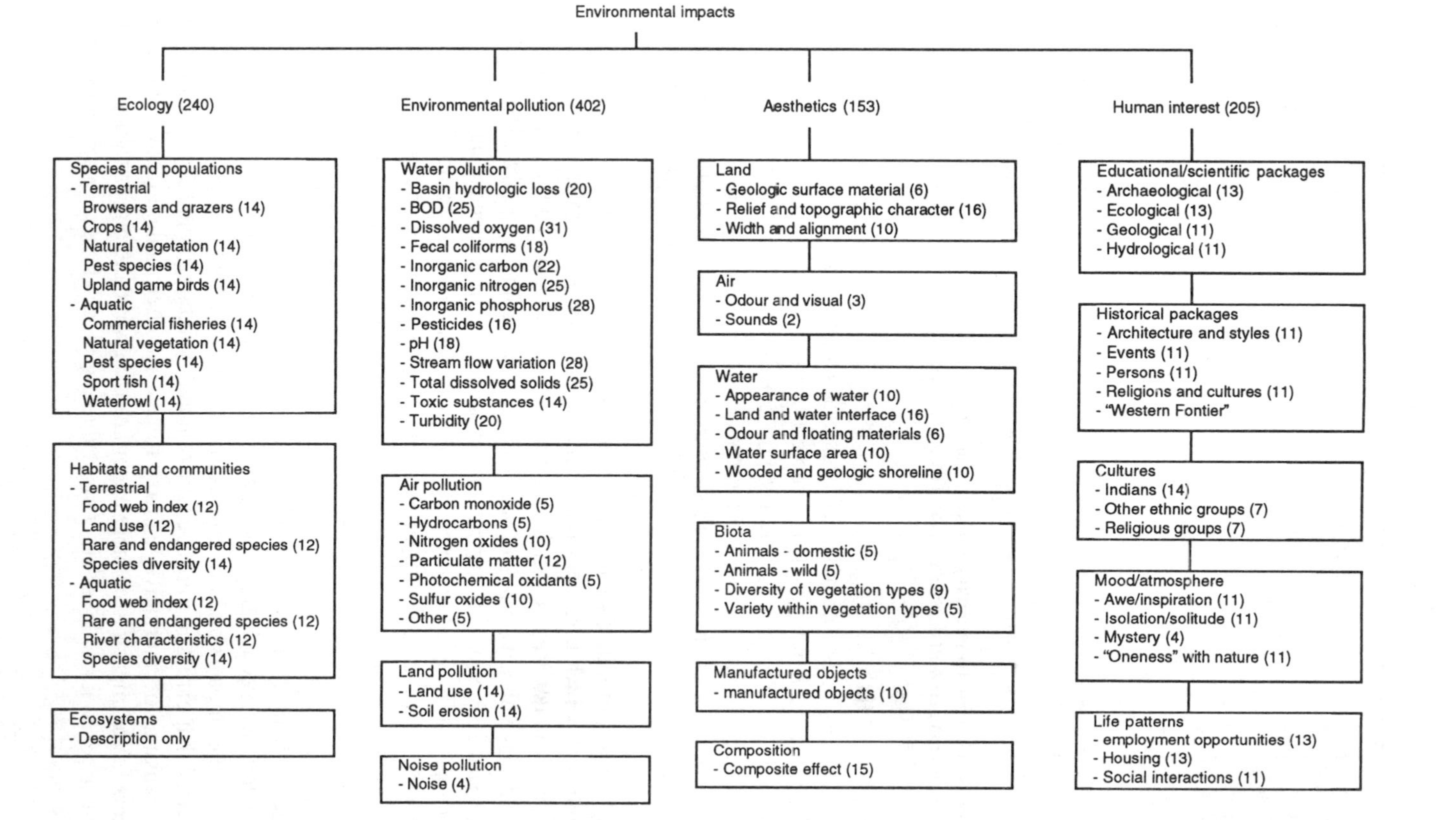

Figure 4.14 Framework for the Battelle Environmental Evaluation system. (*Source:* Dee et al. 1973.)

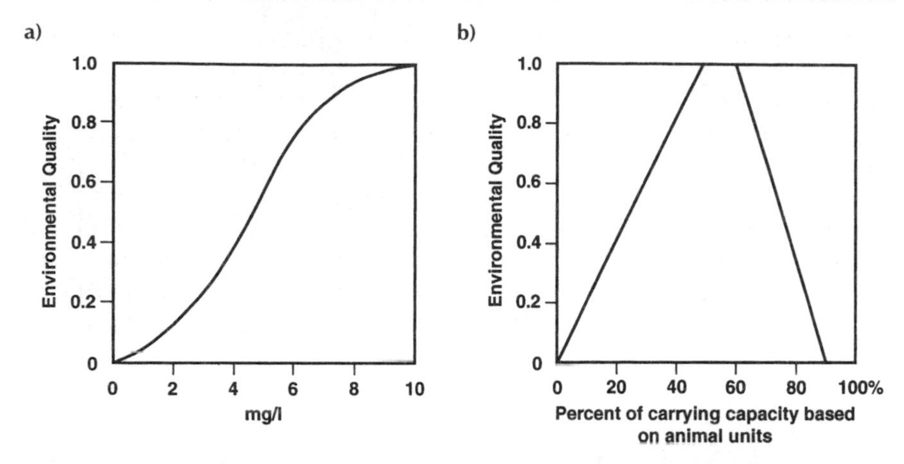

Figure 4.15 Environmental parameter functions for the Environmental Evaluation System: dissolved oxygen and deer : rangeland ratios: a) dissolved oxygen; b) browsers and grazers. *Source:* Dee et al. (1973).

land ratio results in 40% of annual plant production being consumed (environmental quality score 0.8 in Fig. 4.15). A project likely to halve the deer population would cause the score to drop to 0.4. The post-development score would be lower than the pre-development score, so the impact would be negative. This parameter's importance is 14 points out of 1000, so the pre- and post-development scores would be multiplied by 14, and could then be compared with other parameters (Dee et al. 1973).

After examining 54 methods of impact identification, the US Army Corps of Engineers, which is responsible for many water-resource projects requiring EIA, determined that methods such as EES had most potential, and used the principles of EES in forming its *water resources assessment methodology* (WRAM). The WRAM approach assigns project impacts to four accounts: environmental quality, regional development, national economic development and social wellbeing. Factors in each account are weighted and expressed in common terms by the use of functional curves that are similar to EES value functions. Aggregate impact scores are then obtained for each account (Solomon et al. 1977).

Another quantitative method was developed to assess alternative highway proposals (Odum et al. 1975). Unlike EES, this method considers impact duration: long-term irreversible impacts are considered to be more important than short-term reversible impacts and are given ten times more weight. A sensitivity analysis showed that errors in impact estimation and weighting could significantly affect the rankings of alternative highway routes. Another method (Stover 1972) considers future impacts to be more important and gives them higher values than short-term impacts: it multiplies the numerical rating of each future impact by its duration in years.

The attraction of these quantitative methods lies in their ability to "substantiate" numerically that a particular course of action is better than others. This may save decision-makers considerable work and it ensures consistency in assessment and results. However, these methods also have some fundamental weaknesses. They effectively take decisions away from decision-makers (Skutsch & Flowerdew 1976). The methods are difficult for lay-people to understand, and their acceptability depends on the assumptions, especially the weighting schemes, built into them.[3] People carrying out assessments may manipulate results by changing assumptions (Bisset 1978). Quantitative methods also treat the environment as if it were made up of discrete units. Impacts are only related to particular parameters, and much information is lost when impacts are reduced to numbers.

Networks

Network methods explicitly recognize that environmental systems consist of a complex web of relationships, and try to reproduce that web. Impact identification using networks involves following the effects of development through changes in the environmental parameters in the model. The *Sorensen network* was the first network method to be developed; it aimed to help planners to reconcile conflicting land-uses in California. Figure 4.16 shows a section of the network dealing with impacts on water quality. Water is one of the six environmental components, the others being climate, geophysical conditions, biota, access conditions and aesthetics.

The Sorensen method begins by identifying potential causes of environmental change associated with a proposed development action, using a matrix format; for instance, forestry is shown potentially to result in the clearing of vegetation and the use of herbicides and fertilizers. These environmental changes in turn result in specific environmental impacts; in the example, the clearing of vegetation could result in an increased flow of fresh water, which in turn could imperil cliff structures. The analyst stops following the network when an initial cause of change has been traced through all subsequent impacts and changes in environmental conditions, to its final impacts. Environmental impacts can result either directly from the development action, or indirectly through induced changes in environmental conditions. A change in environmental conditions may result in several different types of impact. Sorensen argues that the method should lead to the identification of remedial measures and monitoring schemes (Sorensen 1971).

The Sorensen network does not establish the magnitude or significance of interrelationships between environmental components, nor the extent of change. It requires much time and considerable knowledge of the environment under consideration to construct the network, and it is time-consuming to use manually. Its main advantage is its ability to trace the higher-order impacts of proposed developments. A similar network method, the computerized *IMPACT network*, was designed to assess the impacts of developments on forests and rangelands controlled by the US Forest Service (Thor et al. 1978).

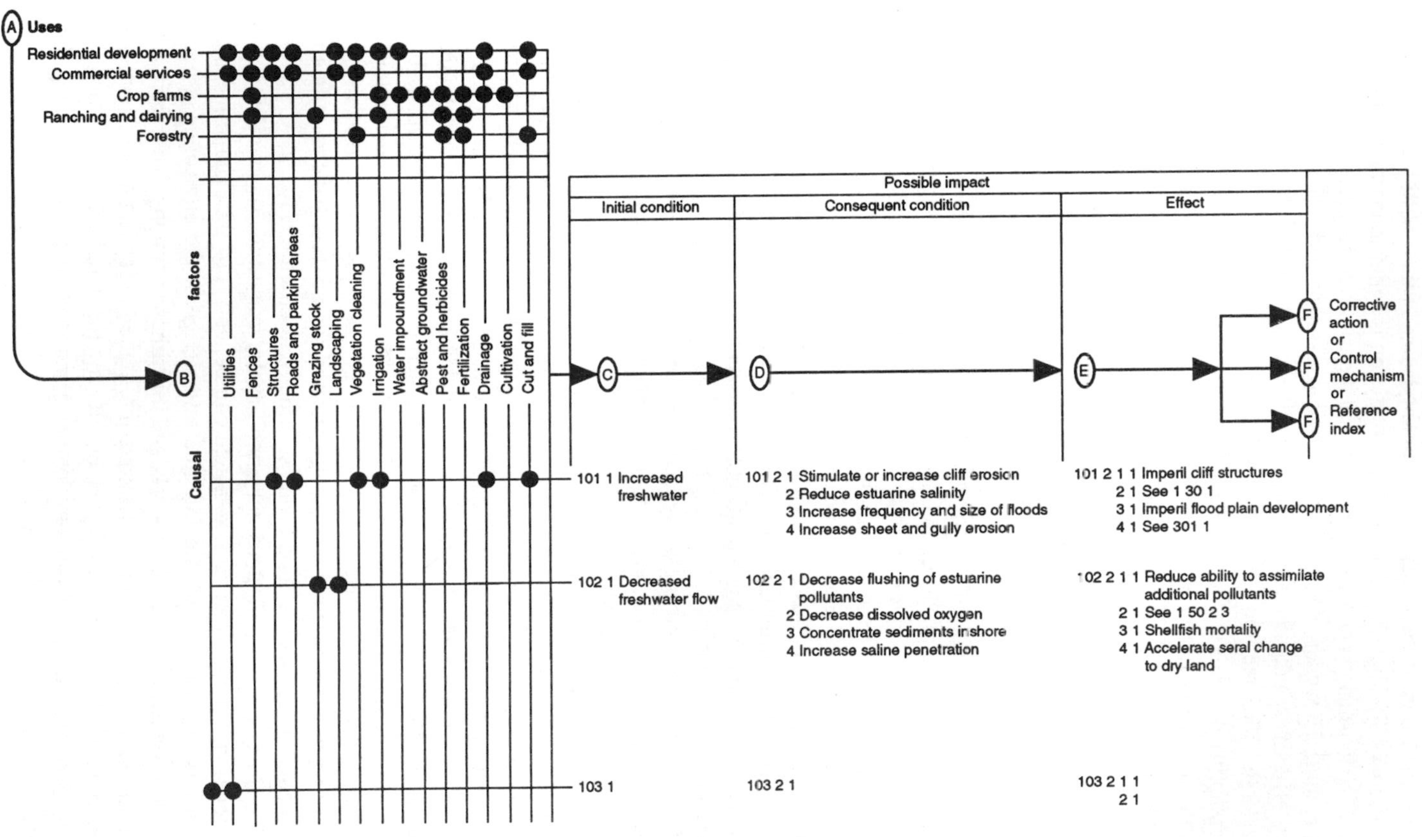

Figure 4.16 Part of the Sorensen Network. *(Source:* Sorensen 1971.)

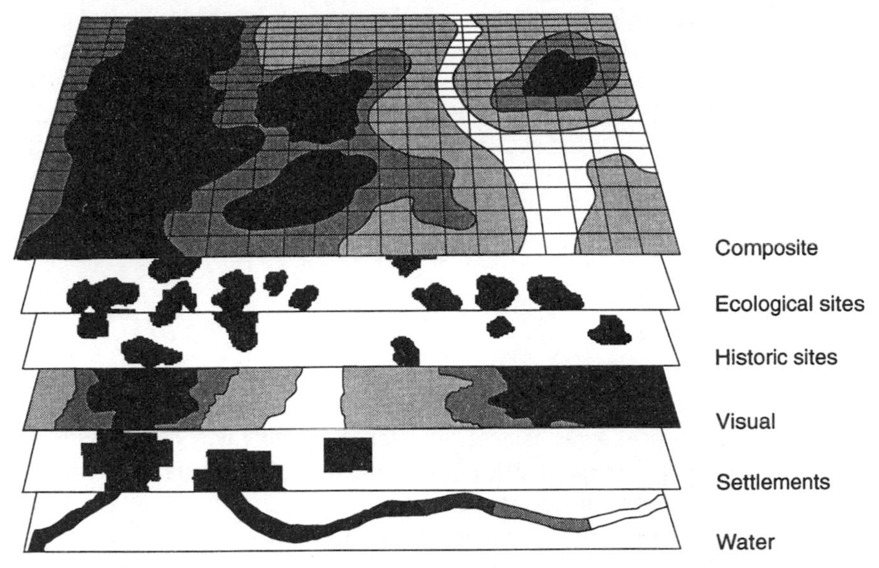

Figure 4.17 Example of overlay maps.

Overlay maps

Overlay maps have been used in environmental planning since the 1960s (McHarg 1968), before the NEPA was enacted. A series of transparencies is used to identify, predict, assign relative significance to and communicate impacts, normally at a scale larger than local. A base map is prepared, showing the general area within which the project may be located. Successive transparent overlay maps are then prepared for individual environmental components that, in the opinion of experts, are likely to be affected by the project (e.g. agriculture, woodland, noise). The project's degree of impact on the environmental feature is shown by the intensity of shading, with darker shading representing a greater impact. The composite impact of the project is found by superimposing the overlay maps and noting the relative intensity of the total shading. Unshaded areas are those where a development project would not have a significant impact. Figure 4.17 shows an example of this technique.

Overlay maps are easy to use and understand, and are popular in practice. They are an excellent way of showing the spatial distribution of impacts. They also lead intrinsically to a low-impact decision. The overlay maps method is particularly useful for identifying optimum corridors for developments such as electricity lines and roads, for comparisons between alternatives, and for assessing large regional developments. The development of computer mapping, and in particular geographical information systems, allows more information to be handled. It also allows different importance weightings to be assigned to the impacts: this enables a sensitivity analysis to be carried out, to see whether changing assumptions about impact importance would alter the decision. However, the method is limited in that it does not consider factors such as the likelihood of the impact occurring, secondary impacts, or the difference between reversible and irreversible impacts. It requires the clear classification of often indeterminate boundaries (such as that between forest and field), and so is not a true representation of conditions on the ground. It relies on the user to identify likely impacts before it can be used. Manual use of a large number of overlays is also often difficult, and is usually limited to about ten transparencies.

Summary

Table 4.4 summarizes the respective advantages of the major impact identification methods discussed in this section.[4] In summary, given the complexity of many impact identification techniques, it is understandable why most EIAs in the UK use checklists or simple matrices, or some hybrid combination including elements from several of the methods discussed. However, as more EIAs are carried out, as legislation concerning indirect and cumulative impacts (see Ch. 13) is enacted, and especially as large developers begin to establish patterns in preparing EIAs, the use of more sophisticated methods for impact identification may

Table 4.4 Comparison of impact identification methods.

	Criterion										
	1	2	3	4	5	6	7	8	9	10	11
Checklists											
Simple/ descriptive/ question	✓	✓						✓	✓	✓	✓
Threshold	✓	✓	✓		✓	✓	✓		✓	✓	✓
Matrices											
Simple	✓	✓						✓	✓	✓	✓
Magnitude/ time-dependent	✓	✓	✓					✓	✓	✓	✓
Leopold	✓	✓	✓		✓			✓	✓		✓
Weighted	✓	✓			✓	✓		✓			✓
Quantitative											
EES/WRAM	✓		✓		✓	✓	✓				
Network											
Sorensen	✓			✓		✓		✓		✓	
Overlay maps		✓	✓		✓	✓		✓	✓	✓	✓

1. Compliance with regulations; 2. comprehensive coverage (social, economic and physical impacts); 3. positive v. negative, reversible v. irreversible impacts, etc.; 4. secondary, indirect, cumulative impacts; 5. significant v. insignificant impacts; 6. compare alternative options; 7. compare against carrying capacity; 8. uses qualitative and quantitative information; 9. easy to use; 10. unbiased, consistent; 11. summarizes impacts for use in EIS.

increase. This may of course not be wholly positive in the EIA process. EIA methods are not politically neutral, and the more sophisticated the method becomes, often the more difficult becomes clear communication and effective participation (see Ch. 6 for more discussion).

4.8 Summary

The early stages of the EIA process are typified by several interacting steps. These include deciding whether an EIA is needed at all ("screening"), consulting with the various parties involved to seek to produce an initial focus on some of the key impacts ("scoping"), and an outline of possible alternative approaches to the project, including alternative locations, scales and processes. Scoping and alternatives are not mandatory in an EIA in the UK, but both can greatly improve the quality of the process. Early in the process the analyst will also wish to understand the nature of the project concerned, and the environmental baseline conditions in the likely impacted area. Projects have several dimensions (e.g. purpose, physical presence, processes and policies, over several stages in the project life-cycle); a consideration of the environmental baseline also involves several dimensions. For both projects and the impacted environment, obtaining relevant data may present challenges.

Impact identification includes most of the activities already discussed. It also usually involves the use of impact identification methods, ranging from simple checklists and matrices to complex computerized models and networks. The simpler methods are generally easier to use, more consistent and more effective in presenting information in the EIS, but their coverage of impact significance, indirect impacts or alternatives is either very limited or non-existent. The more complex models incorporate these aspects, but at the cost of immediacy. In the UK, if any formal impact identification methods are used, they are normally the simpler types. The methods discussed here have relevance also to the prediction, assessment/evaluation, communication and mitigation of environmental impacts, which are discussed in the next chapter.

References

Aspinwall and Company Ltd 1991. *Environmental Audit of Oxfordshire*. Oxford: Oxfordshire County Council.

Bisset, R. 1978. Quantification, decision-making and environmental impact assessment in the United Kingdom. *Journal of Environmental Management* 7(1), 43–58.

Bisset, R. 1983. A critical survey of methods for environmental impact assessment. In *An annotated reader in environmental planning and management,* T. O'Riordan & R. K. Turner (eds), 168–85. Oxford: Pergamon Press.

Bisset, R. 1989. Introduction to EIA methods. Paper presented at the 10th International Seminar on Environmental Impact Assessment, University of Aberdeen, 9–22 July.

Bregman, J. I. & K. M. Mackenthun 1992. *Environmental impact statements*. Chelsea, Michigan: Lewis.

CEQ (Council on Environmental Quality) 1978. National Environmental Policy Act. Implementation of procedural provision: final regulations. *Federal Register* **43**(230), 55977–56007 (29 November 1978).

Clark, B. D., K. Chapman, R. Bisset, P. Wathern 1979. Environmental impact assessment. In D. Lovejoy ed., *Land-use and landscape planning*, D. Lovejoy (ed.), 53–87. Glasgow: Leonard Hill.

Dee, N., J. K. Baker, N. L. Drobny, K. M. Duke, I. Whitman, D. C. Fahringer 1973. An environmental evaluation system for water resources planning. *Water Resources Research* **3**, 523–35.

DoE (Department of the Environment) 1989. *Environmental assessment: a guide to the procedures*. London: HMSO.

DoE 1991. *Policy appraisal and the environment*. London: HMSO.

DoT (Department of Transport) 1983. *Manual of environmental appraisal*. London: HMSO.

Fortlage, C. 1990. *Environmental assessment: a practical guide*. Aldershot: Gower.

Fuller, K. 1992. Working with assessment. In *Environmental assessment and audit, a user's guide 1992–1993* (special supplement to *Planning*), 14–15.

Jain, R. K., L. V. Urban, G. S. Stacey 1977. *Environmental impact assessment*. New York: Van Nostrand Reinhold.

Jones, C. E., N. Lee, C. Wood 1991. *UK environmental statements 1988–1990: an analysis*. Occasional Paper 29, Department of Town and Country Planning, University of Manchester.

Lee, N. 1987. *Environmental impact assessment: a training guide*. Occasional Paper 18, Department of Town and Country Planning, University of Manchester.

Leopold, L. B., F. E. Clarke, B. B. Hanshaw, J. R. Balsley 1971. *A procedure for evaluating environmental impact*. Washington DC: US Geological Survey Circular 645.

McHarg, I. 1968. *A comprehensive route selection method*. Highway Research Record 246. Washington dC: Highway Research Board.

Marstrand, P. K. 1976. Ecological and social evaluation of industrial development. *Environmental Conservation* **3**(4), 303–8.

Mills, J. 1992. *Monitoring the visual impacts of major projects*. MSc dissertation, School of Planning, Oxford Brookes University.

Munn, R. E. 1979. *Environmental impact assessment: principles and procedures*, 2nd edn. New York: John Wiley.

Odum, E. P., J. C. Zieman, H. H. Shugart, A. Ike, J. R. Champlin 1975. In *Environmental impact assessment*, M. Blisset (ed.). Austin: University of Texas Press.

Parker, B. C. & R. V. Howard 1977. The first environmental monitoring and assessment in Antarctica: the Dry Valley drilling project. *Biological Conservation* **12**(2), 163–77.

Rau, J. G. & D. C. Wooten 1980. *Environmental impact analysis handbook*. New York: McGraw-Hill.

Rendel Planning 1990. *Angle Bay energy project environmental statement*. London: Rendel Planning.

Sassaman, R. W. 1981. Threshold of concern: a technique for evaluating environmental impacts and amenity values. *Journal of Forestry* **79**, 84–6.

Schaenman, P. W. 1976. *Using an impact measurement system to evaluate land devel-*

opment. Washington DC: The Urban Institute.

Skutsch, M. M. & R. T. N. Flowerdew 1976. Measurement techniques in environmental impact assessment. *Environmental Conservation* **3**(3), 209–17.

Solomon, R. C., B. K. Colbert, W. J. Hanson, S. E. Richardson, L. W. Canter, E. C. Vlachos 1977. *Water resources assessment methodology (WRAM) – impact assessment and alternative evaluation*. Technical Report no. Y-77-1, Army Corps of Engineers, Vicksburg, Mississippi.

Sorensen, J. C. 1971. *A framework for the identification and control of resource degradation and conflict in multiple use of the coastal zone*. Department of Landscape Architecture, University of California, Berkeley.

Sorensen J. C. & M. L. Moss 1973. *Procedures and programmes to assist in the environmental impact statement process*. Institute of Urban and Regional Development, University of California, Berkeley.

State of California, Governor's Office of Planning and Research 1992. *California Environmental Quality Act: statutes and guidelines*. Sacramento: State of California.

Stover, L. V. 1972. *Environmental impact assessment: a procedure*. Pottstown, Pennsylvania: Sanders & Thomas.

Thor, E. C., G. H. Elsner, M. R. Travis, K. M. O'Loughlin 1978. Forest environmental impact analysis – a new approach. *Journal of Forestry* **76**, 723–5.

TNO 1983. *Handbook of emission factors*. The Hague: Netherlands Ministry of Housing, Physical Planning and Environmental Affairs.

United Nations Environment Programme 1981. *Guidelines for assessing industrial environmental impact and environmental criteria for the siting of industry*. New York: United Nations.

US Agency for International Development 1981. *Environmental design considerations for rural development projects*. Washington DC: US Agency for International Development.

VROM 1985. *Handling uncertainty in environmental impact assessment* (MER Series, vol. 18). The Hague: Netherlands Ministry of Public Housing, Physical Planning and Environmental Affairs.

Wenger, R. B. & C. R. Rhyner 1972. Evaluation of alternatives for solid waste systems. *Journal of Environmental Systems* **2**(2), 89–108.

Wathern, P. 1984. Ecological modelling in impact analysis. In *Planning and ecology*, R. D. Roberts & T. M. Roberts (eds), 80–93. London: Chapman & Hall.

Wood, C. & N. Lee (eds) 1987. *Environmental impact assessment: five training case studies*. Occasional Paper 19, Department of Town and Country Planning, University of Manchester.

Notes

1. Jones et al. (1991) note that, of 100 EISs prepared between 1988 and 1989, 34 discussed alternatives. This was broadly confirmed by a brief review of EISs at Oxford Brookes University.
2. In the US, "agencies should: consider the option of doing nothing; consider alternatives outside the remit of the agency; and consider achieving only a part of their objectives in order to reduce impact".

3. For instance, the EES's assumption that individual indicators of water quality (such as dissolved oxygen at 31 points) are more important than employment opportunities and housing put together (at 26 points) would certainly be challenged by large sectors of the public.
4. Another category of techniques, simulation models, was not discussed because they are still relatively undeveloped and have, to date, only been applied to problems involving a few environmental impacts.

CHAPTER 5
Impact prediction, evaluation and mitigation

5.1 Introduction

The focus of this chapter is on the central steps of impact prediction, evaluation and mitigation. This is the heart of the EIA process, although as already noted the process is not linear. Indeed the whole EIA exercise is about prediction. It is needed at the earliest stages when the project, including alternatives, is being planned and designed, and it continues through to mitigation, monitoring and auditing. Yet, despite the centrality of prediction in EIA, there is a tendency for many studies to underemphasize it at the expense of more descriptive studies. Prediction is often not treated as an explicit stage in the process; clearly defined models are often missing from studies. Even when used, models are not detailed and there is little discussion of limitations. §5.2 examines the dimensions of prediction (what to predict), the methods and models used in prediction (how to predict), and the limitations implicit in such exercises (living with uncertainty).

Evaluation follows from prediction and involves an assessment of the relative significance of the impacts. Methods range from the intuitive to the analytical, from qualitative to quantitative, from formal to informal. Cost–benefit analysis, monetary valuation techniques, and multi-criteria/multi-attribute methods, with their scoring and weighting systems, provide a number of ways into the evaluation issue. The chapter concludes with a discussion of approaches to the mitigation of significant adverse effects. This may involve measures to avoid, reduce, remedy or compensate for the various impacts associated with projects.

5.2 Prediction

Dimensions of prediction (what to predict?)

The objective of prediction is to identify the magnitude and other dimensions of identified change in the environment with a project/action, in comparison with the situation without that project/action. Predictions also provide the basis for the assessment of significance, which is discussed in §5.3.

One starting point to identify the dimensions of prediction in the UK is the *legislative requirements* (see Table 3.5, paras 2c and 3c). These basic specifications are amplified in guidance given in *Environmental assessment: a guide to the procedures* (DoE 1989) as outlined in Table 5.1. As already noted, this listing is limited on the assessment of socio-economic impacts. Table 1.3 provides a broader view of the scope of the environment, and of the environmental receptors that may be affected by a project.

Prediction involves the identification of potential change in *indicators* of such environment receptors. Scoping will have identified the broad categories of impacts in relation to the project under consideration. If a particular environmental indicator (e.g. SO_2 levels in the air) revealed an increasing problem in an area, irrespective of the project/action (e.g. a power station), this should be predicted forwards as the baseline for this particular indicator. These indicators need to be disaggregated and specified to provide variables that are measurable and relevant. For example an economic impact could be progressively specified as

direct employment → local employment → local skilled employment.

In this way, a list of significant impact indicators of policy relevance can be developed.

An important distinction is often made between the prediction of the likely *magnitude* (i.e. size) and the *significance* (i.e. importance for decision-making) of the impacts. Magnitude does not always equate with significance. For example, a large proportionate increase in one pollutant may still result in an outcome within generally accepted standards, whereas a small increase in another may take it above the applicable standards. In terms of the Sassaman checklist (see Fig. 4.7), the latter is crossing the threshold of concern and the former is not. This also highlights the distinction between *objective* and *subjective* approaches. Prediction of the magnitude of impacts should be an objective exercise, although this is not always easy. The determination of significance is a more subjective exercise as it normally involves value judgements.

As noted in Table 1.4, prediction should also identify *direct* and *indirect* impacts (simple cause–effect diagrams may be useful here), the *geographical extent* of impacts (e.g. local, regional, national), whether the impacts are *beneficial* or *adverse*, and the *duration* of the impacts. In addition to prediction over the time horizon of the project (including, for example, the construction, operational and other stages), the analyst should also be alert to the "rate of change" of impacts. A slow build-up in an impact may be more acceptable than a rapid change; the development of tourism projects in formerly remote/undeveloped areas provides

Table 5.1 Assessment of effects, as outlined in UK regulations.

Assessment of effects

(including direct and indirect, secondary, cumulative, short-, medium- and long-term, permanent and temporary, positive and negative effects of project)

Effects on human beings, buildings and man-made features

1 Change in population arising from the development, and consequential environment effects.
2 Visual effects of the development on the surrounding area and landscape.
3 Levels and effects of emissions from the development during normal operation.
4 Levels and effects of noise from the development.
5 Effects of the development on local roads and transport.
6 Effects of the development on buildings, the architectural and historic heritage, archaeological features, and other human artefacts, e.g. through pollutants, visual intrusion, vibration.

Effects on flora, fauna and geology

7 Loss of, and damage to, habitats and plant and animal species.
8 Loss of, and damage to, geological, palaeotological and physiographic features.
9 Other ecological consequences.

Effects on land

10 Physical effects of the development, e.g. change in local topography, effect of earth-moving on stability, soil erosion, etc.
11 Effects of chemical emissions and deposits on soil of site and surrounding land.
12 Land-use/resource effects:
  (a) quality and quantity of agricultural land to be taken;
  (b) sterilization of mineral resources;
  (c) other alternative uses of the site, including the "do nothing" option;
  (d) effect on surrounding land-uses including agriculture;
  (e) waste disposal.

Effects on water

13 Effects of development on drainage pattern in the area.
14 Changes to other hydrographic characteristics, e.g. ground water level, water courses, flow of underground water.
15 Effects on coastal or estuarine hydrology.
16 Effects of pollutants, waste, etc., on water quality.

Effects on air and climate

17 Level and concentration of chemical emissions and their environmental effects.
18 Particulate matter.
19 Offensive odours.
20 Any other climatic effects.

Other indirect and secondary effects associated with the project

21 Effects from traffic (road, rail, air, water) related to the development.
22 Effects arising from the extraction and consumption of materials, water, energy or other resources by the development.
23 Effects of other development associated with the project, eg. new roads, sewers, housing power lines, pipelines, telecommunications, etc.
24 Effects of association of the development with other existing or proposed development.
25 Secondary effects resulting from the interaction of separate direct effects listed above.

Source: DOE 1989.

a contemporary and topical example of the damaging impacts of rapid change. Projects may be characterized by non-linear processes, by time lags between cause and effect, and the intermittent nature of some impacts should be anticipated. The reversibility or otherwise of impacts, their *permanency*, and their *cumulative* and synergistic impacts should also be predicted. Cumulative (or additive) impacts are the collective effects of impacts that may be individually minor but in combination, often over time, may prove major. Such cumulative impacts are difficult to predict, and are often poorly covered or are missing altogether from EIA studies.

Another dimension is the unit of measurement, and the distinction between *quantitative* and *qualitative* impacts. Some indicators are more readily quantifiable than others (e.g. change in drinking water quality, in comparison, for example, with changes in community stress associated with a project). Where possible, predictions should seek to present impacts in explicit units, which can provide a basis for evaluation and trade-off. Quantification can allow predicted impacts to be assessed against various local, national and international standards. Predictions should also include estimates of the *probability* that the impact will occur, which raises the important issue of uncertainty.

Methods and models for prediction (how to predict?)

There are many potential methods to predict impacts; a study undertaken by Environmental Resources Ltd for the Dutch Government in the early 1980s identified 150 different prediction methods used in just 140 EIA studies from the Netherlands and North America (VROM 1984). None provides a magic solution to the prediction problem. "All predictions are based on conceptual models of how the universe functions; they range in complexity from those that are totally intuitive to those based on explicit assumptions concerning the nature of environmental processes . . . the environment is never as well behaved as assumed in models, and the assessor is to be discouraged from accepting off-the-shelf formulae" (Munn 1979).

Predictive methods can be classified in many ways, which are not mutually exclusive. In terms of *scope*, all methods are *partial* in their coverage of impacts, but some seek to be more *holistic* than others. Partial methods may be classified according to type of project (e.g. retail impact assessment), and type of impact (e.g. wider economic impacts). Some may be *extrapolative*, others may be more *normative*. For extrapolative methods, predictions are made that are consistent with past and present data. Extrapolative methods include, for example: trend analysis (extrapolating present trends, modified to take account of changes caused by the project), scenarios (common-sense forecasts of future state based on a variety of assumptions), analogies (transferring experience from elsewhere to the study in hand) and intuitive forecasting (e.g. use of the Delphi technique to seek to achieve group consensus on the impacts of a project) (Green et al. 1989). Normative approaches work backwards from desired outcomes to assess whether the project, in its environmental context, is adequate to achieve them. For example, a desired socio-economic outcome from the construction stage of a major project may be 50% local employment. The achievement of this outcome may necessitate modifications to the project and/or to associated employment policies (e.g. on

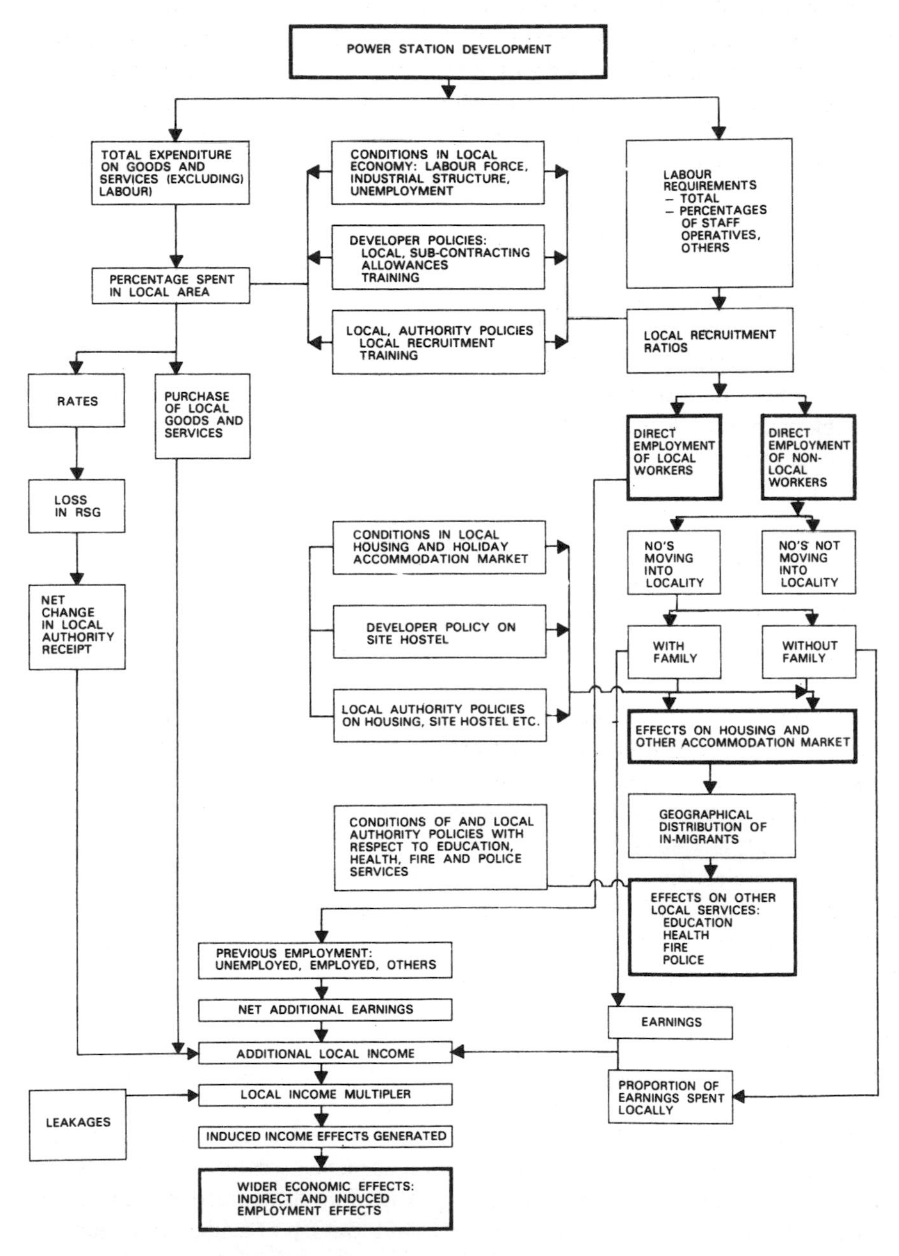

Figure 5.1 A cause–effect flow diagram for the local socio-economic impacts of a power station proposal. *Source:* Glasson et al. 1987.

training). Various scenarios may be tested to determine that most likely to achieve the desired outcomes.

Methods can also be classified according to their *form* as the following six types of models illustrate.

Mechanistic or mathematical models

Mechanistic or mathematical models describe cause–effect relationships in the form of flow charts or mathematical functions. The latter can range from simple direct "input–output" relationships to more complex dynamic mathematical models with a wide array of interrelationships. Mathematical models can be spatially aggregated (e.g. a model to predict the survival rate of a cohort population, or an economic multiplier for a particular area), or more locationally based, predicting net changes in detailed locations throughout a study area. Of the latter, retail impact models, which predict the distribution of retail expenditure using gravity model principles, provide a simple example; the comprehensive land-use locational models of Harris, Lowry, Cripps et al., provide more holistic examples (Journal of American Institute of Planners 1965). Mathematical models can also be divided into deterministic and stochastic models. Deterministic models, like the gravity model, depend on fixed relationships. In contrast, a stochastic model is probabilistic, and indicates "the degree of probability of the occurrence of a certain event by specifying the statistical probability that a certain number of events will take place in a given area and/or time interval" (Loewenstein 1966).

There are many mathematical models available for particular impacts. Reference to various EISs, especially from the USA, and to the literature (e.g. Bregman & Mackenthun 1992, Hansen & Jorgensen 1991, Rau & Wooten 1980, Suter 1993, US Environmental Protection Agency 1993, Westman 1985) reveals the availability of a rich array. For instance, Kristensen et al. (1990) list 21 mathematical models for phosphorus retention in lakes alone. Figure 5.1 provides a simple flow diagram for the prediction of the local socio-economic impacts of a power station development. Key determinants in the model are the details of the labour requirements for the project, the conditions in the local economy, and relevant local authority and developer policies on topics such as training, local recruitment and travel allowances. The local recruitment ratio is a key factor in the determination of subsequent impacts.

An example of a deterministic mathematical model, often used in socio-economic impact predictions, is the multiplier (Lewis 1988), an example of which is shown in Figure 5.2. The injection of money into an economy - local, regional or national - will increase income in the economy by some multiple of the original injection. Modification of the basic model allows it to be used to predict income and employment impacts for various groups over the various stages of the life of a project (Glasson et al. 1988). The more disaggregated (by industry type) input–output member of the multiplier family provides a particularly sophisticated method for predicting economic impacts, but with major data requirements.

$$Y_r = \frac{1}{1-(1-s)(1-t-u)(1-m)} J$$

where

Y_r = change in level of income (Y) in region (r), in £
J = initial income injection (or multiplicand)
t = proportion of additional income paid in direct taxation and National Insurance contributions
s = proportion of income saved (and therefore not spent locally)
u = decline in transfer payments (e.g. unemployment benefits) which result from the rise in local income and employment
m = proportion of additional income spent on imported consumer goods

Figure 5.2 A simple multiplier model for the prediction of local economic impacts.

Mass balance models

Mass balance models establish a mass balance equation for a given "compartment", namely a defined physical entity such as the water in a stream, a volume of soil, or an organism. Inputs to the compartment could be, for instance, water, energy, food or chemicals; outputs could be outflowing water, wastes, or diffusion to another compartment. Changes in the contents of the compartment equal the sum of the inputs minus the sum of the outputs, as illustrated in Figure 5.3. Mass balance models are particularly effective for describing physical changes such as the flow of water in a river basin or the flow of energy through an ecosystem.

Statistical models

Statistical models use statistical techniques such as regression or principal components analysis to describe the relationship between data, test hypotheses, or extrapolate data. For instance they can be used in a pollution-monitoring study

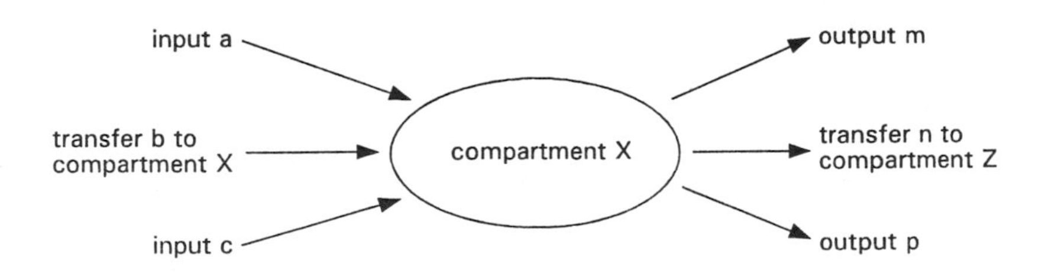

change in compartment X = a + b + c − m − n − p

at steady state, a + b + c = m + n + p

Figure 5.3 Example of a mass balance model.

to describe the concentration of a pollutant as a function of the stream-flow rates and the distance down stream. They can compare conditions at a contaminated and a control site to determine the significance of any differences in monitoring data. They can extrapolate a model to conditions outside the data range used to derive the model - e.g. from toxicity at high doses of a pollutant to toxicity at low doses - or from data that are available to data that are unavailable - e.g. from toxicity in rats to toxicity in humans.

Physical, image, or architectural models

Physical, image, or architectural models are illustrative or scale models that replicate some element of the project–environment interaction. For example a scale model (or computer graphics) could be used to predict the impacts of a development on the landscape or built environment.

Field and laboratory experimental methods

Field and laboratory experimental methods use existing data inventories, often supplemented by special surveys, to predict impacts on receptors. Field tests are carried out in unconfined conditions, usually at approximately the same scale as the predicted impact; an example would be the testing of a pesticide in an outdoor pond. Laboratory tests, such as the testing of a pollutant on seedlings raised in a hydroponic solution, are usually cheaper to run but may not extrapolate well to conditions in natural systems.

Analogue models

Analogue models make predictions based on analogous situations. They include comparing the impacts of a proposed development to a similar existing development; comparing the environmental conditions at one site with those at similar sites elsewhere; or comparing an unknown environmental impact (e.g. of wind turbines on radio reception) with a known environmental impact (e.g. of other forms of development on radio reception). Analogue models can be developed from site visits, literature searches, or monitoring of similar projects. Expert opinion, based, it is to be hoped, on previous relevant experience, may also be used.

Other methods for prediction

The various impact identification methods of Chapter 4 may also be of value in impact prediction. The Sassaman threshold of concern checklist approach has already been noted; the Leopold matrix also builds in magnitude predictions, although the objectivity of a system where each analyst is allowed to develop a ranking system on a scale of 1 to 10 is somewhat doubtful. Overlays can be used to predict spatial impacts and the Sorensen network is useful in tracing through indirect impacts.

Choice of prediction methods

In choosing prediction methods, an assessor should be concerned about their appropriateness for the task involved, in the context of the resources available (Lee 1987). Will the methods produce what is wanted (e.g. a range of impacts, for the appropriate geographical area, over various stages), from the resources available (including time, data, range of expertise)? In addition, criteria of replicability (method is free from analyst bias), consistency (method can be applied to different projects to enable comparisons of predictions to be made) and adaptability, should also be considered in the choice of methods. In many cases more than one method may be appropriate. For instance the range of methods available for predicting impacts on air quality is apparent from the 165 closely typed pages on the subject in Rau & Wooten (1980). Table 5.2 provides an overview of some of the possible methods of predicting the initial emissions of pollutants, which, with atmospheric interaction, may degrade air quality, which may then have effects on key receptors, including humans.

Table 5.2 Examples of methods used in predicting air quality impacts.

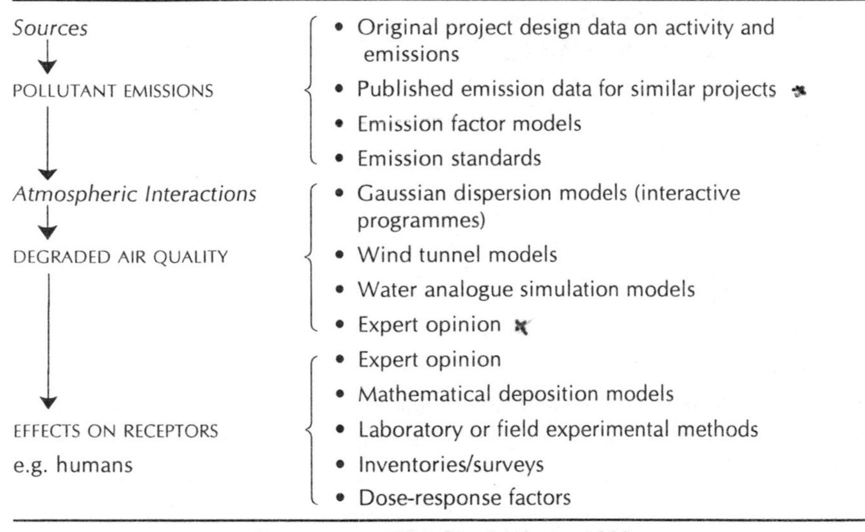

Sources ↓ POLLUTANT EMISSIONS	• Original project design data on activity and emissions • Published emission data for similar projects • Emission factor models • Emission standards
↓ *Atmospheric Interactions* ↓ DEGRADED AIR QUALITY	• Gaussian dispersion models (interactive programmes) • Wind tunnel models • Water analogue simulation models • Expert opinion
↓ EFFECTS ON RECEPTORS e.g. humans	• Expert opinion • Mathematical deposition models • Laboratory or field experimental methods • Inventories/surveys • Dose-response factors

Source: VROM 1984, Rau & Wooten 1980.

In practice, there has been a tendency to use the less formal predictive methods, and especially expert opinion (VROM 1984). Even where more formal methods have been used, they have tended to be simple, for example the use of photomontages for visual impacts, or the use of simple dilution and steady-state dispersion models for water quality. However, simple methods need not be inappropriate, especially for early stages in the EIA process, nor need they be applied uncritically or in a simplistic way. Lee (1987) provides the following illustration:

(a) a single expert may be asked for a brief, qualitative opinion; or

(b) the expert may also be asked to justify that opinion (i) by verbal or mathematical description of the relationships he has taken into account and/or (ii) by indicating the empirical evidence which supports that opinion; or
(c) as in (b), except that opinions are also sought from other experts; or
(d) as in (c), except that the experts are also required to reach a common opinion, with supporting reasons, qualifications, etc.; or
(e) as in (d), except that the experts are expected to reach a common opinion using an agreed process of consensus building (e.g. based on "Delphi" techniques (Golden et al. 1979)).

The development of more complex methods can be very time-consuming and expensive, especially since many of these models are limited to specific environmental components and physical processes, and may only be justified when a number of relatively similar projects are proposed. However, notwithstanding the emphasis on the simple informal methods, there is scope for mathematical simulation models in the prediction stage. Munn (1979) identifies a number of criteria for situations in which computer-based simulation or mathematical models would be useful. The following are some of the most relevant:

- the assessment requires the handling of large numbers of simple calculations;
- there are many complex links between the elements of the EIA;
- the affected processes are time-dependent;
- increased definitions of assumptions and elements will be valuable in drawing together the many disciplines involved in the assessment;
- some or all of the relationships of the assessment can only be defined in terms of statistical probabilities.

Living with uncertainty

Environmental impact statements often appear more certain in their predictions than they should. This may reflect a concern not to undermine credibility and/or an unwillingness to attempt to allow for uncertainty. All predictions have an element of uncertainty, but it is only in recent years that such uncertainty has begun to be acknowledged in the EIA process (De Jongh 1988). There are many sources of uncertainty of relevance to the EIA process as a whole. In their classic works on strategic choice, Friend & Jessop (1977) and Friend & Hickling (1987) identified three broad classes of uncertainty: uncertainties about the physical, social and economic environment (UE), uncertainties about guiding values (UV), and uncertainties about related decisions (UR) (see Fig. 5.4). All three classes of uncertainty may affect the accuracy of predictions, but the focus in an EIA study is usually on uncertainty about the environment. This may include the use of inaccurate and/or partial information on the project and on baseline environmental conditions, unanticipated changes in the project during one or more of the stages of the life-cycle, and oversimplification and errors in the application of methods and mod-

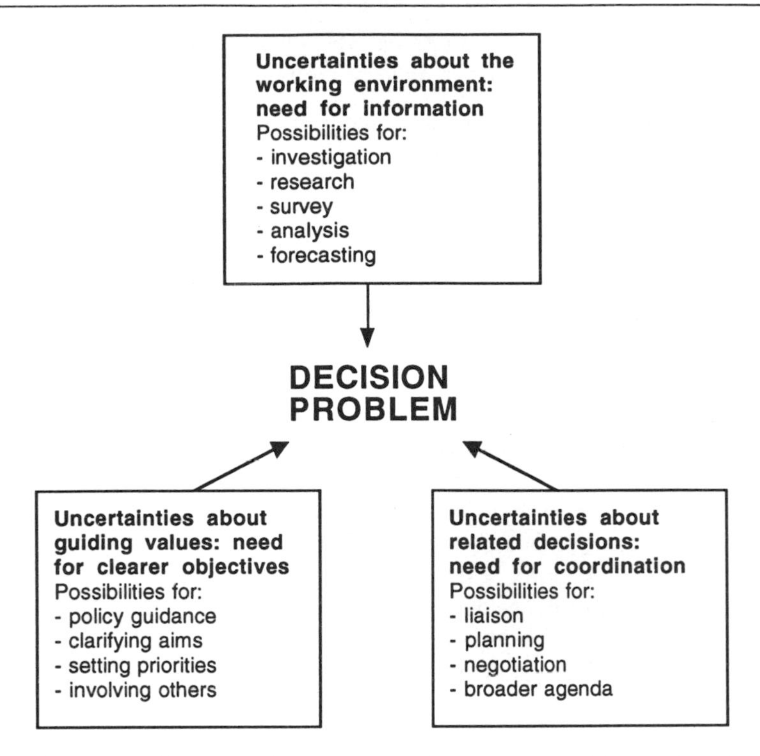

Figure 5.4 The types of uncertainty in decision-making. (*Source:* Friend & Hickling 1987.)

els. Socio-economic conditions may be particularly difficult to predict, as underlying societal values may change quite dramatically over the life, say 30–40 years, of a project.

Uncertainty in EIA predictive exercises can be handled in several ways. The *assumptions underpinning predictions should be clearly stated* (Voogd 1983). Issues of probability and confidence in predictions should be addressed, and ranges may be attached to predictions within which the analyst is "*n*"% confident that the actual outcome will lie. For example, scientific research may conclude that the 95% confidence interval for the noise associated with a new industrial project is 65–70 dBA, which means that only 5 times out of 100 would the dBA be expected to be outside this range. Tomlinson (1989) draws attention to the twin issues of probability and confidence involved in predictions.

> These twin factors are generally expressed through the same word. For example in the prediction "a major oil spill would have major ecological consequences", a high degree of both probability and confidence exists. Situations may arise, however, where a low probability event based upon a low level of confidence is predicted. This is potentially more serious than a higher probability event with high confidence, since low levels of confidence may

preclude expenditure on mitigating measures, ignoring issues of significance. Monitoring measures may be an appropriate response in such situations.

It may also be useful to show impacts under "peak" as well as "average" conditions for a particular stage of a project; this may be very relevant in the construction stage of major projects.

Sensitivity analysis may be used to assess the consistency in relationships between variables. If the relationship between input A and output B is such that whatever the changes in A, there is little change in B, then no further information may be needed. However, where the effect is much more variable, there may be a need for further information. Of course, the best check on the accuracy of predictions is to check on the outcomes of the implementation of a project after the decision. This is too late for the project under consideration, but could be useful for future projects. Conversely, monitoring of outcomes of similar projects may provide useful information for the project in hand. Holling (1978), who believes that the "core issue of EIA is how to cope with decision-making under uncertainty", recommends a policy of adaptive environmental impact assessment, with periodic reviews of the EIA through the project life-cycle. Another procedural approach would be to require an *uncertainty report* as one step in the process; such a report would bring together the various sources of uncertainty associated with a project and the means by which they might be reduced (uncertainties are rarely eliminated).

5.3 Evaluation

Evaluation in the EIA process

Once impacts have been predicted, there is a need to assess their relative significance. Criteria for significance include the magnitude and likelihood of the impact and its spatial and temporal extent, the likely degree of recovery of the affected environment, the value of the affected environment, the level of public concern, and political repercussions. As with prediction, the choice of evaluation method should be related to the task in hand and to the resources available. Evaluation should feed into most stages of the EIA process, but the nature of the methods used may vary, for example, according to the number of alternatives under consideration, according to the level of aggregation of information, and according to the number and type of parties involved (e.g. "in house" and/or "external" consultation).

Evaluation methods can be of various types: including formal or informal, quantitative or qualitative, aggregated or disaggregated (see Voogd 1983, Maclaren & Whitney 1985). The most formal is the *comparison of likely impacts against legal requirements and standards* (e.g. air quality standards, building regulations). Beyond this, all assessments of significance either implicitly or explicitly apply weights to the various impacts (i.e. some are assessed as more important than others). This involves interpretation and the application of judgement. Such judgement can be rationalized in various ways and a range of methods are available, but all involve values and all are subjective. Parkin (1992) sees judgements as being on a continuum between an analytical mode and an intuitive mode. In practice, many fall into the intuitive end of the continuum, but such judgements, made without the benefit of analysis, are likely to be flawed, inconsistent and biased. "Social effects of resource allocation decisions are too extensive to allow the decision to 'emerge' from some opaque procedure free of overt political scrutiny" (Parkin 1992). Analytical methods seek to introduce a rational approach to evaluation.

An important distinction is between two sets of methods: those that assume a common utilitarian ethic with a single evaluation criterion (money), and those based on the measurement of personal utilities, including multiple criteria. The *cost–benefit analysis* (CBA) approach, which seeks to express impacts in monetary units, falls into the former category. A variety of methods, including *multi-criteria analysis*, *decision analysis* and *goals achievement*, fall into the latter. The very growth of EIA is partly a response to the limitations of CBA and to the problems of monetary valuation of environmental impacts. Yet, after two decades of limited concern, there is renewed interest in the monetizing of environmental costs and benefits (DoE 1991). The multi-criteria/multi-attribute methods involve scoring and weighting systems that are also not problem-free. The various approaches are now outlined. In practice, there are many hybrid variations between these two main categories, and these are referred to in both categories.

Cost–benefit analysis and monetary valuation techniques

Cost–benefit analysis itself lies in a range of project and plan appraisal methods that seek to apply monetary values to costs and benefits (Lichfield et al. 1975). At one extreme are *partial* approaches, such as financial-appraisal, cost-minimization and cost-effectiveness methods, which consider only a subsection of the relevant population or only a subsection of the full range of consequences of a plan or project. *Financial appraisal* is limited to a narrow concern, usually of the developer, with the stream of financial costs and returns associated with an investment. *Cost-effectiveness* involves selecting an option that achieves a goal at least cost (for example, devising a least-cost approach to produce coastal bathing waters that meet the CEC Blue Flag criteria). The cost-effectiveness approach is more problematic where there are a number of goals and where some actions achieve certain goals more fully than others (Winpenny 1991).

Cost–benefit analysis is more *comprehensive* in scope. It takes a long view of projects (farther as well as nearer future) and a wide view (in the sense of allowing for side-effects). It is based in welfare economics, and seeks to include all the relevant costs and benefits to evaluate the net social benefit of a project. It was used extensively in the UK in the 1960s and early 1970s for public sector projects, the most famous being the Third London Airport (HMSO 1971). The methodology of CBA has several stages: project definition, identification and enumeration of costs and benefits, evaluation of costs and benefits, discounting and presentation of results. Several of the stages are similar to those in EIA. The basic evaluation principle is to measure in monetary terms where possible – as money is the common measure of value and monetary values are best understood by the community and decision-makers – and then reduce all costs and benefits to the same capital or annual basis. Future annual flows of costs and benefits are usually discounted to a net present value (see Table 5.3). A range of interest rates may be used to show the sensitivity of the analysis to changes. If the net social benefit minus cost is positive, then there may be a presumption in favour of a project. However, the final outcome may not always be that clear. The presentation of results should distinguish between tangible and intangible costs and benefits, as relevant, allowing the decision-maker to consider the trade-offs involved in the choice of one option or another.

Table 5.3 Cost–benefit analysis: presentation of results: tangibles and intangibles.

Category	Alternative 1	Alternative 2
Tangibles		
Annual benefits	£ B1	£ b1
	£ B2	£ b2
	£ B3	£ b3
Total annual benefit	£ B1 + B2 + B3	£ b1 + b2 + b3
Annual costs	£ C1	£ c1
	£ C2	£ c2
	£ C3	£ c3
Total annual costs	£ C1 + C2 + C3	£ c1 + c2 + c3
Net discounted present value (NDPV) of benefits and costs over 'n' years at X%*	£ D	£ E
Intangibles		
Intangibles are likely to include costs and benefits	I1	i1
	I2	i2
	I3	i3
	I4	i4
Intangibles summation (undiscounted)	I1 + I2 + I3 + I4	i1 + i2 + i3 + i4

* e.g. NPDV (Alt1)

$$D = \sum\left[\frac{B1}{(1+X)^1} + \frac{B1}{(1+X)^2} + \cdots + \frac{B1}{(1+X)^n} + \frac{B2}{(1+X)^1} + \cdots + \frac{B2}{(1+X)^n} + \frac{B3}{(1+X)^1} + \cdots + \frac{B3}{(1+X)^n}\right]$$

$$-\sum\left[\frac{C1}{(1+X)^1} + \frac{C1}{(1+X)^2} + \cdots \frac{C1}{(1+X)^n} + \frac{C2}{(1+X)^1} + \cdots + \frac{C2}{(1+X)^n} + \frac{C3}{(1+X)^1} + \cdots + \frac{C3}{(1+X)^n}\right]$$

CBA has excited both advocates (e.g. Dasgupta & Pearce 1978, Pearce et al. 1989, Pearce 1989) and opponents (e.g. Bowers 1990). It does have many problems, including the difficulties of identifying, enumerating and monetizing intangibles. Many environmental impacts fall into the intangible category: for example, the loss of a rare species, the urbanization of a rural landscape, and the saving of a human life. The incompatibility of monetary and non-monetary units makes decision-making problematic (Bateman 1991). Another problem is the choice of discount rate: for example, should a very low rate be used to prevent the rapid erosion of future costs and benefits in the analysis? This choice of rate has profound implications for the evaluation of resources for future generations. There is also the underlying and fundamental problem of the use of the single evaluation criterion of money, and the assumption that £1 is worth the same to any person, whether s/he be a tramp or a millionaire, a resident of a rich commuter belt or of a poor and remote rural community. CBA also ignores distributional effects and aggregates costs and benefits to estimate the change in the welfare of society as a whole.

The *planning balance sheet* (PBS) is a variation on the theme of CBA, and it goes beyond CBA in its attempts to identify, enumerate and evaluate the distribution of costs and benefits between the affected parties. It also acknowledges the difficulty of attempts to monetize the more intangible impacts. PBS was developed by Lichfield et al. (1975) to compare alternative town plans. It is basically a set of social accounts structured into sets of "producers" and "consumers" engaged in various transactions. The transaction could, for example, be an adverse impact such as noise from an airport (the producer) on the local community (the consumer), or a beneficial impact such as the time savings resulting from a new motorway development (the producer) for users of the motorway (the consumers). For each producer and consumer group, costs and benefits are quantified per transaction, in monetary terms or otherwise, and weighted according to the numbers involved. The findings are presented in tabular form, leaving the decision-maker to consider the trade-offs, but this time with some guidance on the distributional impacts of the options under consideration (Fig. 5.5). More recently, Lichfield (1988) has sought to integrate EIA and PBS further in an approach he calls community impact evaluation (CIE).

Partly in response to the "intangibles" problem in CBA, there has also been considerable interest in the development of *monetary valuation techniques* to improve the economic measurement of the more intangible environmental impacts (DoE 1991, Winpenny 1991, Barde & Pearce 1991). The techniques can be broadly classified into direct and indirect and they are concerned with the measurement of preferences about the environment rather than intrinsic values of the environment. The direct approaches seek to measure directly the monetary value of environmental gains – for example, better air quality or an improved scenic view. Indirect approaches measure preferences for a particular effect via the establishment of a "dose–response" type relationship. The various techniques found under the direct and indirect categories are summarized in Table 5.4. Such techniques can contribute to the assessment of the total economic value of an

	Plan A				Plan B			
	Benefits		Costs		Benefits		Costs	
	Capital	Annual	Capital	Annual	Capital	Annual	Capital	Annual
Producers								
X	£a	£b	–	£d	–	–	£b	£c
Y	i_1	i_2	–	–	i_3	i_4	–	–
Z	M_1	–	M_2	–	M_3	–	M_4	–
Consumers								
X′	–	£e	–	£f	–	£g	–	£h
Y′	i_5	i_6	–	–	i_7	i_8	–	–
Z′	M_1	–	M_3	–	M_2	–	M_4	–

£ = benefits and costs that can be monetized
M = where only a ranking of monetary values can be estimated
i = intangibles

Figure 5.5 Example of structure of a planning balance sheet.

action/project, which should not only include user values (preferences people have for using an environmental asset, such as a river for fishing) but also non-user values (where people value an asset but do not use it, although some may wish to do so some day). Of course such techniques have their problems, for example the potential bias in people's replies in the contingent valuation method (CVM) approach. However, simply through the act of seeking a value for various environmental features, such techniques help to reinforce the understanding that such features are not "free" goods and should not be treated as such.

Scoring and weighting and multi-criteria methods

Multi-criteria and multi-attribute methods seek to overcome some of the deficiencies of CBA; in particular they seek to allow for a pluralist view of society, composed of diverse "stakeholders" with diverse goals and with differing values concerning environmental changes. Most of the methods use – and sometimes misuse – some kind of simple scoring and weighting system; such systems generate considerable debate. Here some key elements of good practice are discussed, followed by a brief overview of the range of multi-criteria/multi-attribute methods available to the analyst.

Scoring may use quantitative or qualitative scales according to the availability of information on the impact under consideration. Lee (1987) provides an example (see Table 5.5) of how different levels of impact (in this example noise, where the unit of measurement is in units of $L_{10}dB_A$) can be scored in different systems. These systems seek to standardize the impact scores for purposes of comparison. Where quantitative data are not available, ranking of alternatives may use other approaches, for example using letters (A, B, C, etc.) or words (not significant,

Table 5.4 Summary of environmental monetary valuation techniques.

Direct household production function (HPF)

HPF methods seek to determine expenditure on commodities that are substitutes or complements for an environmental characteristic to value changes in that characteristic. Subtypes include:

- Avertive expenditures: expenditure on various substitutes for environmental change (e.g. noise insulation as an estimate of the value of peace and quiet).
- Travel cost method: expenditure, in terms of cost and time, incurred in travelling to a particular location (e.g. a recreation site) is taken as an estimate of the value placed on the environmental good at that location (e.g. benefit arising from use of the site).

Direct hedonic price methods (HPM)

HPM methods seek to estimate the implicit price for environmental attributes by examining the real markets in which those attributes are traded. Again there are two main subtypes:

- Hedonic house/land prices: these prices are used to value characteristics such as "clean air" and "peace and quiet", through cross sectional data analysis (e.g. on house price sales in different locations).
- Wage risk premia: the extra payments associated with certain higher risk occupations are used to value changes in morbidity and mortality (and implicitly human life) associated with such occupations.

Direct experimental markets

Survey methods are used to elicit individual values for non-market goods. Experimental markets are created to discover how people would value certain environmental changes. Two kinds of questioning, of a sample of the population, may be used:

- Contingent valuation method (CVM): people are asked what they are willing to pay (WTP) for keeping X (e.g. a good view, an historic building) or preventing Y, or what they are willing to accept (WTA) for losing A, or tolerating B.
- Contingent ranking method (CRM) or stated preference: people are asked to rank their preferences for various environmental goods, which may then be valued by linking the preferences to the real price of something traded in the market (e.g. house prices).

Indirect methods

Indirect methods seek to establish preferences through the estimation of relationships between a "dose" (e.g. reduction in air pollution) and an effect (e.g. health improvement). Approaches include:

- Indirect market price approach: the dose–response approach seeks to measure the effect (e.g. value of loss of fish stock) resulting from an environmental change (e.g. oil pollution of a fish farm), by using the market value of the output involved. The replacement-cost approach uses the cost of replacing or restoring a damaged asset as a measure of the benefit of restoration (e.g. of an old stone bridge eroded by pollution and wear and tear).
- Effect on production approach: where a market exists for the goods and services involved, the environmental impact can be represented by the value of the change in output that it causes. It is widely used in developing countries, and is a continuation of the dose–response approach.

Adapted from: DoE 1991, Winpenny 1991, Pearce & Markandya 1990, Barde & Pearce 1991.

Table 5.5 A comparison of different scoring systems.

Method	Alternatives				Basis of score
	A (no action)	B	C	D	
Ratio	65	62	71	75	Absolute $L_{10}dB_A$ measure
Interval	0	−3	+6	+10	Difference in $L_{10}dB_A$ using alternative A as base
Ordinal	B	A	C	D	Ranking according to ascending value of $L_{10}dB_A$
Binary	0	0	1	1	0 = less than $70L_{10}dB_A$ 1 = $70L_{10}dB_A$ or more

Based on Lee (1987).

significant, very significant).

Weighting seeks to identify the relative importance of the various impact types for which scores of some sort may be available (for example, the relative importance of a water pollution impact; the impact on a rare component of flora). Different impacts may be allocated weights (normally numbers) out of a total budget (e.g. 10 points to be allocated between 3 impacts), but by whom? Multi-criteria/multi-attribute methods seek to recognize the plurality of views and weights in their methods; the Delphi approach also uses individuals' weights, from which group weights are then derived. In many studies, however, the weights are those produced by the technical team. Indeed the decision-makers may be unwilling to reveal all their personal preferences, for fear of undermining their negotiating positions. This internalization of the weighting exercise does not destroy the use of weights, but it does emphasize the need for clarification of scoring and weighting systems and, in particular, for the identification of the origin of the weightings used in an EIA. Wherever possible, scoring and weighting should be used to reveal the trade-offs in impacts involved in particular projects or in alternatives. For example, Table 5.6 shows that the main issue is the trade-off between the impact on flora of one scheme and the impact on noise of the other scheme.

Several approaches to the scoring and weighting of impacts have already been introduced in the outline of impact identification methods in Chapter 4. The Leopold matrix includes measures of significance (on a scale of 1 to 10) as well as of magnitude of impacts. The matrix approach can also be usefully modified to identify the distribution of impacts between geographical areas and/or between various affected parties (Fig. 5.6). The quantitative EES and WRAM methods generate weights for different environmental parameters, drawn up by panels of experts. Weightings can also be built into overlay maps to identify areas with the most development potential according to various combinations of weightings. Some of the limitations of such approaches have already been noted in Chapter 4.

Table 5.6 Weighting, scoring and trade-offs.

Impact type	Weight	Scheme A		Scheme B	
	(w)	Score (a)	(aw)	Score (b)	(bw)
Noise	3	5	15	1	3
Loss of flora	4	1	4	3	12
Air pollution	3	2	6	2	6
Total			25		21

Other methods in the multi-criteria/multi-attribute category include, *inter alia*, decision analysis, the goals achievement matrix, multi-attribute utility theory, and judgement analysis. *Decision analysis* is the operational form of decision theory, a theory of how individuals make decisions in the face of uncertainty, which owes its modern origins to von Neuman & Morgenstern (1953). Decision analysis usually involves the construction of a decision tree, an example of which is shown in Figure 5.7. Each branch represents a potential action, with a probability of achievement attached to it.

The *goals achievement matrix* (GAM) was developed as a planning tool by Hill (1968) to overcome the perceived weaknesses of the planning balance sheet approach. GAM makes the goals and objectives of a project/plan explicit, and the evaluation of alternatives is accomplished by measuring the extent to which they achieve the stated goals. The existence of many diverse goals leads to a system of weights. Since all interested parties are not politically equal, the identified groups should also be weighted. The end result is a matrix of weighted objectives and

Group environmental component	Project Action							
	Construction stage actions				Operational stage actions			
	A	B	C	D	a	b	c	d
Group 1 (e.g. indigenous population $\geq$ 45 years old) various								
• Social								
• Physical								
• Economic components								
Group 2 (e.g. indigenous population $<$ 45 years old) various								
• Social								
• Physical								
• Economic components								

Figure 5.6 Simple matrix identification of distribution of impacts.

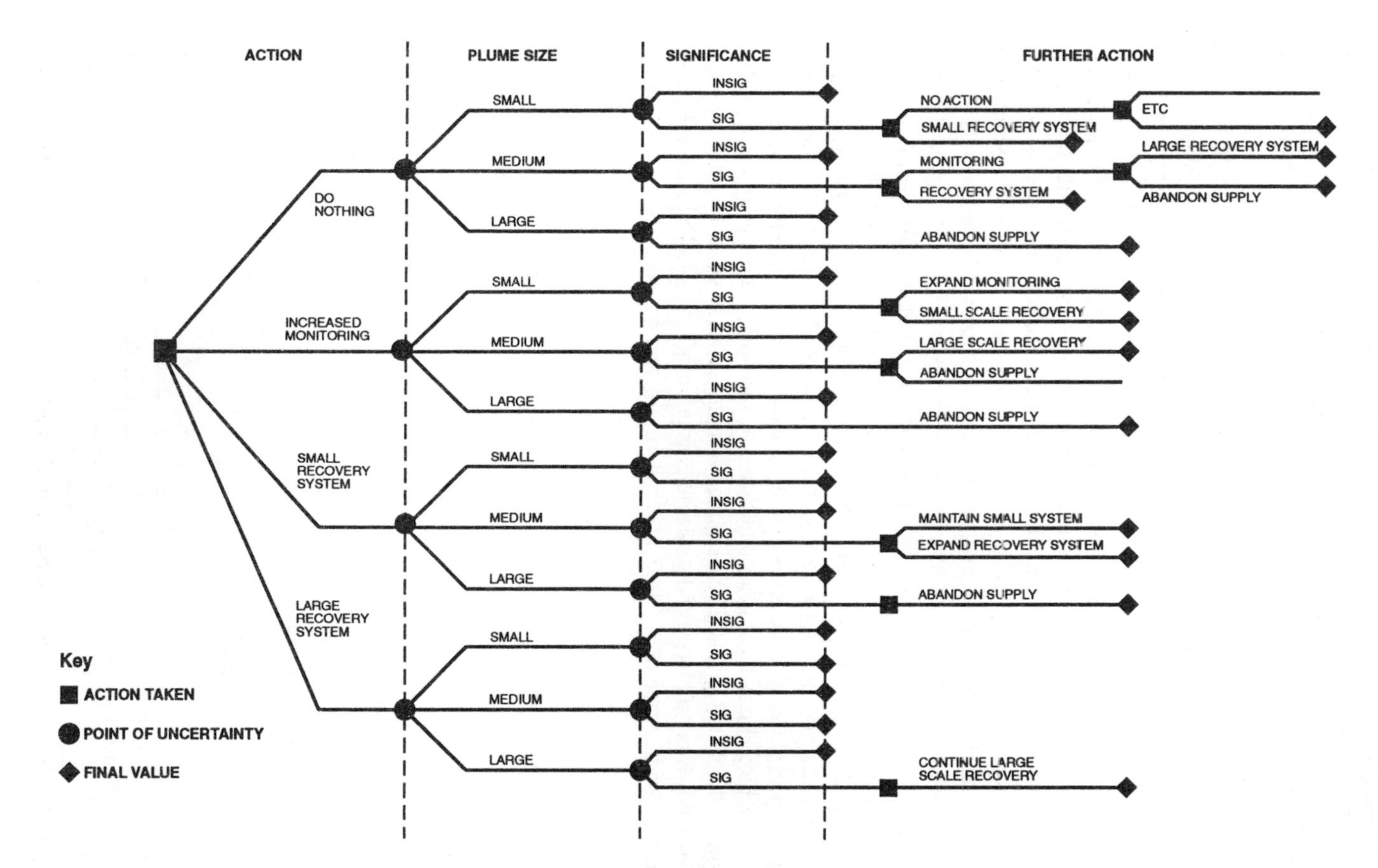

Figure 5.7 A decision tree: problem of groundwater contamination. (*Source:* De Jongh 1988)

weighted interests/agencies (Fig. 5.8). The use of goals and value weights to evaluate plans in the interests of the community, and not just for economic efficiency, has much to commend it. The approach also provides an opportunity for public participation. Unfortunately, the complexity of the approach has limited its use, and the weights and goals used may often reflect the views of the analyst more than those of the interests/agencies involved.

Multi-attribute utility theory (MAUT) has gained a certain prominence in recent years as an evaluation method that can incorporate the values of the key interests involved (Edwards & Newman 1982, Bisset 1988, Parkin 1992). MAUT involves a number of steps, including the identification of the entities (alternatives, objects) to be evaluated, and the identification and structuring of environmental attributes (e.g. noise level) to be measured. The latter may include a "value tree" with general objectives (values) at the top and specific attributes at the bottom. The ranking of attributes is by the key stakeholders/expert group whose values are to be maximized. Attributes are scaled and formal value or utility models are developed to quantify trade-offs among attribute scales and attributes. For further reference, see Parkin (1992) for an outline of the key steps and an application of a "relatively" simple and well proven version of MAUT known as the SMART method.

Finally brief reference is made to the *Delphi method* which can be used to build views of various stakeholders into the evaluation process. The Delphi method is an established means of collecting expert opinion and of gaining consensus among experts on various issues under consideration. It has the advantage of obtaining expert opinion from the individual, with guaranteed anonymity, avoiding the potential distortion caused by peer pressure in group situations. Compared with other evaluation methods it can also be undertaken in a short time period and at relatively low cost.

There have been a number of interesting applications of the Delphi method in EIA (Richey et al. 1985, Green et al. 1989, 1990). Green et al. used the approach to assess the environmental impacts of the redevelopment and re-orientation of

Goal description:			α			β
Relative weight:			2			3
Incidence	Relative weight	Costs	Benefits	Relative weight	Costs	Benefits
Group a	1	A	D	5	E	1
Group b	3	H	J	4	M	2
Group c	1	L	J	3	M	3
Group d	2	–	J	2	V	4
Group e	1	–	K	1	T	5
		Σ	Σ	Σ	Σ	

Figure 5.8 Goals achievement matrix (section of). *Source:* Adapted from Hill 1968.

Bradford's famous Salt Mill. The method involves drawing up a Delphi panel. In the Salt Mill case, the initial panel of 40 included experts with a working knowledge of the project (e.g. planners, tourism officers), councillors, employees, academics, local residents and traders. This was designed to provide a balanced view of interests and expertise. The Delphi exercise usually has a three-stage approach: (1) a general questionnaire asking panel members to identify important impacts (positive and negative); (2) a first-round questionnaire asking panel members to rate the importance of a list of impacts identified from the first stage; and (3) a second-round questionnaire, asking panel members to re-evaluate the importance of each impact in the light of the overall panel response to the first round. However, the method is not without its limitations. The potential user should be aware of the difficulties of drawing up the "balanced" panel in the first place, and of the difficulty of avoiding distortion of the assessment by the differential fall-out of panel members between stages of the exercise, and by overzealous structuring of the exercise by the organizers.

5.4 Mitigation

Types of mitigation measures

Mitigation is defined in EC Directive 85/337 as "measures envisaged in order to avoid, reduce and, if possible remedy significant adverse effects" (CEC 1985). In similar vein, the US Council on Environmental Quality, in its regulations implementing the National Environmental Policy Act, defines mitigation as including: "not taking certain actions; limiting the proposed action and its implementation; repairing, rehabilitating, or restoring the affected environment; presentation and maintenance actions during the life of the action; and replacing or providing substitute resources or environments" (CEQ 1978). The guidance on mitigation measures provided by the UK Government is set out in Table 5.7. It is not possible to specify here all the types of mitigation measures that could be used. Instead the following subsections provide a few examples, relating to biophysical and socio-economic impacts. The reader is also referred to Fortlage (1990) for useful coverage of mitigation measures. A review of EISs for developments similar to the development under consideration may also suggest useful mitigation measures.

At one extreme, the prediction and evaluation of impacts may reveal an array of impacts with such significant adverse effects that the only effective mitigation measure may be to abandon a proposal altogether. A less draconian, and more normal, situation would be to modify aspects of the development action to *avoid* various impacts. Examples of methods to avoid impacts include:

- the control of solid and liquid wastes by recycling on site or by removing them from the site for environmentally sensitive treatment elsewhere;
- the use of a designated lorry route, and day-time working only, to avoid

Table 5.7 Mitigation measures, as outlined in *UK guide to procedures.*

Where significant adverse effects are identified, [describe] the measures to be taken to avoid, reduce or remedy those effects, e.g.:

(a) Site planning

(b) Technical measures, e.g.:
  (i) process selection
  (ii) recycling
  (iii) pollution control and treatment
  (iv) containment (e.g. bunding of storage vessels)

(c) Aesthetic and ecological measures, e.g.:
  (i) mounding
  (ii) design, colour, etc.
  (iii) landscaping
  (iv) tree plantings
  (v) measure to preserve particular habitats or create alternative habitats
  (vi) recording of archaeological sites
  (vii) measures to safeguard historic building or sites

[Assess] the likely effectiveness of mitigating measures.

Source: DOE 1989.

disturbance to village communities from construction lorry traffic and from night construction work; and

- the establishment of buffer zones and the minimal use of toxic substances to avoid impacts on local ecosystems.

Some adverse effects may be less easily avoided; there may also be less need to avoid them completely. Examples of methods to *reduce* adverse effects include:

- the sensitive design of structures, using simple profiles, local materials, and muted colours, to reduce the visual impact of a development, and landscaping to hide and/or blend it into the local environment;
- use of construction site hostels, and coaches for journeys to work, to reduce the impact on the local housing market, and on the roads, of a project with major construction stage employment;
- use of silting basins or traps, planting of temporary cover crops, and scheduling of activities during the dry months to reduce erosion and sedimentation.

During one or more stages of the life of a project, certain environmental components may be temporarily lost or damaged. It may be possible to *repair, rehabilitate or restore* the affected component to varying degrees. For example:

- agricultural land used for the storage of materials during construction may be fully rehabilitated; land used for gravel extraction may be restored to agricultural use, but over a much longer period, and with associated impacts according to the nature of the landfill material used;
- a river or stream diverted by a road project can be unculverted and re-established with similar flow patterns as far as is possible;

- a local community astride a route to a new tourism facility could be relieved from much of the adverse traffic effects through the construction of a bypass (which of course, introduces a new flow of impacts).

There will invariably be some adverse effects that cannot be reduced. In such cases, it may be necessary to *compensate* for adverse effects. For example:

- for loss of public recreational space, or a wildlife habitat, the provision of land with recreation facilities, or creation of a nature reserve, elsewhere;
- for loss of privacy, quietness and safety in houses adjacent to a new road, the provision of sound insulation and/or the purchase by the developer of badly affected properties.

Mitigation measures can become linked with discussions between the developer and the local planning authority on what is known in the UK as "planning gain". Fortlage (1990) advises of some of the potential complications associated with such discussions, and the need to distinguish between mitigation measures and planning gain:

> Before any mitigating measures are put forward, the developer and the local planning authority must agree as to which effects are to be regarded as adverse, or sufficiently adverse to warrant the expense of remedial work, otherwise the whole exercise becomes a bargaining game which is likely to be unprofitable to both parties . . .
>
> Planning permission often includes conditions requiring the provision of planning gains by the developer to offset some deterioration of the area caused by the development, but it is essential to distinguish very clearly between those benefits offered by way of compensation for adverse environmental effects, and those which are a formal part of planning consent. The local planning authority may decide to formulate the compensation proposals as a planning condition in order to ensure that they are carried out, so the developer should beware of putting forward proposals that he does not really intend to implement.

Mitigation measures must be planned in an integrated and coherent fashion to ensure that they really are effective, that they do not conflict with each other, and that they do not merely shift a problem from one medium to another. There may also be beneficial effects of a project on an impacted area, often of a socio-economic nature; where such effects are identified, there should be a concern to ensure that they do happen and do not become diluted. For example, the potential local employment benefits of a project can be encouraged through appropriate skills training programmes for local people; various tenure arrangements can be used to direct houses in new major housing schemes to local people in need.

Mitigation in the EIA process

Like many elements in the EIA process, mitigation is not limited to one point in the assessment. Although it may follow logically from the prediction and assessment of relative significance of impacts, it is in fact inherent in all aspects of the process. The original project design may have already been modified, possibly in the light of mitigation changes made to earlier comparable projects or perhaps as a result of early consultation with the LPA or with the local community. The consideration of alternatives, initial scoping activities, baseline studies and impact identification studies may suggest further mitigation measures. Although more in-depth studies may identify new impacts, mitigation measures may alleviate others. The prediction and evaluation exercise can thus focus on a limited range of potential impacts.

Mitigation measures are normally discussed and documented in each topic section of the EIS (e.g. air quality, visual quality, transport, employment). Those discussions should clarify the extent to which the significance of each adverse impact has been offset by the mitigation measures proposed. A summary chart (see Table 5.8) can provide a clear and very useful overview of the envisaged outcomes, and may be a useful basis for agreement on planning consents. *Residual unmitigated or only partially mitigated impacts* should be identified. These could be divided according to degree of severity: for example, into "less than significant impacts" and "significant unavoidable impacts".

Table 5.8 Example of a section of a summary table for impacts and mitigation measures.

	Impact	Mitigation measure(s)	Level of significance after mitigation[1]
1.	400 acres of prime agricultural land would be lost from the County to accommodate the petrochemical plant.	The only *full* mitigating measure for this impact would be to abandon the project.	SU
2.	Additional lorry and car traffic on the adjacent hilly section of the motorway will increase traffic volumes by 10–20% above those predicted on the basis of current trends.	A lorry crawler lane on the motorway, funded by the developer, will help to spread the volume, but effects may be *partial* and short lived.	SU
3.	The project would block the movement of most terrestrial species from the hilly areas to the east of the site to the wetlands to the west of the site.	A wildlife corridor should be developed and maintained along the entire length of the existing stream which runs through the site. The width of the corridor should be a minimum of 75 ft. The stream bed should be cleaned of silt and enhanced through the construction of occasional pools. The buffer zone should be planted with native riparian vegetation, including sycamore and willow.	LS

Note: SU = Significant unavoidable impact; LS = Less than significant impact

Mitigation measures are of little or no value unless they are implemented. Hence there is a clear link between mitigation and monitoring of outcomes, if and when a project is approved and moves to the construction and operational stages. Monitoring, which is discussed in Chapter 7, must include the effectiveness or otherwise of mitigation measures. The latter must therefore be devised with monitoring in mind; they must be clear enough to allow checking of effectiveness. The use of particular mitigation measures may also draw on previous experience of relative effectiveness, from previous monitoring activity in other relevant and comparable cases.

5.5 Summary

Impact prediction and the evaluation of the significance of impacts often constitute a "black box" in EIA studies. Intuition, often wrapped up as "expert opinion", cannot provide a firm and defensible foundation for this key stage of the process. Various methods, ranging from simple to complex, are available to the analyst, and these can help to underpin analysis. Mitigation measures also come into play particularly at this stage. However, the increasing sophistication of some methods does run the risk of cutting out key actors, and especially the public, from the EIA process. Chapter 6 discusses the important, but currently weak, rôle of public participation, the value of good presentation, and approaches to EIS review and decision-making.

References

Barde, J. P. & D. W. Pearce 1991. *Valuing the environment: six case studies*. London: Earthscan.

Bateman, I. 1991. Social discounting, monetary evaluation and practical sustainability. *Town and Country Planning* **60**(6), 174–6.

Bisset, R. 1988. Developments in EIA methods. In *Environmental impact assessment: theory and practice*, P. Wathern (ed.), 47–60. London: Unwin Hyman.

Bowers, J. 1990. *Economics of the environment: the conservationists' response to the Pearce Report*. Wellington, Telford, England: British Association of Nature Conservationists.

Bregman, J. I. & K. M. Mackenthun 1992. *Environmental impact statements*. Boca Raton, Florida: Lewis.

CEC (Commission of the European Communities 1985). *On the assessment of effects of certain public and private projects on the environment*, OJ L 175, 5 July 1985. Brussels: CEC.

CEQ (Council on Environmental Quality) 1978. *National Environmental Policy Act*, Code of Federal Regulations, Title 40, Section 1508. 20.

Dasgupta A. K. & D. W. Pearce 1978. *Cost–benefit analysis: theory and practice*.

London: Macmillan.
De Jongh, P. E. 1988. Uncertainty in EIA. In *Environmental impact assessment: theory and practice*, P. Wathern (ed.), 62–83. London: Unwin Hyman.
DoE (Department of the Environment) 1989. *Environmental appraisal: a guide to the procedures*. London: HMSO.
DoE (Department of the Environment) 1991. *Policy appraisal and the environment*. London: HMSO.
Edwards, W. & J. R. Newman 1982. *Multiattribute evaluation*. Los Angeles: Sage.
Fortlage, C. 1990. *Environmental assessment: a practical guide* Aldershot, England: Gower.
Friend, J. K. & A. Hickling 1987. *Planning under pressure: the strategic choice approach*. Oxford: Pergamon.
Friend, J. K. & W. N. Jessop 1977. *Local government and strategic choice: an operational research approach to the processes of public planning*, 2nd edn. Oxford: Pergamon.
Glasson, J., M. J. Elson, M. Van der Wee, B. Barrett 1987. *Socio-economic impact assessment of the proposed Hinkley Point C power station*. Impacts Assessment Unit, Oxford Polytechnic.
Glasson, J., M. Van der Wee, B. Barrett 1988. A local income and employment multiplier analysis of a proposed nuclear power station development at Hinkley Point in Somerset. *Urban Studies* **25**, 248–61.
Golden, J., R. P. Duellette, S. Saari, P. N. Cheremisinoff 1979. *Environmental impact data book*. Ann Arbor, Michigan: Ann Arbor Science Publishers.
Green, H., C. Hunter, B. Moore 1989. Assessing the environmental impact of tourism development: the use of the Delphi technique. *International Journal of Environmental Studies* **35**, 51–62.
Green, H., C. Hunter, B. Moore 1990. Assessing the environmental impact of tourism development *Tourism Management* (June 1990), 111–20.
Hansen, P. E. & S. E. Jorgensen (eds) 1991. *Introduction to environmental management*. New York: Elsevier.
Hill, M. 1968. A goals-achievement matrix for evaluating alternative plans. *Journal of the American Institute of Planners* **34**, 19.
HMSO 1971. *Report of the Roskill Commission on the Third London Airport*. London: HMSO
Holling, C. S. (ed.) 1978. *Adaptive environmental assessment and management*. New York: John Wiley.
Journal of American Institute of Planners 1965. Theme issue on urban development models – new tools for planners". May 1965.
Kristensen, P., J. P. Jensen, E. Jeppesen 1990. *Eutrophication models for lakes*. Research Report C9. Copenhagen: National Agency of Environmental Protection.
Lee, N. 1987. *Environmental impact assessment: a training guide*. Occasional Paper 18, Department of Town and Country Planning, University of Manchester.
Lewis, J. A. 1988. Economic impact analysis: a UK literature survey and bibliography. *Progress in Planning* **30**(3), 161–209.
Lichfield, N. 1988. *Economics in urban conservation*. Cambridge: Cambridge University Press.
Lichfield, N., P. Kettle, M. Whitbread 1975. *Evaluation in the planning process*. Oxford: Pergamon.
Loewenstein, L. K. 1966. On the nature of analytical models. *Urban Studies* **3**.
Maclaren V. W. & J. B. Whitney (eds) 1985. *New directions in environmental impact*

assessment in Canada. London: Methuen.

Munn, R. E. 1979. *Environmental impact assessment: principles and procedures*. New York: John Wiley.

Parkin, J. 1992. *Judging plans and projects*. Aldershot, England: Avebury.

Pearce, D. 1989. Keynote speech at the 10th International Seminar on Environmental Impact Assessment and Management, University of Aberdeen, 9–22 July 1989.

Pearce, D. & A. Markandya 1990. *Environmental policy benefits: monetary valuation*. Paris: OECD.

Pearce, D., A. Markandya, E. B. Barbier 1989. *Blueprint for a Green economy*. London: Earthscan.

Rau J. G. & D. C. Wooten 1980. *Environmental impact analysis handbook*. New York: McGraw-Hill.

Richey, J. S., B. W. Mar, R. Horner 1985. The Delphi technique in environmental assessment. *Journal of Environmental Management* **21**(1), 135–46.

Suter II, G. W. 1993. *Ecological risk assessment*. Chelsea, Michigan: Lewis.

Tomlinson, P. 1989. Environmental statements: guidance for review and audit. *The Planner* **75**(28), 12–15.

US Environmental Protection Agency 1993. *Sourcebook for the environmental assessment (EA) process*. In preparation.

Von Neuman, J. & O. Morgenstern 1953. *Theory of games and economic behaviour*. Princeton, New Jersey: Princeton University Press.

Voogd, J. H. 1983. *Multicriteria evaluation for urban and regional planning*. London: Pion.

VROM 1984. *Prediction in environmental impact assessment*. The Hague: Netherlands Ministry of Public Housing, Physical Planning and Environmental Affairs.

Westman, W. E. 1985. *Ecology, impact assessment and environmental planning*. New York: John Wiley.

Winpenny, J. T. 1991. *Values for the environment: a guide to economic appraisal* (Overseas Development Institute). London: HMSO.

CHAPTER 6
Participation, presentation and review

6.1 Introduction

The aim of the EIA process is to provide information about a proposal's likely environmental impacts to the developer, public and decision-makers so that a better decision may be made. As such, how the information is presented, how the various interested parties use that information, and how the final decision incorporates the results of the EIA and the views of the various parties, are essential components in the EIA process.

Traditionally, the British system of decision-making has been characterized by administrative discretion and secrecy, with limited input by the public (McCormick 1991). However, there have been recent moves towards greater public participation in decision-making, and especially towards greater public access to information. In the environmental arena, the Environmental Protection Act of 1990 requires Her Majesty's Inspectorate of Pollution (HMIP) and local authorities to establish public registers of information on potentially polluting processes; the White Paper on the Environment (*This common inheritance*) and its annual updates compile environmental data and set forth an environmental agenda in a publicly available form; and EC Directive 85/337's requirements for EIA allow greater public access to information previously not compiled, or considered to be confidential. These trends can be expected to continue in the future: EC Directive 90/313 (Commission of the European Communities 1990) requires Member States to make provisions for freedom of access to information on the environment by the end of 1992, and the UK government is setting up suitable arrangements.

Past experience shows that the overall benefits of openness can exceed its costs, despite the expenditure and delays associated with full-scale public participation in the project planning process. The case of British Gas has already been noted (House of Lords 1981). More recently, speaking about the EIA process in Ireland, the conservation manager of Europe's largest zinc/lead mine noted that:

> . . . properly defined and widely used, [EIA is] an advantage rather than a deterrent. It is a mechanism for ensuring the early and orderly consideration of all relevant issues and for the involvement of affected communities. It is in this last area that its true benefit lies. We have entered an era when the people decide. It is therefore in the interests of developers to ensure that they, the people, are equipped to do so with the confidence that their concern is recognized and their future life-style protected. (Dallas 1984)

Despite the positive trends towards greater public participation in the EIA process and improved communication of EIA findings, both are still underdeveloped in the UK. Few developers make a real effort to gain a sense of the public's views before presenting their planning applications and EISs to local authorities. Few local authorities have the time or resources to gauge public opinion adequately before making their decisions. Few EISs are truly well presented, although standards of presentation have improved rapidly since mid-1988.

This chapter discusses how public consultation and participation can be fostered, and how the results of participation can be used to improve local relations and speed up the application process. The effective presentation of the EIS is then discussed. Guidelines for reviewing EISs and assessing their accuracy and comprehensiveness are considered and the chapter concludes with a discussion of the process of decision-making and post-decision legal challenges.

6.2 Public consultation and participation[1]

UK procedures

The rôle of public consultation and participation in the EIA process is to assure the quality, comprehensiveness, and effectiveness of the EIA, as well as to ensure that the public's views are adequately taken into consideration in the decision-making process. Public participation can be useful at most stages of the process:

- in determining the scope of the EIA
- in evaluating the relative significance of the likely impacts
- in providing specialist knowledge about the site
- in proposing mitigation measures
- in ensuring that the EIS is objective, truthful and complete, and
- in monitoring any conditions set on the development agreement.

Article 6 of EC Directive 85/337 requires that:

> Member States shall ensure that:
> - any request for development consent and any information gathered pursuant to Article 5 are made available to the public
> - the public concerned is given the opportunity to express an opinion before

> the project is initiated.
> The detailed arrangements for such information and consultation shall be determined by the Member States. . .

In the UK, this has been translated by the various EIA regulations into the following general requirements. Notices must be published in two local newspapers and posted at the proposed site at least seven days before the submission of the development application and EIS. These notices must describe the proposed development, state that a copy of the EIS is available for public inspection with other documents relating to the development application for at least 21 days, give an address where copies of the EIS may be obtained and the charge for the EIS, and state that written representations on the application may be made to the competent authority for at least 28 days after the notice is published. Finally, when a charge is made for an EIS, it must be "reasonable", taking into account printing and distribution costs. The details of these provisions differ between regulations, e.g. the Electricity and Pipe-line Works Regulations make no provisions for the public to make representations, and the Harbour Works Regulations require the EIS to be publicly available for 42 days.

Environmental assessment: a guide to the procedures (DOE 1989), the government manual to developers, notes:

> Developers should also consider whether to consult non-statutory bodies concerned with environmental issues, and the general public, during the preparation of the environmental statement. Bodies of these kinds may have particular knowledge and expertise to offer . . . While developers are under no obligation to publicize their proposals before submitting a planning application, consultation with local amenity groups and with the general public can be useful in identifying key environmental issues, and may put the developer in a better position to modify the project in ways which would mitigate adverse effects and recognize local environmental concerns. It will also give the developer an early indication of the issues which are likely to be important issues at the formal application stage if, for instance, the proposal goes to public inquiry.

From this it is clear that in the UK the requirements for public participation have been implemented half-heartedly at best and, in turn, developers and the competent authorities have generally limited themselves to only the minimum legal requirements. As such the potential benefits of public participation are not being achieved in the UK. This section discusses how "best practice" public participation can be encouraged. It begins by considering what is meant by "the public". The advantages and disadvantages of public participation are then discussed, and requirements for effective participation are established. Finally, methods of public participation are reviewed and compared. The reader is also referred to Jain et al. (1977), Mollison (1992), O'Riordan & Turner (1983), and Westman (1985) for further information.

Who are "the public"?

The simple term "the public" actually refers to a complex amalgam of interest groups, which changes over time and from project to project. "The public" can be broadly classified into two main groups. The first consists of the voluntary groups, quasi-statutory bodies, or issues-based pressure groups, which are usually concerned with a specific aspect of the environment or with the environment as a whole. The second group consists of the people living near the proposed development who may be directly affected by it. These two groups can have very different interests and resources. The organized groups may have extensive financial and professional resources at their disposal, may concentrate on specific aspects of the development, and may see their participation as a way of gaining political points or national publicity. People living locally, instead, may lack the technical, educational, or financial resources, and familiarity with relevant procedures, to put their points across effectively, yet they are the ones who will be the most directly affected by the development (Mollison 1992). The individual members of the two groups, in turn, come from a wide range of backgrounds and have a wide variety of opinions. A multiplicity of "publics" exist, each of which has specific views, which may well conflict with those of other groups and of EIA "experts".

Participation can be tightly controlled by regulations specifying which groups and organizations are eligible to participate or by criteria identifying those considered to be directly affected by a development (e.g. living within a certain distance of the development). The design of any public participation scheme should attempt to identify the "publics" concerned and take the steps required to ensure their fullest involvement.

Advantages and disadvantages of public participation

Developers do not usually favour public participation. It may upset a good relationship with the local planning authority. It carries the risk of giving the project a high profile, with attendant costs in time and money. It may not lead to a conclusive decision on a project, as diverse interest groups have different concerns and priorities; the decision may also represent the views of the most vocal interest groups rather than of the overall public. Most developers' contact with the public comes only at the stage of planning appeals and inquiries; by this time, "participation" has often evolved into a systematic attempt to stop the project. Thus, many developers never see the positive side of public participation, because they do not give it a chance.

Historically, public participation has also had connotations of extremism, confrontation, delays, and blocked development. In the USA, NEPA-related lawsuits have stopped major development projects, including oil and gas developments in Wyoming, a ski resort in California, and clear-cut logging in Alaska (Turner 1988). In Japan, riots (so violent that six people died) delayed construction of the Narita

Airport near Tokyo by five years in the late 1960s and early 1970s. In the UK, perhaps the most visible forms of public "participation" have involved protesters wearing gas masks at nuclear waste disposal and power station sites, and people threatening to lie down in front of the bulldozers working on the M3 motorway at Twyford Down. Public participation may provide the legal means for intentionally obstructing development; protracted delay of a project can be an effective method of defeating it.

On the other hand, public participation can be used positively to convey information about a development, clear up misunderstandings, allow a better understanding of relevant issues and how they will be dealt with, and identify and deal with areas of controversy while a project is still in its early planning phases. The process of considering and responding to the unique contributions of local people or special interest groups may suggest measures the developer could take to avoid local opposition and environmental problems. These measures are likely to be more innovative, viable, and publicly acceptable than those proposed solely by the developer. Project modifications made early in the planning process, before plans have been fully developed, are easier and cheaper to accommodate than those made later. Projects that do not have to go to inquiry are considerably cheaper than those that do. Early public participation also prevents an escalation of frustration and anger, thus helping to avoid the possibility of more forceful "participation". Finally, implementation of a project generally proceeds more cheaply and smoothly if local residents agree with the proposal, with fewer protests, a more willing labour force, and fewer complaints about impacts such as noise and traffic. The developers of a motor-racing circuit note:

> The environmental statement was the single most significant factor in convincing local members, residents and interested parties that measures designed to reduce existing environmental impacts of motor racing had been uppermost in the formulation of the new proposals. The extensive environmental studies which formed the basis of the statement proved to be a robust defence against the claims of objectors and provided reassurance to independent bodies such as the Countryside Commission and the Department of the Environment. Had this not been the case, the project would undoubtedly have needed to be considered at a public inquiry.
>
> (Hancock 1992)

Objectives and methods of public participation

Lack of information, or misinformation about the nature of a proposed development, prevents adequate public participation, and causes resentment and criticism of the project. One objective of public participation is thus to *provide information* about the development and its likely impacts. Before an EIS is prepared, information can be provided at public meetings and exhibitions, through presentations in schools, and on telephone hotlines. This information should be as candid and truthful as possible: people will be on their guard against evasions or biased information, and will look for confirmation of their fears. The EIS, when it is prepared, will also provide information about the project.

The way in which information is conveyed can influence public participation. Highly technical information can be understood by only a small proportion of the public. Information in different media (e.g. newspapers, television) will reach different sectors of the public. Ideally, a variety of methods for conveying information should be used, catering to different interest groups. EISs should also be presented in such a way as to be meaningful to both specialists and lay-people; this will be discussed further in §6.3.

Another objective of public participation is to help to *identify issues* that concern local residents. These issues are often not the same as those of concern to the developer or outside experts. Public participation exercises should thus

Table 6.1 Advantages and disadvantages of levels of increasing public influence.

Approaches	Extent of public power in decision-making	Advantages	Disadvantages
Information feedback			
Slide or film presentation, information kit, newspaper account, notices, etc.	Nil	Informative, quick	No feedback; presentation subject to bias
Consultation			
Public hearing, ombudsperson or representative, etc.	Low	Allows two-way information transfer; allows limited discussion	Does not permit ongoing communication; somewhat time-consuming
Joint planning			
Advisory committee, structured workshop, etc.	Moderate	Permits continuing input and feedback; increases education and involvement of citizens	Very time-consuming; dependent on what information is provided by planners
Delegated authority			
Citizens' review board, citizens' planning commission, etc.	High	Permits better access to relevant information; permits greater control over options and timing of decision	Long-term time commitment; difficult to include wide representation on small board

Source: Westman 1985.

achieve a two-way flow of information to allow residents to voice their views. The public participation exercises may well identify conflicts between the needs of the developer and those of various sectors of the public. This should ideally *lead to solutions* to these conflicts, and to agreement on future courses of action that reflect the joint objectives of all parties. Public participation is likely to be greatest where public comments are most likely to *influence decisions*. Westman (1985) has identified four levels of increasing public power in participation methods: information-feedback approaches, consultation, joint planning, and delegated authority. Table 6.1 lists advantages and disadvantages of these levels. Several criteria for good public participation can thus be identified: it should provide information, cater for different levels of technical sophistication and for special interests, achieve a two-way flow of information, and have an impact on decision-making.

There are many different ways of initiating public participation. A few are listed in Table 6.2, along with an indication of how well they fulfil the objectives and criteria listed above. These methods provide examples of several of the "eight rungs on a ladder of citizen participation", ranging from non-participation

Table 6.2 Methods of public participation and their effectiveness.

	Provide information	Cater for special interests	Two-way communication	Impact on decision-making
Explanatory meeting, slide/film presentation	✓	½	½	–
Presentation to small groups	✓	✓	✓	½
Public display, exhibit, models	✓	–	–	–
Press release, legal notice	½	–	–	–
Written comment	–	½	½	½
Poll	½	–	✓	✓
Field office	✓	✓	½	–
Site visit	✓	✓	–	–
Advisory committee, task force, community representative	½	½	✓	✓
Working groups of key actors	✓	½	✓	✓
Citizen review board	½	½	✓	✓
Public inquiry	✓	½	½	✓/–
Litigation	½	–	½	✓/–
Demonstration protests, riots	–	–	½	✓/–

Adapted from Westman 1985.

(manipulation, therapy), through tokenism (informing, consultation, placation) to citizen power (partnership, delegated power, citizen control) identified by Arnstein (1971).

6.3 EIS presentation

Although the EIA regulations specify the minimum contents required in an EIS (in Appendix 3 of the Town and Country Planning (Assessment of Environmental Effects) Regulations 1988, they do not give any standard for the presentation of this information. EISs range from a three-page typed and stapled EIS, to glossy brochures with computer graphics, and multi-volume documents in purpose-designed binders. This section discusses the contents, organization, clarity of communication and presentation of an EIS.

Contents and organization

An EIS should be *comprehensive*. It must, at minimum, fulfil the requirements of the relevant EIA legislation concerning contents. As will be discussed in Chapter 8, past EISs have not all fulfilled these requirements. This is, however, improving rapidly and local planning authorities are increasingly likely to require information on topics that they feel have not been adequately discussed in an EIS. A good EIS will also go further than the minimum requirements if other significant impacts are identified. Most EISs are broadly organized into four sections: a non-technical summary, a discussion of relevant methods and issues, a description of the project and of environmental baseline conditions, and a discussion of the project's likely environmental impacts (which may include the discussion of baseline environmental conditions and predicted impacts, proposed mitigation measures and residual impacts). Ideally, an EIS should also include the main alternatives that were considered, and proposals for monitoring. It could include much or all of the information given in Appendix 4 of *Environmental assessment: a guide to the procedures* (DOE 1989); Table 1.1 provides an example of a good EIS outline.

An EIS should *explain why some impacts are not addressed*. The introductory chapter, or an appendix, should include a "finding of no significant impact" section that explains why some impacts may be considered to be insignificant. If, for instance, the development is unlikely to affect the climate, a reason for this conclusion should be given. An EIS should *emphasize key points*. These should have been identified during the scoping exercise, but additional issues may arise during the course of the EIA. The EIS should *set the context* of the issues. The names of the developer, relevant consultants, relevant local planning authorities and consultees should be listed, along with a contact person who can supply further information. The main relevant planning issues and legislation should be explained. The EIS should also indicate any references used, and include a bibliography. The cost of the EIS should be given.

The preparation of a *non-technical summary* is particularly important in an EIS, as this is often the only part of the document that the public and decision-makers will read. The Dutch suggest that this summary "be such that a lay member of the public can read it and then be able to pass a considered opinion on the alternatives described and their environmental impact" (Government of the Netherlands 1991). It should thus briefly cover all relevant impacts and highlight key impacts, and should ideally contain a list or a table that allows the reader to identify key points at a glance. Chapter 4 gave examples of techniques for identifying and summarizing impacts.

An EIS should preferably be one *unified document*, with perhaps a second volume of appendices. A common problem with the organization of EISs stems from the way in which environmental impacts are assessed. The developer (or the consultants co-ordinating the EIA) often subcontracts parts of the EIA to consultancies that specialize in a particular field (e.g. ecological specialists, landscape consultants). These in turn prepare reports in a wide range of lengths and styles, making a variety of (possibly different) assumptions about the project and likely future environmental conditions, and proposing different and possibly conflicting mitigation measures. One way in which developers have attempted to circumvent this problem has been to summarize the impact predictions in a main text, and add the full reports as appendices to the main body of the EIS. Another has been to put a "company cover" on each report, and present the EIS as a multi-volume document, with each volume discussing a single type of impact. Both of these methods are problematic: the "appendix method" in essence discounts the great majority of findings, and the "multi-volume method" is cumbersome to read and carry, and often lacks integration. Neither method attempts to present findings in a cohesive manner, emphasize key impacts, or propose a coherent package of mitigation and monitoring measures. A good EIS would incorporate the information from subcontractors' reports into one coherent document that uses consistent assumptions and proposes consistent mitigation measures.

The EIS should be kept as *brief* as possible while still *presenting the necessary information*. The main text should include all relevant discussion of impacts, and appendices should present only additional data and documentation. In the USA, EISs are generally expected to be less than 150 pages long, while in the UK about 15% of EISs exceed 150 pages (Jones et al. 1991). On the other hand, about half of the EISs prepared in the UK to date are less than 50 pages long. Although length is not in itself an indicator of quality, commentators agree that statements of less than 50 pages are unlikely to be adequate (Coles et al. 1992, Lee 1992).

Clarity of communication

Weiss (1989) nicely notes that an unreadable EIS is an environmental hazard: "The issue is the quality of the document, its usefulness in support of the goals of environmental legislation, and, by implication, the quality of the environmental stewardship entrusted to the scientific community . . . An unreadable EIS not only hurts the environmental protection laws and, thus, the environment. It also turns the sincere environmental engineer into a kind of 'polluter'". Weiss identifies three classes of errors that reduce how well EISs communicate information:

- strategic errors, namely "mistakes of planning, failure to understand why the EIS is written and for whom"
- structural errors related to the EIS's organization, and
- tactical errors of poor editing.

The EIS has to communicate information to a wide range of audiences, from the decision-maker, to environmental experts, to the lay person. Although it cannot fulfil all the expectations of all its readers, it can go a long way towards being a useful document for a wide audience. At a minimum it should be *well written*, with good spelling and punctuation. It should have a clear structure, with easily visible titles and a logical flow of information. A table of contents, with page numbers marked, should be included in the main text, allowing easy access to information. Key points should be clearly indicated, perhaps in a table at the front or back of the report.

The EIS should include a *minimum of technical jargon*. Any jargon it does include should be explained in the text or in footnotes:

Wrong It is believed that the aquiclude properties of the Brithdir seams have been reduced and there is a degree of groundwater communication between the Brithdir and the underlying Rhondda beds, although . . . numerous seepages do occur on the valley flanks with the retention regime dependent upon the nature of the superficial deposits.

Right The accepted method for evaluating the importance of a site for waterfowl (i.e. waders and wildfowl) is the "1% criterion". A site is considered to be of national importance if it regularly holds at least 1% of the estimated British population of a species of waterfowl. "Regularly" in this context means counts (usually expressed as annual peak figures), averaged over the past 5 years.

The EIS should clearly *state any assumptions* on which impact predictions are based:

Wrong As the proposed development will extend below any potential [archaeological] remains, it should be possible to establish a method of working which could allow adequate archaeological examinations to take place.

Right For each operation an assumption has been made of the type and number of plant involved. These are:
Demolition – 2 pneumatic breakers, tracked loader
Excavation – backacter excavator, tracked shovel . . .

The EIS should be *specific*. Although it is easier and more defensible to claim that an impact is "significant" or "likely", the resulting EIS will be little more than a vague collection of possible future trends.

Wrong The landscape will be protected by the flexibility of the proposed [monorail] to be positioned and designed to merge in both location and scale into and with the existing environment.

Right From these [specified] sections of road, large numbers of proposed wind turbines would be visible on the skyline, where the towers would appear as either small or indistinct objects and the movement of rotors would attract the attention of road users. The change in the scenery caused by the proposals would constitute a major visual impact, mainly due to the density of visible wind turbine rotors.

Predicted impacts should be *quantified* if possible, perhaps with a range, and the use of non-quantified descriptors such as "severe" or "minimal" should be explained:

Wrong The effect on residential properties will be minimal with the nearest properties . . . at least 200 m from the closest area of filling.

Right Without the bypass, traffic in the town centre can be expected to increase by about 50–75% by the year 2008. With the bypass, however, the overall reduction to 65–75% of the 1986 level can be achieved.

Even better, predictions should give an indication of the probability of an impact occurring, and the degree of confidence with which the prediction can be made (see Ch. 9 for a good example). In cases of uncertainty, the EIS should propose worst-case scenarios:

Right In terms of traffic generation, the "worst-case" scenario would be for 100% usage of the car park. . . . For a more realistic analysis, a redistribution of 50% has been assumed.

Finally, the EIS should be *honest and unbiased.* A review of local authorities noted that "[a] number of respondents felt that the Environmental Statement concentrated too much on supporting the proposal rather than focusing on its impacts and was therefore not sufficiently objective" (Kenyan 1991). Developers cannot be expected to conclude that their projects have such major environmental impacts that they should be stopped. However, it is unlikely that all major environmental issues will have been resolved by the time the statement is written.

Wrong The proposed site lies adjacent to lagoons, mud and sands which form four regional Special Sites of Scientific Interest [authors' comment: note the misnomer]. The loss of habitat for birds, is unlikely to be significant, owing to the availability of similar habitats in the vicinity.

Table 6.3 provides a simple example of a clear presentation of the environmental effects of a road development on adjacent areas.

Presentation

Although it would be good to report that EISs are read only for their contents and clarity, in reality, as for prime ministers and presidents, presentation can have a

Table 6.3 Presentation of environmental effects.

Feature	Effects	Affected	Timescale	Magnitude	Controversial	Probability	Mitigation	Significance
Pigeon House Road	Reduced risk of HGV traffic	Residents	Short term permanent	Local	No	High	None	Minor beneficial
	Perceived severance due to elevated structure	Residents	Short term reducing with time	Local	Potentially	Low	None	Minor adverse
Bremen Grove	Reductions in traffic flow by about 80%	Residents and children using the park	Short term permanent	Local	No	High	None	Minor beneficial
Beach Road	Reductions in traffic flow by about 80%	School children	Short term permanent	Local	No	High	None	Minor beneficial

Source: P. Tomlinson (Ove Arup).

major influence on how they are received. EIAs are, indirectly, public relations exercises, and EISs can be seen as publicity documents for developers. Good presentation can convey a concern for the environment, a rigorous approach to the impact analysis, and a positive attitude to the public. Bad presentation, in turn, suggests a lack of care, and perhaps a lack of financial backing. Similarly, good presentation can help to convey information clearly, whereas bad presentation can negatively affect even a well organized EIS.

The presentation of an EIS will say much about the developer. The type of paper used – recycled or not, glossy or not, heavy- or lightweight – will affect the image projected, as will the choice of colour or black-and-white diagrams, and the use of dividers (if any) between chapters. The "ultra-Green" company will opt for double-sided printing on recycled paper, and the "luxury developer" will use glossy, heavyweight paper with a distinctive binder. Generally, a relatively strong binder that stands up well under heavy handling is most suitable for EISs. Unless the document is very thin, a spiral binder is likely to snap or bend open with continued handling and stapled documents are likely to tear. Multivolume documents are difficult to keep together unless a box is provided.

Finally, the use of maps, graphs, photomontages, diagrams and other forms of visual communication can greatly help the EIS presentation. As noted in Chapter 4, a location map, a site layout of the project, and a process diagram, are virtually essential to the proper description of a development. Maps showing, for example, the extent of visual impacts, the location of designated areas, or classes of agricultural land, are a succinct and clear way of presenting such information. Graphs are often much more effective than tables or figures in conveying numerical information. Forms of visual communication break up the page, and add interest to an EIS.

6.4 Review of EISs

The comprehensiveness and accuracy of EISs are matters of concern. As will be shown in Chapter 8, many EISs do not meet even the minimum regulatory requirements, much less provide adequate information on which to base decisions. In some countries (for example, the Netherlands, Canada, Malaysia and Indonesia) EIA commissions have been established to review EISs and act as a quality assurance process. However, in the UK there are no mandatory requirements regarding the pre-decision review of EISs to ensure that they are comprehensive and accurate. A planning application cannot be judged to be invalid if it is accompanied by an inadequate or incomplete EIS; a competent authority may only request further information, or refuse permission and risk an appeal. In addition, many competent authorities do not have the full range of technical expertise needed to assess the adequacy and comprehensiveness of an EIS. Some authorities, especially those that receive few EISs, have consequently had difficulties in dealing with the technical complexities of EISs. In about 10–20% of cases, consultants have been brought in to review EISs. Other authorities have joined the Institute of Environmental Assessment, which reviews one EIS at no cost to member organizations. Others have been reluctant to buy outside expertise, especially at a time of restrictions on local spending (McDonic 1992, Fuller 1992). At the time of writing, the DoE is in the process of dealing with this problem by commissioning research on the review and use of environmental information in decision-making.

In an attempt to fill the void previously left by the national government, several non-mandatory review criteria have been established. Effective review criteria should allow a competent authority to:

- ensure that all relevant information has been analyzed and presented
- assess the validity and accuracy of information contained in the EIS
- quickly become familiar with the proposed project and consider whether additional information is needed
- assess the significance of the project's environmental effects
- evaluate the need for mitigation and monitoring of environmental impacts, and
- advise on whether a project should be allowed to proceed (Tomlinson 1989).

To fulfil these criteria, Tomlinson (1989) proposed review criteria in the form of yes/no questions concerning nine main issues: administration/procedural requirements, effective communication, impact identification, alternatives, information assembly, baseline description, impact prediction, mitigation measures, and monitoring/audits.[2]

Lee & Colley (1990) in turn proposed an hierarchical review framework. At the top of the hierarchy is a comprehensive mark (A = well performed and complete, through to F = very unsatisfactory) for the entire report. This mark is based on marks given to four broad subheadings: description of the development, local environment and baseline conditions; identification and evaluation of key impacts; alternatives and mitigation of impacts; and communication of results. Each of these, in turn, is based on two further layers of increasingly specific topics/questions. Lee & Colley's criteria have been used either directly, or in a modified form (e.g. by the Institute of Environmental Assessment) to review a range of EISs in the UK. It is the most commonly used review method in the UK.

The review criteria given in Table 6.4 are an amalgamation and extension of Tomlinson's and Lee & Colley's criteria. It is unlikely that any EIS will fulfil all of the criteria. Similarly, some criteria may not apply to all projects. However, they should act as a checklist of good practice, both for those preparing and those reviewing EISs. Table 6.5 shows possible ways of using these criteria. Example (a), which relates to minimum requirements, amplifies the presence or otherwise of key information. Example (b) includes a simple grading, which could be on the A to F scale used by Lee & Colley, for each criterion (only one of which is shown here). Example (c) is a variation of example (b).

Table 6.4 EIS review criteria.

Minimum requirements (T&CP Regulations 1988 and other UK EIA regulations)

1. Description of the proposed development
   - information about the site
   - information about the design and size/scale of the development
2. Data necessary to identify and assess the main effects that the development is likely to have on the environment.
3. Description of the likely significant impacts, direct and indirect, of the development on the environment:
   - human beings; flora; fauna; soil; water; air; climate; landscape; interactions between the above; material assets; cultural heritage
4. Measures to avoid, reduce or remedy all significant adverse effects.
5. Non-technical summary

Suggestions for best practice

1. Purpose, location and physical presence of proposed development. Discuss
   - purpose and rationale of project
   - location of proposed project on map
   - land area and uses (construction, operation, decommissioning) on map
   - design and size of the development in figures
   - duration of construction, operation, close-down
   - proposed access and transport arrangements (construction, operation, close-down)
2. Main physical characteristics of production processes. Describe
   - overall operation of process involved
   - type and quantity of resources used e.g. minerals, energy, water, other
   - transportation requirements (of inputs and outputs)
   - estimated type, quantity (rate of production), composition, strength of residues/emissions (construction, operation):
     - aqueous wastes
     - gaseous and particulate emissions
     - soil/land pollutants, solid wastes
     - noise and vibration
     - heat and light
     - radiation
     - others
   - potential for accidents, hazards and emergencies
   - processes for the containment, treatment and disposal of wastes
3. Main socio-economic characteristics of production process. Describe
   - labour requirements of project, including size, duration, sources, skills categories and training
   - provision of developer facilities (e.g. housing)
   - local service requirements (e.g. catering)
   - expenditure flows
   - flow of social activity (service demands, community participation, etc.)
4. Description of environment and determination of "baseline" conditions, i. e. the probable future state of the environment, in the absence of the project, taking into account natural fluctuations and human activities. Describe
   - environment affected with map
   - important components (physical and socio-economic) of affected environment
   - potentially significant effects away from immediate site
   - the existing and likely future status of valued environmental resources and land-use
   - any variability in natural systems (e.g. seasonal change in habitat use), and human use (e.g. arising from polluting activities)
5. Planning framework
   - explain relevant national, regional, county, district and/or subject planning policies
   - if the proposal does not conform to guidance, justify the departure
6. Alternatives. Discuss
   - the need for the project
   - the "no action" option

- alternative locations
- alternative scales
- alternative processes
- alternative site layouts
- alternative operating conditions
- alternative decisions
- main reason for choosing the proposed development

7. Impact identification and scoping

- identify individuals, communities and statutory agencies likely to be affected by the project
- discuss issues considered important by:
  - local authority(-ies)
  - statutory consultees
  - experts consulted
  - voluntary organizations
  - local residents
  - general public
- use an impact identification method (e.g. matrix, checklist)
- carry out a scoping exercise: identify all activities with significant impacts on valued environmental attributes; state why particular environmental attributes and activities are valued
- define temporal and spatial boundaries for specific environmental impacts

8. Impact prediction

- discuss likely significant impacts of development on environment that may result from:
  - use of natural resources
  - elimination of wastes
  - emission of pollutants and creation of other nuisances
  - use of labour and services
  - flow of expenditure
  - flow of social activities
- for each impact, discuss:
  - variable affected
  - deviations from baseline condition
  - impact magnitude, including geographical extent
  - adverse or beneficial
  - direct and indirect
  - duration, over project life-cycle
  - (ir)reversibility
  - permanency
  - probability
  - confidence in prediction
  - cumulative impacts
  - relation to national and international standards (and justify standards)
  - relation to relevant statutory designations
  - relation to planning and environmental policies
  - impact in both standard and non-standard operating conditions
  - quantify impacts where possible
- explain basis for predictions, e.g. case studies, models
- identify and justify key variables and assumptions
- attempt to isolate project-generated impacts from other changes resulting from non-project activities and variables
- ensure that predictions are logically linked to baseline conditions, mitigation measures and monitoring proposals

9. Assessment of significance/evaluation. Outline

- approach to evaluation of relative significance of impacts used
- monetary valuation approaches used (e.g. CBA, monetary valuation techniques)
- multi-criteria evaluation approaches used, including
  - scoring systems
  - weighting systems
- distribution of impacts

10. Mitigation measures

- for all major adverse impacts, consider the following broad approaches:
  - avoid impacts

- reduce impacts
  - compensate for impacts
  - remedy impacts (repair, rehabilitate or restore affected components)
- more specifically consider the following measures:
  - project modification, including design, layout and timing
  - technical measures (e.g. recycling, pollution control, containment)
  - aesthetic and ecological measures (e.g. mounding, plantings, measures to preserve particular habitats, recording of archaeological sites)
  - replacement/compensation for affected resources
  - provision of alternative facilities
- explain rationale for choosing mitigation measures
- discuss effectiveness of proposed mitigation measures
- explain commitment to mitigation measures (e.g. in a proposed Sec. 106 agreement)
- indicate significant residual or unmitigated impacts, and justify why these impacts should not be mitigated

11. Consultation
- discuss attempts to contact public and obtain public opinion
- note statutory consultees contacted
- include a copy (or summary) of main comments from consultees and the public
- explain measures taken to respond to these comments, including changing development proposal

12. Difficulties. Indicate
- and account for any gaps in data
- any difficulties, including technical deficiencies and lack of know-how
- any basis for questioning assumptions, data and information

13. Monitoring and auditing
- consider all important impacts for monitoring, especially where uncertainty exists
- outline a monitoring programme, including:
  - who is responsible for financing, undertaking, and reporting the results of the monitoring
  - the frequency of monitoring (this may vary by impact)
  - who reviews and acts on monitoring information
  - actions to be taken if monitoring identified problems
  - plans for making monitoring information publicly available
  - distinctions between project stages (especially between construction and operation)
- consider all important impacts for monitoring, especially where uncertainty exists
- detail an audit programme to check on predictions, mitigation measures and to improve future predictions

14. Presentation
- mention:
  - the relevant EIA legislation
  - the name of the developer
  - the name of the competent authority(ies)
  - the name of the organization preparing the EIS
  - the name, address and telephone number of a contact person
- arrange the EIS logically into sections/chapters
- include a table of contents with page numbers
- summarize chapters except where they are short
- summarize the main impacts, e.g. in a table
- acknowledge external sources in full
- explain technical terms
- use tables, graphs and figures as appropriate
- present the document as an integrated whole, including consistent standards of analysis
- give prominence to severe impacts
- minimize bias
- give no undue emphasis on construction, operation, close-down

15. Non-technical summary. Ensure that
- there is a non-technical summary
- the summary is comprehensive
- it is appropriate for decision-makers
- it is appropriate for public participation

Table 6.5 Examples of possible uses for EIS review criteria.

(a)

Criterion	Presence/absence (page number)	Information provided	Key information absent
1. Description of the proposed development:			
Site	✓ (p. 5)	Location (in plans), existing operations, access	Working method, vehicle movements, restoration plans
Design	X		Site buildings (location, size), employment, restoration
. . .			

(b)

Criterion	Presence/absence (page number)	Comments	Grade
1. Purpose, location and physical presence of proposed development:			
Purpose and rationale of project	✓ (p. 11)	Briefly in introduction, more details in Sec. 2	A
Location of proposed project on map	✓ (p. 12)		B
. . .			

(c)

Criterion	A	B	C	None	
5. Alternatives:					
Need for the project	✓				Discusses need for more landfill in Sec. 3. However, no estimates are made for the proposed landfill's capacity
"No action" option				✓	not discussed
. . .					

6.5 Decisions on projects

EIA and planning permission

Where required by the local authority, as competent authority, an EIS must be submitted with a planning application for a project under consideration. The determination of applications with an EIS must be made within a 16-week period, unless the developer agrees to a longer period. As noted in §6.4 it is at this stage that the review of a statement is undertaken. The LPA is required to have regard to the EIS, as well as to other material considerations, in determining the application. In reaching its decision, the LPA will also consider the views it has received from statutory consultees and from the public, plus any further information it requested from the developer. The range of decision options are as for any planning application: the LPA may grant permission for the project (with or without conditions) or refuse permission. The LPA may suggest further mitigation measures following consultations, and will seek to negotiate these with the developer. The Town and Country Planning (Assessment of Environmental Effects) Regulations 1988 state that:

> The local planning authority or the Secretary of State or an Inspector shall not grant planning permission pursuant to an application to which this regulation applies, unless they have first taken the environmental information into consideration.

Circular 15/88 (App. B, para. 49) elaborates:

> In considering the planning application the local planning authority is required to have regard to all of the environmental information, that is the information contained in the environmental statement and any comment made by the statutory consultees and representations from members of the public, as well as to other material considerations.

As Atkinson & Ainsworth (1992) point out:

> The system operates on the basis that, since EA exists for the benefit of the public, the LPA will act to impose the principal sanction of the system where the developer fails to comply with the procedures: a refusal of planning permission. To that end, any concerned member of the public may make representations to the LPA during the planning process, drawing its attention to the relevant procedures of the 1988 Regulations. If these representations are ignored, a request may be made to the Secretary of State to exercise his extensive powers of call-in to determine the application himself.

If the development is refused, the developer may wish to appeal against the decision. If the development is permitted, individuals or organizations may want to challenge the permission. The Secretary of State may also "call-in" an application, for a variety of reasons. A public inquiry may result.

EIA and public inquiries

Public inquiries for projects with an EIS raise important issues, the first of which concerns the rôle of the Planning Inspectorate. The EIA Regulations allow inquiry inspectors and the Secretary of State (a) to require the submission of an EIS before a public inquiry, if they regard this as appropriate, and (b) to request further information from a developer where they consider that the EIS is inadequate as it stands. The information contained in an EIS will be among the material considerations to be taken into account by an Inspector in arriving at his recommendations. To date, there is little evidence to indicate how inquiry inspectors have exercised these powers or the extent to which they have taken EISs into account.

A second issue concerns the use made of the EIS by the various parties involved in a public inquiry, either before or during the inquiry. The environmental impact of proposals (especially highways, traffic and landscape issues) will certainly be examined in detail during any inquiry, and this examination will draw upon the information and analysis contained in the EIS. In distinguishing between matters of fact and matters of interpretation, and in identifying the most significant impacts of the development, an EIS may assist the developers and local planning authorities in arriving at a list of agreed matters before the start of an inquiry. This should avoid unnecessary delays during the inquiry itself. Again, to date, there is little evidence to indicate (a) how EISs have been used during public inquiries and (b) whether the consideration of environmental issues during such inquiries has been any different from what would have occurred before the introduction of the EIA regulations. There is clearly a need for more research into this important area. A recent report commissioned by the DoE (1991) made the following recommendations:

> The DoE . . . should make the results of the Secretary of State's or inspectors' decisions relating to called-in planning applications or appeals involving EIA available for publication in the Journal of Planning and Environment Law; . . . The use made in public inquiries of EISs by the various parties should be investigated to determine whether EIA is being integrated appropriately in the call-in and appeal procedures.

Challenging a decision: judicial review

The UK planning system has no official provisions for an appeal against development consent. However, if permission is granted, a third party may wish to challenge that decision on the grounds, for example, that no EIA was prepared when it should have been, or that the competent authority did not adequately consider the relevant environmental information. The only way to do this is through judicial review proceedings in the UK High Court, or through the EC.

Judicial review proceedings in the UK courts first require that the third party shows that it has "standing" to bring in the application, namely sufficient interest in the project by virtue, for example, of potential personal injury if the project proceeds. Establishing standing is one of the main difficulties in applying for judicial review in the UK, and procedures for establishing standing are being reconsidered. If standing is established, the third party must then convince the court that the competent authority did not act according to the relevant EIA procedures. The court does not make its own decision about the merits of the case, but only reviews the way in which the competent authority arrived at its decision:

> [T]he court will only quash a decision of the [competent authority] where it acted without jurisdiction or exceeded its jurisdiction or failed to comply with the rules of natural justice in a case where those rules apply or where there is an error of law on the face of the record or the decision is so unreasonable that no [competent authority] could have made it. (Atkinson & Ainsworth 1992)

Various possible scenarios emerge. A competent authority may fail to require an EIA for a Schedule 1 project, or may grant permission for such a project without considering the environmental information. In such a case, its decision would be void.

A competent authority may decide that a project does not require EIA because it is not in Schedule 1 or in Schedule 2 with significant environmental effects. This was the issue in the case of *Regina v. Swale Borough Council and Medways Port Authority ex parte RSPB* (1991, 1 *PLR* 6) concerning the construction of a storage area for cargo, which would have required the infilling of Lappel Bank, a mudflat important for its wading birds. The appellants argued that the project fell either within Schedule 1 or within Schedule 2 with significant environmental effects; the local planning authority felt that an EIA was unnecessary. The courts held that the decision about whether or not a project falls within a Schedule is a decision for the local authority, open for review only if no reasonable local authority could have made it.

A competent authority may make a decision in the absence of a formal EIA, but with environmental information available in other forms. This was the case in *Regina v. Poole Borough Council, ex parte Beebee and others* (1991, *Journal of Planning Law* 643) concerning a decision to develop part of Canford Heath. In this case the courts ruled that, despite the lack of an EIS and the attendant rigour and publicity, enough environmental information was available for the council to make an informed decision. A different judgement might have been made had the competent authority been shown to have made its decision before it had all of the relevant environmental information.

In several cases (e.g. the M3 at Twyford Down and a large afforestation scheme in Glen Dye (1992, *Journal of Environmental Law* 289)) a competent authority approved a project, in the absence of an EIA, after 3 July 1988 – when Directive 85/337 should have been implemented – but before the relevant UK regulation came into effect. These cases have raised the issue of how far the Directive applies to transitional projects proposed before 3 July 1988, but decided after that date. Macrory (1992) summarizes the courts' judgement: "The Directive is intended to influence the 'process at every state', and in the absence of clear transitional measures it would be against the aims of the Directive to attempt to retrospectively impose such requirements on decision-making procedures already commenced". The cases have also focused on whether Directive 85/337 can have direct effect in the EC Member States. Some legal experts argue that it should have direct effect for Annex I projects but not for Annex II projects; others argue that it should also have direct effect for Annex II projects likely to have significant environmental impacts where "no Member State could reasonably have . . . decided otherwise" (Macrory 1992). No clear resolution has emerged to date.

In summary, judicial review of competent authorities' decisions have to date been limited by the issue of standing, and by the courts' narrow interpretation of the duties of competent authorities under the EIA regulations. Future court cases may widen this interpretation, but it is very unlikely that the UK courts will play as active a rôle as those in the USA did in relation to the NEPA.

Challenging a decision: European Community

Another avenue for third parties to challenge a competent authority's decision to permit development, or not to require EIA, is through the EC. Such cases need to show that the UK failed to fulfil its obligations as a Member State under the Treaty of Rome by not properly implementing EC legislation, in this case, Directive 85/337. In such a case, Article 169 of the Treaty allows a declaration of non-compliance to be sought from the European Court of Justice. The issue of standing is not a problem here, since the Commission of the EC can begin proceedings either on its own initiative or based on the written complaint of any individual. To use this mechanism, the Commission must first state its case to the Member State and seek its observations. The Commission may then issue a "reasoned opinion". If the Member State fails to comply within the specified time, the case proceeds to the European Court of Justice.

This mechanism has been used, although not (yet) to its full conclusion, against the UK government in the case of seven projects:

- extension of the M3 bisecting Twyford Down, near Winchester
- extension of the M11 to link it to the Blackwall Tunnel
- construction of a clinical waste incinerator at South Warwick Hospital in the West Midlands
- construction of a soft drinks and can manufacturing plant at Brackmills in Northamptonshire
- construction of a high-speed train link between the Channel Tunnel and a London rail terminal
- extension of British Petroleum's gas-separation plant at Kinneil near Falkirk
- the East London River Crossing linking Becton in the Docklands to Greenwich and passing through the ancient woodland of Oxleas Wood.

In October 1991, the then EC Commissioner of the Environment, Carlo Ripa di Meana, wrote to the UK Secretary of State for Transport, requesting information on these seven projects.[3] The UK government responded by sending a letter of observations and ten boxes of documents to the EC. The EC dropped legal proceedings against the M3 proposal in late July 1992, after being satisfied with the UK's response that the Directive had been complied with (construction has since begun, amidst a storm of protest). The EC accepted that consent for the M11 had been granted before the Directive came into effect. The waste incinerator has been refused on appeal, a second EIA has been prepared for the Brackmills plant, and the EC was satisfied that the Channel Tunnel rail link and terminal could be considered as separate projects (*Simmons & Simmons Environmental Law Newsletter* 15). However, in late 1992 the EC was still considering issuing a "reasoned opinion" over the case of the Kinneil gas-separation plant, arguing that no public consultation had taken place, and that the competent authority had not considered the information that the developer had prepared (*Planning* 980). In May 1993 the EC issued a "reasoned opinion" on the East London River Crossing, alleging that the procedural requirements of the Directive had not been complied with, and that no non-technical summary had been prepared (*Simmons and Simmons Environmental Law Newsletter* 18). In July 1993 the DOT announced that the project would be redesigned to avoid damage to Oxleas Wood.

Under Article 171 of the Treaty of Rome, if the European Court of Justice finds that a Member State has failed to fulfil an obligation under the Treaty, it may require the Member State to take necessary measures to comply with the Court's judgement. Under Article 186, the EC may take interim measures to require a Member State to desist from certain actions until a decision is taken on the main action. However, to do so the Commission must show the need for urgent relief, and that irreparable damage to Community interests would result if these measures were not taken. Suggested amendments to the Treaty of Maastricht would enable the European Court of Justice to impose fines on Member States in the future. The reader is referred to Atkinson & Ainsworth (1992), Buxton (1992), and Salter (1992a,b,c) for further information.

6.6 Summary

Active public participation and good presentation are key aspects of a successful EIA process. Both have been undervalued. Presentation of environmental information is, happily, improving rapidly, but public participation is likely to remain a weak aspect of EIA in the UK until developers and competent authorities see that benefits outweigh costs.

Formal review of EIA is also rarely carried out, despite the availability of several non-mandatory review guidelines. A recent government initiative is likely to develop, and encourage the use of, such guidelines, even if they are not made mandatory. Such review procedures can contribute to the processing of the EIS as part of the decision-making stage. The impact of the EIS on the quality of the outcome is discussed in Chapter 8.

Several appeals against development consents or against competent authorities' failure to require EIA have been brought to the UK courts or the EC. The UK courts have been unwilling to overturn the decisions of competent authorities, and have generally given a narrow interpretation of the duties of competent authorities under the EIA regulations. The EC, by contrast, has proven willing to challenge the UK government on its implementation of Directive 85/337 and on some specific decisions resulting from this implementation.

More positively, the next step in a good EIA procedure is the monitoring of the development's actual impacts, and the comparison of actual and predicted impacts. This is discussed in Chapter 7.

References

Arnstein 1971. A ladder of public participation in the USA. *Journal of the Royal Town Planning Institute* **57**.

Atkinson, N. & R. Ainsworth 1992. Environmental assessment and the local authority: facing the European imperative. *Environmental Policy and Practice* **2**(2), 111–28.

Buxton, R. 1992. Scope for legal challenge. In *Environmental assessment and audit: a user's guide*, 43–4. Gloucester: Ambit.

Coles, T. F., K. G. Fuller, M. Slater 1992. Practical experience of environmental assessment in the UK. *Proceedings of Advances in Environmental Assessment Conference*. London: IBVC Technical Services.

Council of the European Community 1990. Directive 90/313, *Official Journal* L158, 23 June 1990, 56.

CPRE 1991. *The Environmental Assessment Directive – five years on*. London: Council for the Protection of Rural England.

Dallas, W. G. 1984. Experiences of environmental impact assessment procedures in Ireland. In *Planning and ecology*, R. D. Roberts & T. M. Roberts (eds), 389–95. London: Chapman & Hall.

DoE (Department of the Environment) 1989. *Environmental assessment: a guide to the procedures*. London: HMSO.

DoE (Department of the Environment) 1991. *Monitoring environmental assessment and planning*. A report by the EIA Centre, Department of Planning and Landscape, University of Manchester, HMSO.

Elkin, T. J. & P. G. R. Smith 1988. What is a good environmental impact statement? Reviewing screening reports from Canada's national parks. *Journal of Environmental Management* **26**, 71–89.

Fuller, K. 1992. Working with assessment. In *Environmental assessment and audit: a user's guide*, 14–15. Gloucester: Ambit.

Government of the Netherlands 1991. *Environmental impact assessment*. The Hague: Ministry of Housing, Physical Planning and the Environment.

Hancock, T. 1992. Statement as an aid to consent. In *Environmental assessment and audit: a user's guide*, 34–5. Gloucester: Ambit.

House of Lords 1981. *Environmental assessment of projects* (Select Committee on the European Communities, 11th Report, Session 1980–81). London: HMSO.

Jain, R. K., L. V. Urban, G. S. Stacey 1977. *Environmental impact analysis*. New York: Van Nostrand Reinhold.

Jones, C. E., N. Lee, C. Wood 1991. *UK environmental statements 1988–1990: an analysis*. Occasional Paper 29, EIA Centre, University of Manchester.

Kenyan, R. C. 1991. Environmental assessment: an overview on behalf of the RICS. *Journal*

of Planning and Environment Law, 419–22.
Lee, N. 1992. Improving quality in the assessment process. In *Environmental assessment and audit: a user's guide*, 12–13. Gloucester: Ambit.
Lee, N. & R. Colley 1990. *Reviewing the quality of environmental statements*. Occasional Paper 24, EIA Centre, University of Manchester.
Macrory, R. 1992. Analysis of the petition of the Kincardine and Deeside District Council. *Journal of Environmental Law* **4**(2), 298–304.
McCormick, J. 1991. *British politics and the environment*. London: Earthscan.
McDonic, G. 1992. The first four years. In *Environmental assessment and audit: a user's guide*, 2–3. Gloucester: Ambit.
Mollison, K. 1992. A discussion of public consultation in the EIA process with reference to Holland and Ireland. Unpublished paper, MSc/Diploma course in Environmental Assessment and Management, Oxford Polytechnic.
O'Riordan, T. & R. K. Turner 1983. *An annotated reader in environmental planning and management*. Oxford: Pergamon.
Salter, J. R. 1992a. Environmental assessment: the challenge from Brussels. *Journal of Planning and Environment Law* (January), 14–20.
Salter, J. R. 1992b. Environmental assessment – the need for transparency. *Journal of Planning and Environment Law* (March), 214–21.
Salter, J. R. 1992c. Environmental assessment – the question of implementation. *Journal of Planning and Environment Law* (April), 313–18.
Tomlinson, P. 1989. Environmental statements: guidance for review and audit. *The Planner* **75**(28), 12–15.
Turner, T. 1988. The legal eagles. *The Amicus Journal* (Winter), 25–37.
Weiss, E. H. 1989. An unreadable EIS is an environmental hazard. *The Environmental Professional* **11**, 236–40.
Westman, W. E. 1985. *Ecology, impact assessment and environmental planning*. New York: John Wiley.

Notes

1. Although this section refers to public consultation and participation together as "public participation", the two are in fact separate. Consultation is in essence an exercise concerning a passive audience: views are solicited, but respondents have little active influence over any resulting decisions. In contrast, public participation involves an active rôle for the public, with some influence over any modifications to the project and over the ultimate decision.
2. These are based on a Canadian framework initially established by Elkin & Smith (1988).
3. Further actions by the Commissioner, including the issue of a press release on the matter, provoked the UK Prime Minister to write to EC President Delors, accusing the Commissioner of unwarranted intrusion into national affairs.

CHAPTER 7
Monitoring and auditing: after the decision

7.1 Introduction

Major projects – such as roads, airports, power stations, petrochemicals plants, mineral developments and holiday villages – have a life-cycle, with a number of stages (see Fig. 1.5). The life-cycle may cover a very long period (e.g. 50–60 years for the planning, construction, operation and decommissioning of a fossil-fuelled power station). EIA, as it is currently practised in the UK and in many other countries, relates primarily to the period *before* the decision. At its worst, it is a partial linear exercise related to one site, produced in-house by a developer, without any public participation. However, EIA should not stop at the decision. It should be more than an auxiliary to the procedures to obtain a planning permission; rather it should be a means to obtain good environmental management *over the life* of the project. This involves building monitoring and auditing into the EIA process.

The following section clarifies the definitions of and differences between monitoring and auditing, and outlines their (potentially) important rôles in EIA. An approach to the better integration of monitoring into the process is then outlined, drawing in particular on Californian practice. This is followed by a discussion of approaches to environmental impact auditing, including a review of recent attempts to audit a range of EISs in several countries. The final section draws briefly on detailed monitoring and auditing studies of the local socio-economic impacts of the construction of the Sizewell B PWR nuclear power station in the UK.

7.2 The importance of monitoring and auditing in the EIA process

Monitoring involves the measuring and recording of physical, social and economic variables associated with development impacts (e.g. traffic flows, air quality, noise, employment levels). It seeks to provide information on the characteristics and functioning of variables in time and space, and in particular on the occurrence and magnitude of impacts. Monitoring can improve project management. It can be used, for example, as an early-warning system, to identify harmful trends in a locality before it is too late to take remedial action. It can help to identify and correct unanticipated impacts. Monitoring can also provide an accepted database, which can be useful in mediation between interested parties. Thus, monitoring of the origins, pathways and destinations of, for example, dust in an industrial area, may clarify where the responsibilities lie. Monitoring is also essential for successful environmental impact auditing, and can be one of the most effective guarantees of commitment to undertakings and to mitigation measures.

As noted by Buckley (1991), the term environmental *auditing* is currently used in two main ways. *Environmental impact auditing*, which is covered in this chapter, involves comparing the impacts predicted in the EIS with those that actually occur after implementation, in order to assess whether the impact prediction performs satisfactorily. The audit can be of both impact predictions (How good were the predictions?) and of mitigation measures/conditions attached to the development (Is mitigation effective? Are the conditions being honoured?). This approach to auditing contrasts with *environmental management auditing*, which focuses on public and private corporate structures and programmes for environmental management and the associated risks and liabilities. The latter approach is discussed further in Chapter 12.

In total, monitoring and auditing can make important contributions to the better planning and EIA of future projects (see Fig. 7.1). Sadler (1988) writes of the need to introduce feedback in order to learn from experience; we must avoid the constant "reinventing of the wheel" in EIA. Monitoring and auditing of outcomes can contribute to an improvement in all aspects of the EIA process, from understanding baseline conditions to the framing of effective mitigation measures. Greene et al. (1985) in addition note that monitoring and auditing should reduce time and resource commitments to EIA by allowing all participants to learn from past experience; they should also contribute to a general enhancing of the credibility of proponents, regulatory agencies, and EIA processes. We are on an upward learning curve, and there is a considerable growth of interest in examining the effectiveness of the EIA process in practice. Unfortunately there are some significant issues that have greatly limited the use of monitoring and auditing to date. These and possible ways forwards for monitoring and auditing in practice are now discussed.

7.3 Monitoring in practice

Monitoring implies the systematic collection of a potentially large quantity of information over a long period. Such information should include not only the "traditional" *indicators* (e.g. ambient air quality, noise levels, size of workforce)

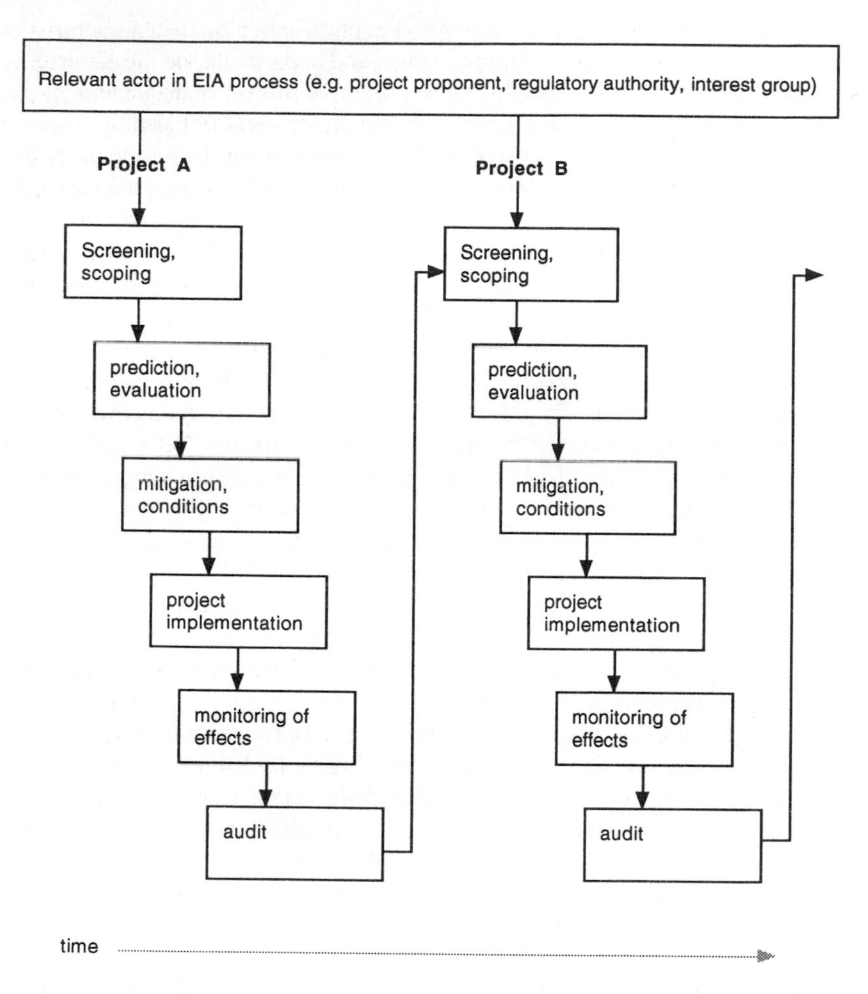

Figure 7.1 Monitoring and auditing and learning from experience in the EIA process. Adapted from: Bisset & Tomlinson 1988, Sadler 1988.

but also causal underlying factors (e.g. local authority and developer *decisions and policies*). The causal factors are the determinants of impacts and these may have to be changed if there is a wish to modify impacts. Opinions held by people about the project and its impacts may also be influential in determining responses to it. A systematic attempt to identify such *opinions* can be an important input into a monitoring study. The information collected needs to be stored, analyzed and communicated to relevant participants in the EIA process. A primary requirement therefore is to focus monitoring activity only on "those environmental parameters expected to experience a significant impact, together with those parameters for which the assessment methodology or basic data were not so well established as desired" (Lee & Wood 1980).

Monitoring is an integral part of EIA; baseline data, project descriptions, impact predictions and mitigation measures should be developed with monitoring implications in mind. An EIS should include a *monitoring programme* which has clear objectives, temporal and spatial controls, adequate duration (e.g. covering major stages in project implementation), practical methodologies, sufficient funding, clear responsibilities, and open and regular reporting. Ideally, the monitoring activity should include a partnership between the parties involved; for example, the collection of information could involve the developer, local authority and local community. Monitoring programmes should also be adapted to the dynamic nature of the environment (Holling 1978).

Unfortunately, however, monitoring is not a mandatory step in many EIA procedures, including those current in the UK. In the *absence of mandatory procedures*, it is usually difficult to persuade developers that it is in their interest to have a continuing approach to EIA. This is particularly the case where the proponent has a one-off project, and has less interest in learning from experience for application to future projects. Fortunately, we can turn to some examples of good practice in a few other countries. A brief summary of monitoring procedures in Canada is included in Chapter 11. The monitoring procedures used in California, for projects which are subject to the California Environmental Quality Act (CEQA), are also of interest (California Resources Agency 1988).

Since January 1989, state and local agencies in California have been required to adopt a monitoring and/or reporting programme for mitigation measures and project changes that have been imposed as conditions to address significant environmental impacts. The aim is to provide a mechanism that will help to ensure that mitigation measures will be implemented in a timely manner in accordance with the terms of the project approval. Monitoring refers to the observation and oversight of mitigation activities at a project site, whereas reporting refers to the communication of the monitoring results to the agency and public. If project implementation is to be phased, the mitigation and subsequent reporting and monitoring may also have to be phased. If monitoring reveals that mitigation measures are ignored or are not completed, sanctions could be imposed; these can include for example, "stop work" orders, fines and restitution. The components of a monitoring programme would normally include the following:

- summary of significant impacts identified in the environmental impact report
- mitigation measures recommended for each significant impact
- monitoring requirements for each mitigation measure
- person or agency responsible for the mitigation measure monitoring
- timing and/or frequency of the monitoring
- agency responsible for ensuring compliance with the monitoring programme
- reporting requirements.

Figure 7.2 provides an extract from a monitoring programme for a wood-waste conversion facility at West Berkeley in California.

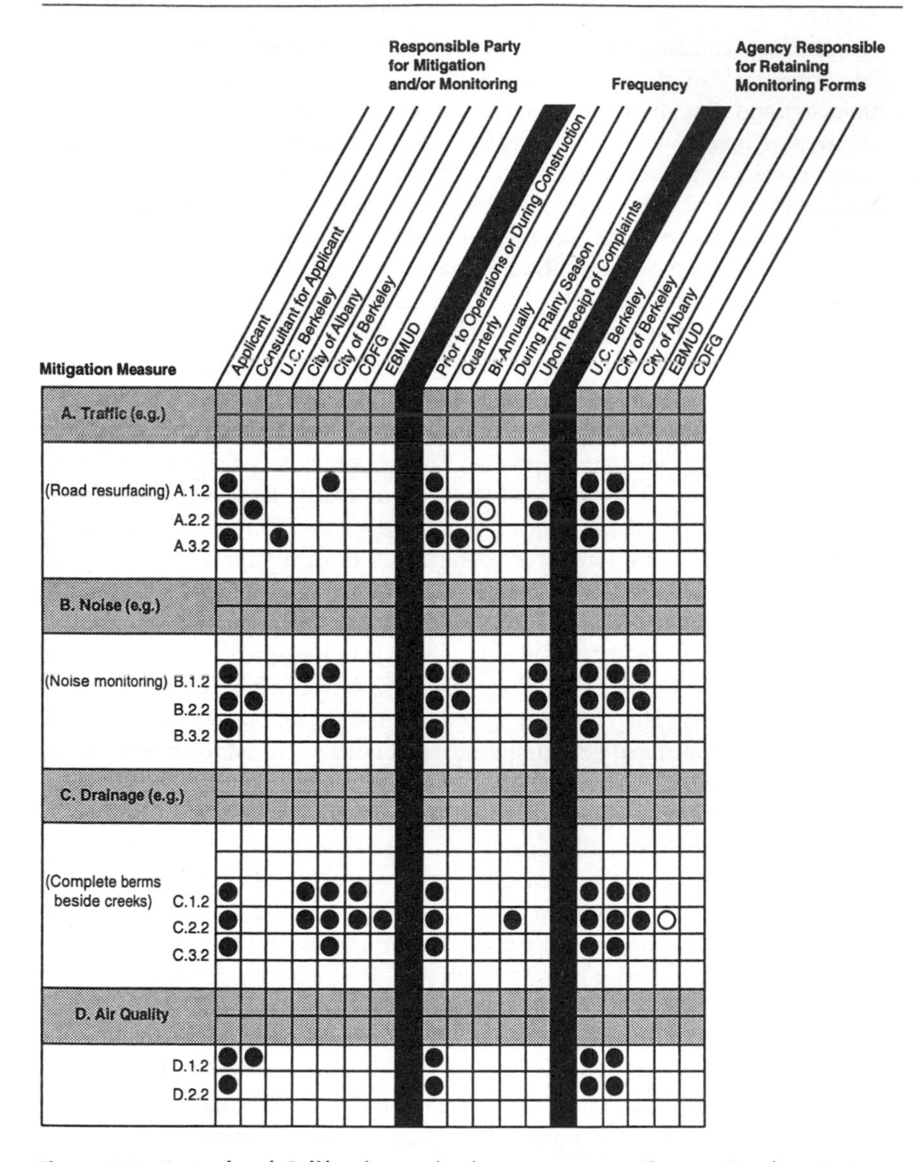

Figure 7.2 Example of California monitoring programme. (*Source:* Baseline Environmental Consulting 1989)

7.4 Auditing in practice

Auditing is already developing a considerable variety of types. Tomlinson & Atkinson (1987a,b) have attempted to standardize *definitions* with a set of terms for seven different points of audit in the "standard" EIA process, as follows:

- decision point audit (draft EIS) – by regulatory authority in the planning approval process
- decision point audit (final EIS) – also by regulatory authority in the planning approval process
- implementation audit – covers start up, and could include scrutiny by government and public, with focus on the proponent's compliance with mitigation and other imposed conditions
- performance audit – covers full operation, and could also include government and public scrutiny
- predictive techniques audit – compares actual with predicted impacts as a means of comparing the value of different predictive techniques
- project impact audits – also compares actual impacts with predicted, and provides feedback for improved project management and for future projects
- procedures audit – external review (e.g. by public) of procedures used by government and industry during EIA processes.

These terms can and do overlap. The focus here is on project performance and implementation audits. Whatever the focus, auditing faces major *problems*. Buckley (1991) identifies the following:

- EISs often contain very few testable predictions, which may only relate to relatively minor impacts
- the environmental parameters that are monitored may not correspond with those for which predictions were made
- monitoring techniques may not enable predictions to be tested, because, *inter alia*, of mismatch of time periods and locations, too few samples, etc.
- projects are almost always modified between the design used for the EIA and in practice
- monitoring data provided by the developers/project operators may possibly be biased towards their interests.

Such problems may partly explain the dismal record of Canadian EISs examined, from an ecological perspective, by Beanlands & Duinker (1983), where accuracy of prediction appeared to be the exception rather than the rule. There are several examples, also from Canada, of situations where the EIA has failed to predict significant impacts. Berkes (1988) indicated how an EIA on the James Bay HEP megaproject (1971–85) failed to pick up a sequence of interlinked impacts, which resulted in a significant increase in the mercury contamination of fish and in mercury poisoning of the Native American peoples. Dickman (1991) identified the failings of an EIA to pick up the impacts of increased lead and zinc mine tailings on the fish population in Garrow Lake, Canada's most northerly hypersaline lake. Such outcomes are not unique to Canada. Canada is a leader in monitoring, and the incidence of such research may result in improved and better predictions than in most countries. Findings from the limited auditing activity in the UK are not too encouraging. A study of four major developments in the UK – the Sullom Voe (Shetlands) and Flotta (Orkneys) oil terminals, Cow Green reservoir and the Redcar steelworks – suggested that 88% of the predictions were not auditable. Of those that were auditable, fewer than half were accurate (Bisset 1984). Mills's (1992) monitoring study of the visual impacts of five recent UK major project developments (trunk road, two wind-farms, power station and opencast coal-mine) revealed that there were often significant differences between what was stated in the EIS and what actually happened on the ground. Project descriptions changed fundamentally in some cases, landscape descriptions were restricted to land immediately surrounding the site, and aesthetic considerations were often omitted. However, mitigation measures were generally carried out well!

One of the more comprehensive nationwide auditing studies of the precision and accuracy of environmental impact predictions has been carried out by Buckley (1991) in Australia. At the time of his study, he found that adequate monitoring data to test predictions were available for only 3% of the up to 1000 EISs produced between 1974 and 1982. In general he found that testable predictions and monitoring data were available only for large complex projects, which had often been the subject of public controversy, and where monitoring was aimed primarily at testing compliance with standards rather than impact predictions. Some examples of the over 300 major and subsidiary predictions tested are illustrated in Table 7.1.

Table 7.1 Examples of auditing of environmental impact predictions.

Component/ parameter	Type of development	Predicted impact	Actual impact	Accuracy/ precision
Surface water quality: salts, pH	Bauxite mine	No detectable increase in stream salinity	None detected	Correct
Noise	Bauxite mine	Blast noise $<115dB_A$	Only 90% $<115dB_A$	Incorrect: 90% accurate, worse
Workforce	Aluminium smelter	1500 during construction	Up to 2500	Incorrect: 60% accurate, worse

Source: Buckley 1991.

Overall, Buckley found the average accuracy of quantified, critical, testable predictions was 44% ±5% standard error. The more severe the impact, the less accurate the predictions. Inaccuracy was highest for predictions of groundwater seepage. Accuracy assessments are of course influenced by the degree of precision applied to a prediction in the first place. In this respect, the use of ranges, reflecting the probabilistic nature of many impact predictions, may be a sensible way forward and would certainly make compliance monitoring more straightforwards and less subject to dispute. Buckley's national survey does not provide grounds for complacency, with less than 50% accuracy. Indeed, as it was based on monitoring data provided by the operating corporations concerned, it may present a better result than would be generated from a wider trawl of EISs. On the other hand, it is to be hoped that we are learning from experience, and that more recent EISs may contain better and more accurate predictions.

7.5 A UK case study: monitoring and auditing the local socio-economic impacts of the Sizewell B PWR construction project

Background to the case study

Although impacts monitoring and auditing are not mandatory in EIA procedures in the UK, physical and socio-economic effects of developments are not completely ignored. For example, several public agencies monitor particular pollutants. Local planning authorities monitor some of the conditions attached to development permissions. However, there is no systematic approach to the monitoring and auditing of impact predictions and mitigation measures. This case study reports on one attempt to introduce a more systematic, although still very partial, approach to the subject.

In the 1970s and early 1980s, the UK had an active programme of nuclear power station construction. This included a commitment, since revised, to build a family of pressurized water reactor (PWR) stations. The first to be approved was Sizewell B in East Anglia. The approval was controversial and it followed the longest public inquiry in UK history. Construction started in 1987, and completion is programmed for 1994. Its continuing development is still controversial. The Impact Assessment Unit (IAU) in the School of Planning at Oxford Brookes University had undertaken a number of studies of power station impacts, including contribution to EISs, with a focus on the socio-economic impacts. A proposal was made to the relevant public utility, the Central Electricity Generating Board (CEGB), that the construction of Sizewell B provided an invaluable opportunity to monitor the project construction stage in detail, and to check on the predictions made in the public inquiry and on the mitigating conditions attached to the approval. Although the predictions were not formally packaged in an EIS but rather as a series of reports based on the Public Inquiry, research was extensive and comprehensive (DoEn 1986). The CEGB supported a monitoring study, which began in 1988. To the credit of the utility (which is now Nuclear Electric, following privatization) there has been a continuing commitment to the monitoring study, despite the uncertainty about further PWR developments in Britain. Monitoring reports for the first five years of construction have now been completed (Glasson et al. 1989–93).

Operational characteristics of the monitoring study

It is important to clarify the *objectives of the monitoring study*, otherwise irrelevant information may be collected and resources may be wasted. Figure 7.3 outlines

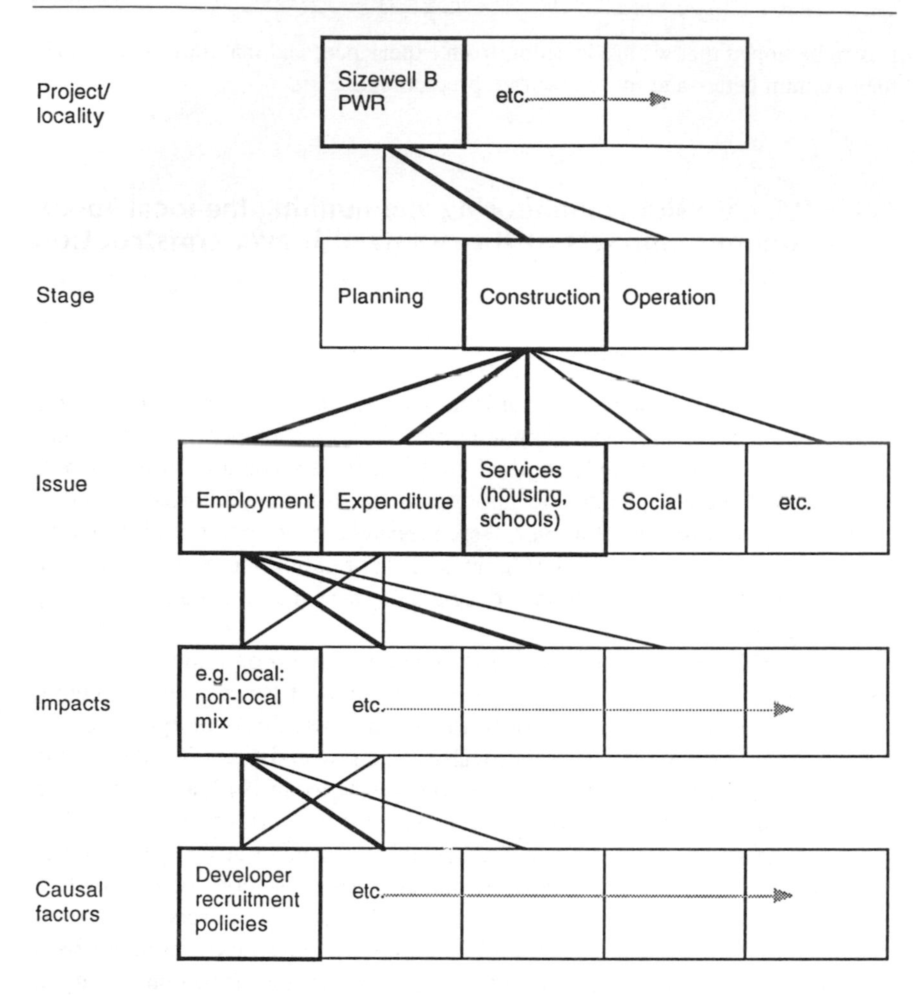

Figure 7.3 Scope of study and data base organization – Sizewell B monitoring study. (*Source:* Glasson et al. 1989–93)

the scope of the study. The development under consideration is the construction stage of the Sizewell B PWR 1200 MW nuclear power station. The focus is on the socio-economic impacts of the development, although with some limited consideration of physical impacts. The socio-economic element of EIA involves "the systematic advanced appraisal of the impacts on the day-to-day quality of life of people and communities when the environment is affected by development or policy change" (Bowles 1981). This involves consideration of impacts on employment, social structure, expenditure, services, etc. Although to date socio-economic studies have often been the poor relation in impact assessment studies, meriting no more than a chapter or two in EISs, they are important, not least because they involve the consideration of the impacts of developments on people, who can answer back and object to developments (Glasson & Heaney 1993).

The highest priority in the study has been the identification of the local employment impacts of the development; this emphasis reflects the key rôle of employment impacts in the generation of other local impacts, particularly upon accommodation and local services. In addition to providing an updated and improved database to inform future assessments, assisting project management of the Sizewell B project in the local community and auditing impact predictions, the study is also monitoring and auditing some of the conditions and undertakings associated with permission to proceed with the construction of the power station. These include undertakings on the use of rail and the routing of road construction traffic, as well as conditions on the use of local labour and local firms, local liaison arrangements and (traffic) noise (DoEn 1986).

The monitoring study includes the collection of a range of information, including statistical data (e.g. the mix of local and non-local construction stage workers, the housing tenure status and expenditure patterns of workers), decisions, opinions and perceptions of impacts. The spatial scope of the study extends to the commuting zone for construction workers (Fig. 7.4). The study includes information from the developer and the main contractors on site, from the relevant local authorities and other public agencies, and from the local community and the construction workforce. Geography A-level students at the local upper school helped to collect data on local perceptions of impacts via biennial major questionnaire surveys in the town of Leiston, which is adjacent to the construction site. A major survey of the socio-economic characteristics and activities of a 20% sample of the project workforce was also carried out every two years. The Impacts Assessment Unit team operated as the catalyst to bring the data together. There has been a high level of support for the study, and the results are openly available in published annual monitoring reports and in summary broadsheets, which are available free of charge to the local community (Glasson et al. 1989–93).

The study has highlighted some *methodological difficulties with monitoring and auditing*. The first relates to the disaggregation of project related impacts from baseline trends. Data are available that indicate local trends in several variables, such as unemployment levels, traffic volumes and crime levels. But problems are encountered in attempting to explain these local trends. To what extent are they attributable to (a) the construction project itself, (b) national and regional factors, or (c) other local changes independent of the construction project? It is relatively straightforward to isolate the rôle of national and regional factors, but the relative rôles of the construction project and other local changes are very difficult to determine. "Controls" are used where possible to isolate the project related impacts.

A second problem relates to the identification of the indirect, knock-on effects of the construction project. Indirect impacts – particularly on employment – may well be significant, but they are not easily observed or measured. For example, indirect employment effects may result from the replacement of employees leaving local employment to take up work on site. Are these local recruits replaced

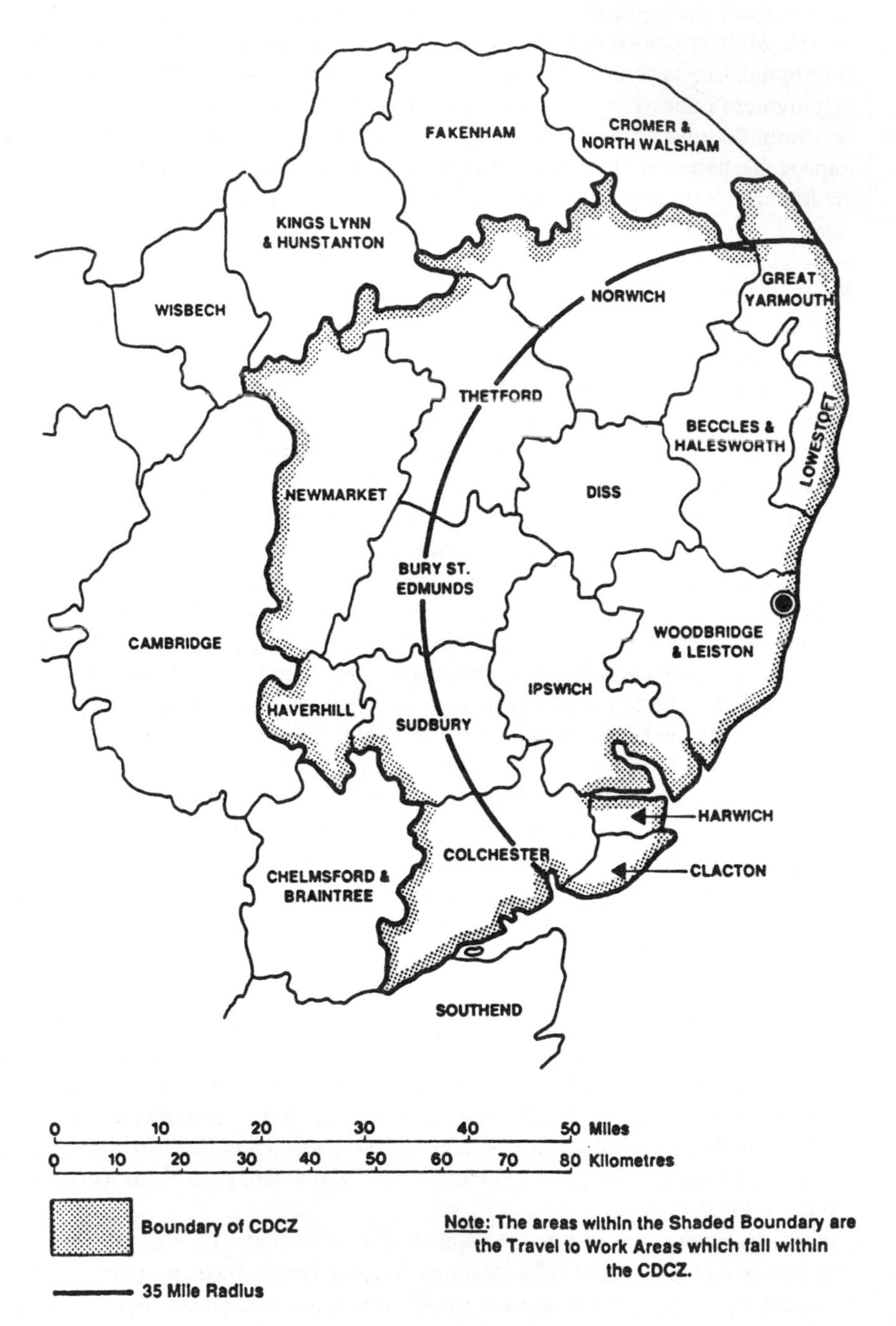

Figure 7.4 Sizewell B commuting zone – monitoring study area. (*Source:* Glasson et al. 1989–93)

by their previous employers? If so, do these replacements come from other local employees, the local unemployed or in-migrant workers? It has not been feasible to obtain this sort of information. Further indirect employment impacts may stem from local businesses gaining work as suppliers or contractors at Sizewell B. They may need to take on additional labour to meet their extra workload. The extent to which this has occurred is again very difficult to estimate, although surveys of local companies have provided some useful information on these issues.

Some findings from the studies

A very brief summary of a number of the findings are outlined below and in Figure 7.5a,b.

Employment

A key prediction and condition was that a large proportion of construction employment should go to local people (within daily commuting distance of the site). This has been the case with a proportion of over 50% for much of the construction period, although, predictably in a rural area, local people do the largely semi-skilled or unskilled jobs. However, as the employment on site has increased, with a shift from civil engineering to mechanical and electrical engineering trades, so the pressure on maintaining the 50% proportion has increased. A training centre was opened in the nearest local town of Leiston, in 1989, to supply between 80 to 120 trainees per annum from the local unemployed.

Local economy

A major project has an economic multiplier effect on a local economy. By the end of 1991, Sizewell B workers were spending about £500,000 per week in Suffolk and Norfolk, Nuclear Electric had placed orders worth over £40 million with local companies, and a "good neighbour" policy was funding a range of community projects (including £1.9 million for a swimming pool in Leiston).

Housing

A major project, with a large in-migrant population, can also distort the local housing market. One mitigating measure at Sizewell B was the requirement on the developer to provide a large site hostel. A 600-bed hostel (subsequently increased to 900) has been provided. It has been very well used, accommodating over 40% of in-migrants to the development at an average occupancy rate of over 85% in 1991, and it has helped to siphon off demand for accommodation in the locality.

Traffic and noise

Traffic impact on local towns and villages can be a major adverse effect of a large construction project. To mitigate it, there is a designated construction route to Sizewell B. Information from the monitoring of traffic flows on designated and non-designated (control) routes indicated that this mitigation measure was work-

Employment impacts

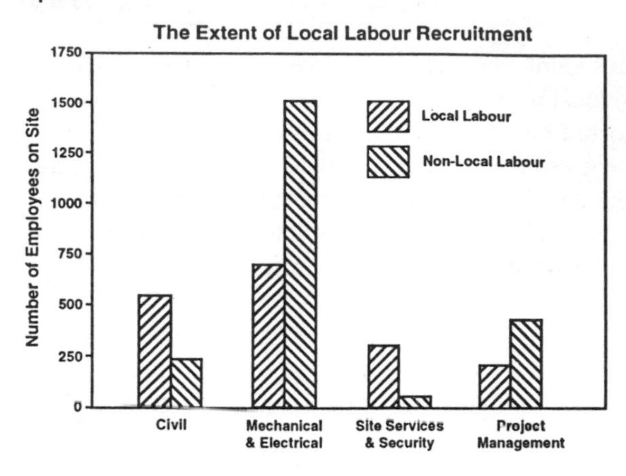

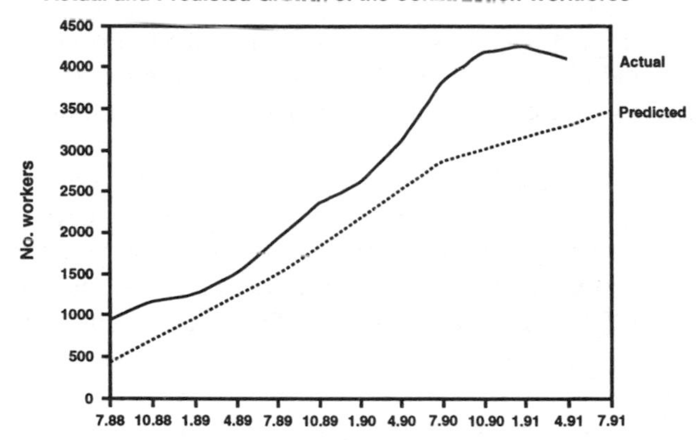

Social impacts

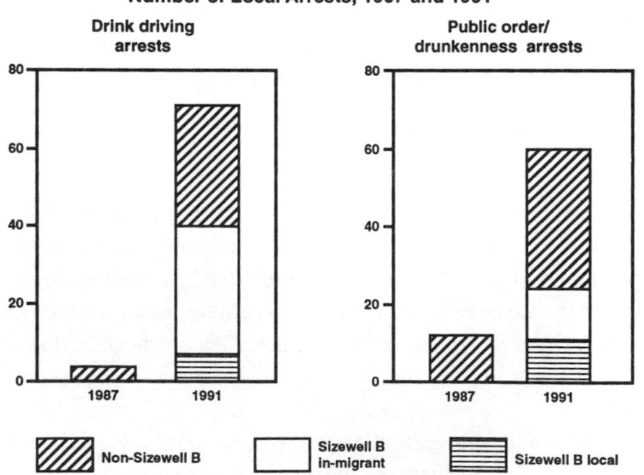

Figure 7.5a Brief summary of some findings from the Sizewell B PWR construction project monitoring and auditing study. (*Source*: Glasson et al. 1989–93)

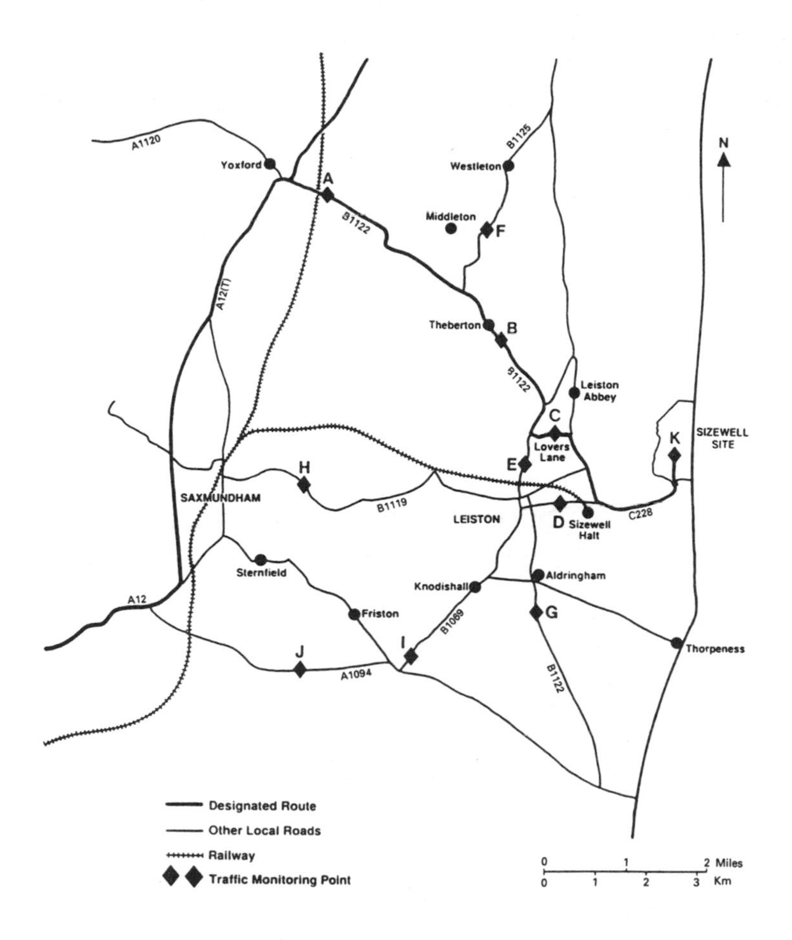

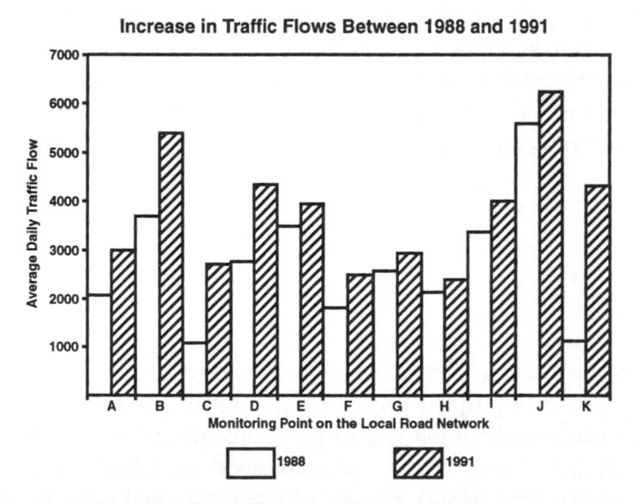

Figure 7.5b More findings from the Sizewell B PWR construction project monitoring and auditing study. (*Source:* Glasson et al. 1989–93)

ing. Between 1988 and 1991, total traffic rose substantially at the four monitoring points on the designated route, but much less so at most of the seven points not on that route. Construction noise on site has been an issue locally. Monitoring has led to modifications in some construction methods, notably improvements to the railway sidings and changes in the piling methods used.

Crime

An increase in local crime is normally associated with the construction stage of major projects. The Leiston police division did see a significant increase in the number of arrests in certain offence categories after the start of the project. However, local people (not employed on the project) were involved in the majority of the arrests, and in the increase in arrests, with the exception of drink/driving, where Sizewell B employees (mainly in-migrants) accounted for most arrests and for most of the increase. Overall, the police considered the project workforce to be relatively trouble-free, with fewer serious offences than anticipated.

Residents' perceptions

Surveys of local residents in 1989 and 1991 revealed more negative than positive perceived impacts, with increased traffic and worker disturbance being seen as the major negative impacts. The main positive impacts of the project were seen to be the employment, additional trade and ameliorative measures associated with it. The monitoring of complaints about the development revealed a substantial fall-off of complaints over time, despite the rapid build-up of the project.

The monitoring of impacts, and auditing of predictions and mitigation measures revealed that many of the predictions used in the Sizewell B public inquiry were reasonably accurate – although there was an underestimate of the build-up of construction employment and an overestimate of the secondary effects on the local economy. Mitigation measures appear to be having some effect. Other local issues have been revealed by monitoring, allowing some modifications to manage the project better in the community. Unfortunately, such systematic monitoring is still discretionary in the UK and is very much dependent on the goodwill of the developer.

7.6 Summary

The mediation of the relationship between a project and its environment is needed throughout the life of a project. Environmental impact assessment is meant to establish the terms and conditions for project implementation, yet there is often relatively little follow-through to this stage and even less follow-up after the fact. Some projects have very long lives and their impacts need to be monitored on a regular basis. Such monitoring can improve project management and contribute to the auditing of both impact predictions and mitigation measures. Monitoring and auditing can provide essential feedback to improve the EIA process, yet this is probably the weakest step of the process in many countries. Discretionary measures are not enough; monitoring and auditing need to be more fully integrated into EIA procedures on a mandatory basis.

References

Baseline Environmental Consulting 1989. *Mitigation monitoring and reporting plan for a wood-waste conversion facility, West Berkeley, California*. Emeryville, California: Baseline Environmental Consulting.

Beanlands, G. E. & P. Duinker 1983. *An ecological framework for environmental impact assessment*. Institute for Resource and Environmental Studies, Dalhousie University, Halifax, Nova Scotia.

Berkes, F. 1988. The intrinsic difficulty of predicting impacts: lessons from the James Bay hydro project. *Environmental Impact Assessment Review* **8**, 201–20.

Bisset, R. 1984. Post development audits to investigate the accuracy of environmental impact predictions. *Zeitschift für Umweltpolitik* **7**, 463–84.

Bisset, R. & P. Tomlinson 1988. Monitoring and auditing of impacts. In *Environmental impact assessment*, P. Wathern (ed.), 117–26. London: Unwin Hyman.

Bowles, R. T. 1981. *Social impact assessment in small communities*. London: Butterworth.

Buckley, R. 1991. Auditing the precision and accuracy of environmental impact predictions in Australia. *Environmental Impact Assessment Review* **11**, 1–23.

California Resources Agency 1988. *California EIA Monitor*. Sacramento: State of California.

Dickman, M. 1991. Failure of environmental impact assessment to predict the impact of mine tailings on Canada's most northerly hypersaline lake. *Environmental Impact Assessment Review* **11**, 171–80.

DoEn (Department of Energy) 1986. *Sizewell B Public Inquiry: report by Sir Frank Layfield*. London: HMSO.

Glasson, J., A. Chadwick, R. Therivel 1989–93. *Local socio-economic impacts of the Sizewell B PWR construction project*. Impacts Assessment Unit, School of Planning, Oxford Brookes University.

Glasson, J. & D. Heaney 1993. Socio-economic impacts: the poor relations in British environmental impact statements. *Journal of Environmental Planning and Management* **36**(3), 335–43.

Greene, G., J. W. MacLaren, B. Sadler 1985. Workshop Summary. In *Audit and evaluation in environmental assessment and management: Canadian and international experience*, 301–21. Banff, Alberta: The Banff Centre.

Holling, C. S. (ed.) 1978. *Adaptive environmental assessment and management*. Chichester, England: John Wiley.

Lee, N. & C. Wood 1980. *Methods of environmental impact assessment for use in project appraisal and physical planning*. Occasional Paper 7, Department of Town and Country Planning, University of Manchester.

Mills, J. 1992. *Monitoring the visual impacts of major projects*. MSc dissertation, School

of Planning, Oxford Brookes University.
Sadler, B. 1988. The evaluation of assessment: post-EIS research and process. In *Environmental impact assessment*, P. Wathern (ed.), 129–42. London: Unwin Hyman.
Tomlinson P. & S. F. Atkinson 1987a. Environmental audits: proposed terminology. *Environmental Monitoring and Assessment* **8**, 187–98.
Tomlinson P. & S. F. Atkinson 1987b. Environmental audits: a literature review. *Environmental Monitoring and Assessment* **8**, 239.

PART 3
Practice

LET'S MAKE SURE I'VE GOT THIS RIGHT. WE GET TO KEEP SOME LIZARDS AND BLUE BUTTERFLIES ON OUR HEATHLAND, AND IN RETURN YOU GET TO BUILD 3,200 NEW HOUSES ON OUR GREEN BELT....

CHAPTER 8
Overview of UK practice to date

8.1 Introduction

Chapters 8 to 11 examine recent EIA practice. Chapter 8 provides an overview of the first four-and-a-half years of UK practice since EC Directive 85/337. This is developed further, with reference to particular sectors and their associated legislation, in Chapters 9 and 10. In Chapter 9 the focus is on new settlements and in Chapter 10 on trunk roads and power stations. Each of these two chapters includes case studies that provide examples of current practice in various aspects of the EIA process. Chapter 11 sets the discussion in the wider context of international practice.

Between the mid-1970s and the mid-1980s, approximately 20 EISs were prepared annually in the UK (Petts & Hills 1982). After the implementation of Directive 85/337 in mid-1988, this number rose dramatically and, despite the recession, more than 300 EISs per annum were produced in the early 1990s. This chapter discusses first where EISs can be found: where collections of EISs are held, and where lists of EISs are published. It considers the number and type of EISs that have been prepared in the UK since mid-1988. It then reviews the achievements of, and problems with, the UK's implementation of Directive 85/337. The information in this section was correct at the time of writing in early 1993; it will change as more EIAs are carried out.

8.2 Sources for EISs

When a local planning authority receives an EIS, it is required to send a copy to the regional office of the Department of the Environment (DoE), which then forwards it to the DoE library in London once the application has been determined. However, this process is a long one; in early 1993, the DoE library in London held about 200 EISs (including a large number of EISs not falling under the town and country planning system), which is less than one-fifth of the EISs that had been prepared until then. In some cases, local authorities are simply not submitting copies of EISs to the DoE: "A total of 16 such cases were unearthed, out of 153 known statements submitted under the planning rules to the end of 1989" (ENDS Report, August 1991, 12–15). The DoE library is open to the public by appointment; photocopies can be made off the premises. Other government agencies such as the Department of Transport (DoT), Ministry of Agriculture, Fisheries and Food (MAFF) and Forestry Authority also hold collections and lists of the EISs that fall under their jurisdiction. For instance the DoT has a collection of approximately 100 EISs in its London office, and the Department of Trade and Industry holds about 75. These collections are, however, generally not publicly available, although access for research purposes may be allowed.

The *Journal of Planning and Environmental Law* lists EISs received by the DoE, Scottish Development Department and Welsh Office on a quarterly basis, giving the project type, local authority and planning decision. It is probably the most consistently up-to-date list available, but is also definitely not comprehensive. It has other problems: it sometimes lists duplicate entries for the same project, and at other times lists projects under names that are quite different from those on the front covers of the EISs.

The Institute of Environmental Assessment has compiled details of approximately 1,100 EISs, which it published as an encyclopedia of EISs in 1993. It has a collection of about 250 EISs, which are available with pre-arrangement with institute staff and can also be mailed on a one-week loan basis. The Institute has also published several reports based on reviews of EISs (e.g. Coles & Tarling 1991, Coles et al. 1992, Fuller 1992).

The EIA Centre at the University of Manchester keeps a large database of EISs and EIA-related literature. It collects a representative selection of UK EISs each year for research purposes (25% of all EISs in 1991/92, although this is likely to decrease). Its collection of approximately 400 UK and 20–30 overseas (mostly Canadian, US, Australian and Malaysian) EISs is, like its database, open to the public by appointment. Researchers at the EIA Centre have published an analysis of 472 EISs prepared before December 1990 (Jones et al. 1991, Wood 1991), and authored the DoE's report *Monitoring environmental assessment and planning* (DoE 1991).

Oxford Brookes University's collection of approximately 500 UK and 15 overseas (mostly US and Canadian) EISs is open to the public by appointment. Photocopies can be made on the premises. The University's Impacts Assessment Unit publishes an annual directory of EISs (Therivel 1990, 1991, Heaney & Therivel 1993). The addresses of these and other organizations are given in Appendix 4.

Various other organizations have collections or databases of EISs. The Royal

Society for the Protection of Birds (RSPB) has approximately 50–100 EISs for projects that have a particular impact on birds (e.g. forestry and barrage projects). Other organizations, such as the Institute of Terrestrial Ecology, English Nature, the Council for the Protection of Rural England (CPRE), and National Rivers Authority, as well as many environmental consultancies, have limited collections of EISs. These collections are generally for in-house use only and are not open to the public.

Gaining access to EISs, listing them, or forming a collection, is complicated by several problems. First, some projects fall under more than one Schedule classification, for example mineral extraction schemes (Sched. 2.2) that are later filled in with waste (Sched. 2.11), or industrial/residential developments (Sched. 2.10) that also have a leisure component (Sched. 2.11). Secondly, the mere description of a project is often not enough to identify the regulations under which its EIA was carried out. For instance, power stations may fall under planning or electricity regulations depending on size. Roads may fall under highways or planning regulations depending on whether they are trunk roads or local highways. Thirdly, many EISs do not mention when, by whom or for whom they were prepared. Finally, the cost of EISs is often prohibitive. For instance, recent DoT statements have regularly cost hundreds of pounds. Problems such as these make the acquisition of EISs very difficult. Various organizations, e.g. the Institute of Environmental Assessment, University of Manchester and Oxford Brookes University have called for one central repository for all EISs in the UK.

8.3 Quantitative analysis

In the absence of a central government lead in maintaining a comprehensive database of EISs, several organizations have begun to establish such databases, notably the Institute of Environmental Assessment (Coles et al. 1992) and Oxford Brookes University (Therivel 1990, 1991, Heaney & Therivel 1993) and, to a lesser extent, the University of Manchester (Jones et al. 1991). This quantitative analysis seeks to establish how many EISs have been produced, for which projects, where, for whom and by whom. This section is based primarily on information from Oxford Brookes University, but these findings are very similar to those of the other organizations.

Number of EISs

Between 1300 and 1500 EIAs were carried out in the UK in the first four-and-a-half years after the implementation of Directive 85/337. A series of three annual surveys of local authorities carried out by Oxford Brookes University showed that, of the 423 (84%) authorities that responded to the surveys by late 1992, 354 (84%) had received at least one EIS, while 69 (16%) had not received any. Figures

8.1 and 8.2 show the number of EISs known to have been received by, respectively, county and regional councils, and district, borough, metropolitan and city councils. Predictably, on average, county and regional councils received considerably more EISs (average 4.5) than the other authorities (average 1.9). Additional EISs were received by other competent authorities not covered by the questionnaire (e.g. in Northern Ireland and by the Forestry Authority).

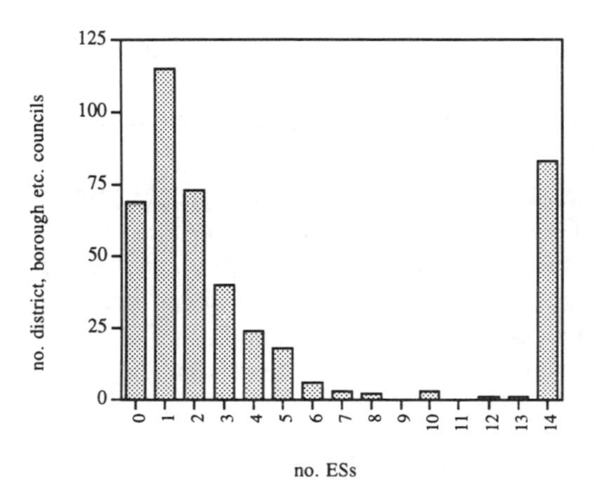

Figure 8.1 **EISs prepared in the UK, July 1988 to October 1992: number received by district, borough, metropolitan and city councils.** (*Source:* Heaney & Therivel 1993)

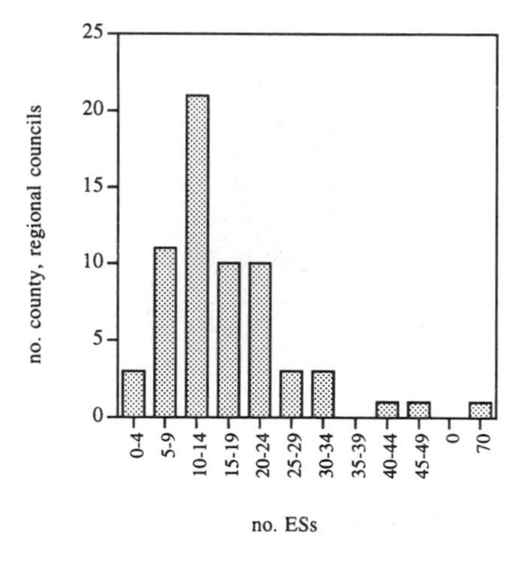

Figure 8.2 **EISs prepared in the UK, July 1988 to October 1992: number received by county and regional councils.** (*Source:* Heaney & Therivel 1993)

These numbers were then extended to cover the 69 authorities for which nothing is known. If none of these received any EISs, still most (70%) authorities would have received at least one EIS by late 1992, and an average of approximately 300 EISs would have been produced in the UK annually since the implementation of the Directive. If all 69 authorities received the district/borough council average of 1.9 EISs, then approximately 1500 EISs would have been produced in the first four and a half years of Directive 85/337's implementation, approximately 330 per year.

Notwithstanding the constraints on development caused by the recession of the early 1990s, the number of EISs being prepared seems to have been rising every year, from less than 200 in the first year of the Directive's implementation to about 400 in the fourth year. Of these, about 70–75% are in England, another 15% in Scotland, and about 10% in Wales.

Types of projects

Figure 8.3 shows the regulations under which the EISs were prepared, and Figure 8.4 shows broadly the types of projects for which EISs have been prepared. About 10% of these projects were Schedule 1 projects, primarily power stations, special roads, and waste treatment facilities. For Schedule 2 projects, the largest category, comprising about one-third of all projects, is that of so-called infrastructure projects (Sched. 2.10). Of these, about half are roads, harbours, and aerodromes (2.10d), about half are business, retail, and residential developments (2.10a,b), and another 4% are for other infrastructure projects. Extraction (Sched. 2.2) accounts for approximately 15% of projects, with coal extraction accounting for 5%. Waste disposal (Sched. 2.11c,d,e) accounts for 15%, and leisure developments (Sched.

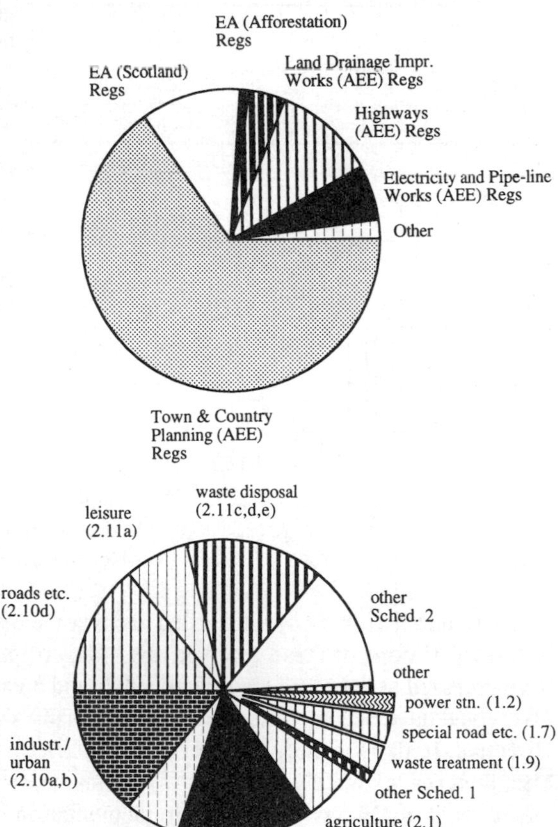

Figure 8.3 EISs prepared in the UK, July 1988 to October 1992: regulations under which EISs were prepared. (*Source:* Heaney & Therivel 1993)

Figure 8.4 EISs prepared in the UK, July 1988 to October 1992: types of projects for which EISs have been prepared. (*Source:* Heaney & Therivel 1993)

2.11a) for about 7%. Despite the doubling of the annual number of known EISs over the 1988–92 period, these proportions remained broadly unchanged. They also correspond with information given in other similar surveys (Jones et al. 1991, Coles et al. 1992).

Location of projects

Figure 8.5 shows the distribution of known EISs in England, Scotland and Wales by county/region. Generally, more is known about English and Scottish EISs than about Welsh ones, so the figure is more likely to underestimate the latter's number. The most EISs to date have been prepared for projects in Kent: approximately 70. London, Strathclyde, Grampian, Cambridgeshire and Northamptonshire are next highest, with EISs prepared for over 30 projects each. In most counties (69%), between 10 and 29 projects requiring EISs have been proposed. Those counties with the fewest EISs, less than 10, are primarily located in northern England/southern Scotland, western Wales, and Somerset/Dorset.

The high number of EISs in the southeast of England can be attributed to the greater level of economic activity currently (or at least until recently) taking place there; the Channel Tunnel in particular has spawned a number of proposals for major secondary projects. In Scotland, the high number of EISs can be attributed to the large area of the regional councils. Lower numbers of EISs can be attributed to lower economic activity, smaller size, less familiarity with the requirements of Directive 85/337, or an unwillingness to answer questionnaires! The types of development vary considerably by region, with southwest England getting (unsurprisingly) many EISs for wind-farms and marinas; southeast England for business/residential developments and transport infrastructure; northern England for heavy industry, agriculture and mineral extraction; and Scotland and Wales for leisure developments and mineral/coal extraction.

Clients and consultants

Studies by Wood (1991) show that approximately 40% of EISs are produced for the public sector and 60% for the private sector. Public-sector projects are largely for national agencies; private-sector projects are mainly one-off projects for a wide range of clients. Approximately 75% of EISs are produced for the clients by consultancies, with 25% produced in-house, although this division is blurred by a range of hybrid arrangements, for example, where external consultants may provide specialist inputs for EISs largely produced in house. A particularly interesting subset is that of the 10% of EIAs where one agency acts as both the project proponent and the competent authority (e.g. the DoT for roads, the Forestry Authority for afforestation, or local authorities for some waste disposal projects). For instance, many roads for which line orders are confirmed are judged by the DoT not to require EIA; in the first 18 months of the regulations, at least 49 such judgements were made (Hansard 1990).

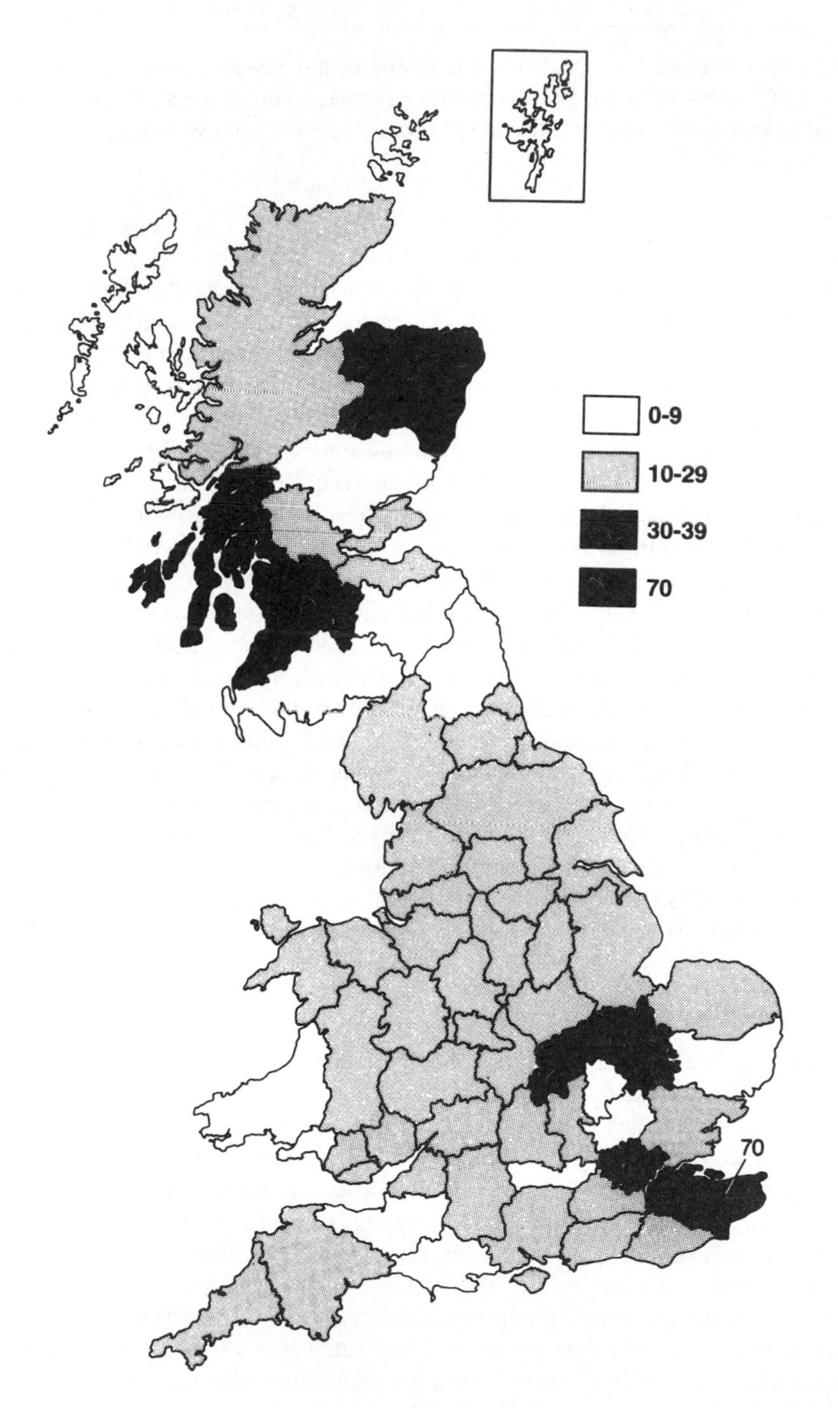

Figure 8.5 EISs prepared in the UK, July 1988 to October 1992: location of projects. (*Source:* Heaney & Therivel 1993)

Over 250 consultancies now claim to have produced at least one EIS. Although several large consultancies have produced substantial numbers of EISs, no single consultancy has more than 5% of the market. So the typical consultant is likely to have very limited experience. Consultancy activity is primarily related to the production of EISs, but there is a growing market in the review of EISs. The large majority of EISs received by local authorities (80%) have been reviewed in house, but 20% have been sent to consultants for this important activity. Local authority experience is of course also limited to date. Although more authorities are now receiving EISs, in most cases they have received only one or two. Procedures and expertise do evolve with practice. Kent County Council, which has received the most EISs, has prepared a handbook for developers on how to carry out EIA (Kent County Council 1991); other handbooks have been prepared for Cheshire and Essex (Cheshire County Council 1989, Essex County Council 1992) and other local authorities are likely to follow suit.

8.4 Qualitative analysis

Underpinning any qualitative analysis of the implementation of EIA in the UK are the requirements of the EC and of UK government legislation. Directive 85/337 has a number of limitations, including its coverage of types of projects (e.g. EIA is not required for defence projects), its failure to make some key steps of the EIA process mandatory (e.g. public participation, consideration of alternatives, and monitoring), its allowing developers to undertake EIAs for their own projects, and of course the narrow focus on the project level only. All of these limitations are reinforced by UK legislation. Indeed, some projects that require EIA under Directive 85/337, including some agriculture projects and afforestation projects that do not require Forestry Authority grants, are not covered in UK EIA legislation. Such legislative and procedural issues, and possible ways forwards, are considered in more detail in Chapters 12 and 13. The focus in this section is on the quality of EIA and the resultant EISs under the operative legislation. This includes discussion of the interpretation of the need for EIA, the scope of EIA, consultation, content of EISs, review, overall quality of EISs and the rôle of EIA in the decision-making process. The latter relates particularly to the over-riding question of whether the EIA system is achieving its objectives.

Need for EIA

Competent authorities in the UK are given wide discretion to determine which Annex II projects require EIA within a framework of varying criteria and thresholds established by the 20 relevant regulations. This is leading to inconsistent application of Directive 85/337's requirements. For instance, a 1991 search (DoE 1991) of 24 local planning authority returns, registers and files revealed 30 projects in 12 authorities for which EIAs could have been required but were not. Most of these types of development (e.g. six mineral extraction schemes, two landfills, 17 mixed-use developments) had been subject to EIA in other local authorities. The decision not to require an EIA had mostly been taken by junior members of staff who had never considered the need for an EIA, or who thought (incorrectly) that no EIA was required if the land was designated for that type of use in the development plan, or if the site was being extended or redeveloped rather than newly developed.

Similarly, different DoE regional offices have given different decisions on appeals for what are essentially very similar developments. For instance:

> . . . there have been two applications for major out-of-town regional shopping centres, both well above the threshold given in the Circular (15/88 on Environmental Assessment) of 10,000 sq. metres. The request for the smaller of the two (The Richings at Ivers, South Buckinghamshire DC) was successfully challenged by the developer. The request for the larger of the two (at Lea Cross, Newham Borough) was also challenged and in this case an assessment was required. The direction letters came from two different DoE Regional offices. (Gosling 1990)

The DoE regional offices' decisions have also differed substantially from the thresholds indicated in Circular 15/88:

> The threshold for sand and gravel cases is set as "sites of more than 50 ha may well require EIA and significantly smaller sites could require EIA if they are in a sensitive area or if subjected to particularly obtrusive operations" . . . Two of the applications in Cambridgeshire are well over the threshold. The Barleycroft Farm (117 ha) proposal is not identified within the minerals local plan . . ., is part in an Area of Best Landscape and is close to residential development. On being challenged a direction was received that assessment was not required. The Fenstanton case (74 ha) is entirely within an Area of Best Landscape and the application has been refused without waiting on the receipt of the environmental statement. (Gosling 1990).

> The surprising result of the challenge by the potential developer of the new settlement of 2000 dwellings at Nine Mile Hill in East Cambridgeshire District was that an assessment was not required. A bald statement was made in a two-paragraph direction letter that "The Secretary of State has considered the matter and concludes that the proposed development is not of the type listed in schedules 1 or 2 of the above regulations and that an environmental statement is not required. (Gosling 1990)

> In another [case], a very substantial new housing area capable of accommodating more than 10,000 people together with associated development similarly evaded the net. (McDonic 1992)

This confusion is not made any the less by the fact that if the Secretary of State for the Environment decides that an EIA is needed, he must publish the reasons, but if he decides that it is not, then no such explanation is required. Finally, many developers voluntarily submit environmental documents without specifying whether these documents are EISs or not. In some cases, competent authorities choose to treat them as EISs, with the attendant requirements for consultation and publicity. In other cases, however, the authorities simply treat them as additional information (DoE 1991). To improve this system, clearer screening criteria are needed to allow the competent authority to determine which projects are likely to have a significant environmental impact (see Ch. 12 for further discussion).

Scope of EIA

Competent authorities also have a great deal of discretion in determining the scope of EIAs. As was discussed in Chapter 3, Directive 85/337's Annex III was interpreted in UK legislation as being in part mandatory and in part discretionary. Table 8.1 shows the type of information included in EISs, based on a survey of 100 EISs prepared before the end of 1989 (Jones et al. 1991). It shows that, although the mandatory requirements of the legislation are generally carried out, the discretionary elements (e.g. consideration of alternatives, forecasting methods, secondary and indirect impacts, scoping) have, understandably, been carried out less often. Since EIS quality is improving, it is likely that a more up-to-date analysis would show higher proportions of EISs including the relevant information. One explanation for this limited coverage may be that local authorities are consulted by developers in the scoping stage in only about half of all cases, and that in about one-third the local authorities are not consulted at all before submission of the EIS. In the USA this problem has been partly alleviated by the requirement that a Finding of No Significant Impact (FONSI) must be prepared for those aspects of the EIS determined during (mandatory) scoping not to require further analysis. In its five-year review, the EC Commission is considering making all of Annex III mandatory, which would go a long way towards solving this problem.

More specifically concerning EISs for road schemes, DoT standard HD 18/88 (DoT 1989), which sets out the requirements for EIS contents, has been described as being "totally inadequate in defining even the minimal requirements to be provided. Moreover, they omit even a discretionary requirement for further information, for instance on forecasting methods or alternatives" (CPRE 1991). The Standard omits several potential environmental impacts that could result from a road scheme, and its statement that individual road schemes do not affect climatic factors and are unlikely to affect soil or water is questionable. In addition, the *Manual of environmental appraisal* (DoT 1983) was not amended between 1983 and mid-1993 despite its inadequacy as an instrument of EIA and its central rôle in DoT decision-making (CPRE 1991). This is discussed further in Chapter 10.

Table 8.1 EISs prepared in the UK July 1988 to December 1989: comprehensiveness.

Type of information	EISs including information (%)
Specified information	
Description of proposed development	93
Data to identify and assess the main environmental effects	76
Description of likely significant effects	88
With reference to:	
• human beings	75
• flora/fauna	85
• soil/geology	51
• water	65
• air	54
• climate	24
• landscape	91
• interaction between them	14
• material assets/cultural heritage	48
Description of mitigating measures	75
Non-technical summary	67
Additional information	
Physical characteristics of proposed development:	
• construction	51
• operation	74
Residues and emissions from the development	
With reference to:	
• water	63
• air	54
• soil	29
• noise	68
• vibration	17
• light	9
• heat	2
• radiation	1
Outline of main alternatives studied	34
Forecasting methods used	45
Difficulties in compiling information	4

Source: Jones et al. 1991.

Consultation with statutory consultees and the public

Despite legal requirements for consultation with statutory consultees, often this does not happen. A study of 24 local planning authorities that had received EISs showed that in one-third of the cases local planning authorities distributed EISs to statutory consultees, and in another one-third developers or consultants forwarded EISs to the consultees. Of the two statutory consultees that must be consulted for every EIA, the Countryside Commission was not consulted in 25% of cases, and the Nature Conservancy Council (now English Nature) was not consulted in one case (DoE 1991, Lee & Brown 1992). One solution to this problem may be for the DoE periodically to send a list of completed EISs to statutory consultees, so that they can ensure that they have received copies.

Similarly, public consultation in the EIA process is often limited. Developers and competent authorities shy away from a process that is felt to be cumbersome, contentious, and often counterproductive. One environmental consultant's comments about public meetings are:

> In a phrase, "don't do it". Judging from experience elsewhere, we have serious reservations about the value of holding presentations or debates in public forum, and believe that these should only be considered in exceptional circumstances . . . One-to-one consultations are time consuming and laborious, but can provide very worthwhile results. (Milne 1991)

However, part of this problem is undoubtedly attributable to unfamiliarity with, and perhaps a defensive approach to, public participation.

Review of EISs

The review of EISs is often ad hoc, without the use of formal review criteria or outside expertise. Despite the ready availability of the Lee & Colley (1992) review criteria, only about onc-third of local authorities use any form of review methods at all, and then usually as indicative criteria, to pinpoint areas for further investigation, rather than in a formal way (Coles et al. 1992, Greek 1992, Robinson 1991, Wood & Jones 1992). Environmental consultancies seem to use the criteria, but not so much to review EISs as to check that the EISs they prepare are complete (Greek 1992). About 10–20% of EISs are sent for review by external consultants or by the Institute of Environmental Assessment; but even where outside consultants are hired to appraise an EIS, it is doubtful whether the appraisal will be wholly unbiased if the consultants might otherwise be in competition with each other. There are also problems involved in getting feedback from the reviewing consultants quickly enough, given the tight timetable for making a project determination. A novel approach being used by some developers requires consultants that are bidding to carry out an EIA to include as part of their bid an "independent" reviewer who will assure the quality of the consultants' work.

Overall quality of EISs

The overall quality of EISs is improving, but is still not particularly good. Using the review criteria developed by Lee & Colley (1992) discussed in Chapter 6, the EIA Centre at the University of Manchester has reviewed 83 EISs prepared between July 1988 and early 1991. Table 8.2 compares the quality of these EISs over time. The proportion of satisfactory or just satisfactory EISs almost doubled between 1988/89 and 1990/91, indicating that some of the early teething problems associated with Directive 85/337 have been resolved. However, the proportion of clearly unsatisfactory EISs (about a quarter) fell only slightly during that time. Some EISs are patently poor, such as the three-page EIS for an asbestos landfill site, or the abattoir study written by the butcher (McDonic 1992). A survey carried out for the DoE showed that local planning authorities and statutory consultees found about one-third of EISs to be satisfactory, about one-third to be unsatisfactory, and about one-third to be neither satisfactory nor unsatisfactory (DoE 1991).

Table 8.2 Quality of EISs (%).

	year 1 (1988/89)	year 2 (1989/90)	year 3 (1990/91)
Satisfactory (A,B)	14	22	30
Just satisfactory (C)	20	26	30
Just unsatisfactory (D)	40	28	17
Unsatisfactory (E,F)	26	24	23

Source: Lee & Brown 1992.

Lee & Brown (1992) identify several trends linked to EIS quality. EISs produced under different legislations are of different quality. Of the EISs prepared under the electricity regulations, all were judged to be satisfactory (categories A, B or C) under the Lee & Colley criteria. 50% of those under the Scottish regulations, 41% under the town and country planning regulations, 25% under the highways regulations, and 14% under the drainage regulations were satisfactory; the rest were unsatisfactory (categories D, E or F). More recent EISs prepared by the DoT are of much better quality than the earlier ones, however, and hopefully this trend will continue with the publication of the new *Manual of environmental appraisal*. EISs prepared under other regulations are also likely to improve.

EISs for large-scale developments are usually better than those for smaller projects: 65% of EISs for large projects, 35% for medium projects, and 20% for small projects were satisfactory. Similarly, longer EISs were generally better than shorter ones: whereas only 10% of EISs of 1–25 pages were satisfactory, this rose to 31% for 26–50 pages, 64% for 51–100 pages, and 78% for more than 100 pages. Finally, EISs prepared by inexperienced developers or consultants were generally of poorer quality than those prepared by experienced developers or consultants; this is shown in Table 8.3 (Lee & Brown 1992).

Table 8.3 Satisfactory EISs: developers' and consultants' experience (%).

Consultant	Developer No experience	Developer Some experience
None used	8%	27%
No experience	30%	50%
Some experience	69%	69%

Source: Lee & Brown 1992.

The quality of EISs may also be influenced by cost considerations. The cost of EISs in the UK has been shown to be between 0.000025% and 5% of project costs (Coles et al. 1992), which is a considerable range. If a developer accepts the lowest bid for preparing an EIS, the adequacy of the ensuing EIS may be limited by the consultants' lack of time, expertise, or equipment. Fuller (1992) argues that this may not be helpful to a developer in the long run:

> A poor-quality statement is often a major contributory factor to delays in the system, as additional information has to be sought on issues not addressed, or only poorly addressed, in the original . . . Therefore, reducing the cost of an environmental assessment below the level required for a thorough job is often a false economy.

Finally, quality seems to vary between sections of EISs, with site descriptions, the description of the development, baseline conditions, mitigation measures, and the general layout and presentation of the EIS generally being carried out considerably better than, for example, scoping and the identification of impacts and the consideration of alternatives (Coles et al. 1992, Lee & Brown 1992). Predictions and the assessment of significance are particularly poorly covered, and the coverage of monitoring is rare.

EIA *in decision-making*

As noted in Chapter 1, EIA has several purposes, including acting as an aid to decision-making, to the better formulation of projects and, fundamentally, to the achievement of sustainable development. An assessment of the effects of EIA on the latter aim is beyond the scope of this book, and may be premature. Some comments can be made on the other aims. Local and central government in the UK were very wary about the introduction of EIA into the UK, primarily via the planning system. One achievement appears to have been that the introduction has been made with minimal disruption to the planning system (Jones et al. 1991), although there have been problems of delay.

Only about half of the planning applications accompanied by an EIS are determined within the required 16-week period (Wood & Jones 1991, Tarling 1991). The average time is about 36 weeks, with about one application in five taking more than a year to determine (Tarling 1991). "The entire [EIA] process, from the notification of intent for the project to the decision took on average just under a year and a quarter (62 weeks) with environmental statement preparation taking 25 weeks" (Coles et al. 1992). Of the planning officers questioned in the study by Lee & Brown, about half felt that the EIS had not influenced how long it took to reach a decision; the rest were about evenly split between those who felt that the EIA had speeded up or slowed down the process. About two-thirds of developers and planning officers felt that the EIA had slightly increased the cost to the developer of obtaining planning permission.

However, the problems imposed by these delays seem to be counterbalanced by improvements in the decision-making process. Although "there are indications that the statement and the consultations based on it may not yet have been fully integrated into the later stages of the development control system" (Lee 1992), the EIA system does seem to be influencing decision-making. A study of 24 pre-1990 planning EISs showed that in two-thirds of cases the proposal was modified as a result of suggestions made during the EIA process; most of the modifications were made after the EIS was submitted. Another study of 41 EISs (Tarling 1991) showed that modifications to the project as a result of the EIA process were required in almost half the cases. Most modifications were regarded as significant. The case studies in the following chapters provide examples of how projects change during the EIA process, leading to the reduction or elimination altogether of some adverse environmental impacts.

8.5 Summary

The statutory EIA system, so contested in conception, was born in 1988 and has grown rapidly since. Although there is no consistent UK government source for information on EISs, there are several other useful sources that facilitate a quantitative analysis. This shows the production of over 300 EISs per annum, mainly for Schedule 2 projects, and with a growing spread to most local authority areas. A qualitative analysis, drawing on various recent studies, reveals some improvement in the overall quality of EISs, but also a number of important issues relating, inter alia, to need, scope, consultation, and review. The following chapters now take a more focused view of UK practice.

References

Cheshire County Council 1989. *Planning practice notes 2: the Cheshire environmental assessment handbook*. Chester: CCC.

Coles, T. F., K. G. Fuller, M. Slater 1992. *Practical experience of environmental assessment in the UK*. Proceedings of Advances in Environmental Assessment Conference,

IBVC Technical Services, London.

Coles, T. F. & J. P. Tarling 1991. *Environmental assessment: experience to date.* Horncastle, England: Institute of Environmental Assessment.

CPRE 1991. *The Environmental Assessment Directive – five years on.* London: Council for the Protection of Rural England.

DOE (Department of the Environment) 1991. *Monitoring environmental assessment and planning.* A report by the EIA Centre, Department of Planning and Landscape, University of Manchester. London: HMSO.

DOT (Department of Transport) 1983. *Manual of environmental appraisal.* London: HMSO.

DOT 1989. *Departmental standard HD 18/88. Environmental assessment under the EC Directive 85/337.* London:DOT.

Essex County Council 1992. *The Essex guide to environmental assessments.* Chelmsford: ECC.

Fuller, K. 1992. Working with assessment. In *Environmental assessment and audit: a user's guide*, 14–15. Gloucester: Ambit.

Gosling, J. 1990. *The Town and Country Planning (Assessment of Environmental Effects) Regulations 1988: the first year of application.* Proposal for a working paper, Department of Land Management and Development, University of Reading.

Greek, M. 1992. *Evaluation of environmental assessment review techniques and criteria in the context of the environmental assessment process.* Bachelor's dissertation, School of Planning, Oxford Brookes University.

Hansard 1990. Parliamentary Written Answer on trunk road schemes and EA, 3 December 1990.

Heaney, D. & R. Therivel 1993. *Directory of environmental statements 1988–1992.* Oxford: School of Planning, Oxford Polytechnic (now Oxford Brookes University).

Jones, C. E., N. Lee, C. Wood 1991. *UK environmental statements 1988–1990: an analysis*, Occasional Paper 29, EIA Centre, Department of Planning and Landscape, University of Manchester.

Kent County Council 1991. *Handbook of environmental assessment.* Maidstone: Kent County Council.

Lee, N. 1992. Improving quality in the assessment process. *Environmental assessment and audit: a user's guide*, 12–13. Gloucester: Ambit.

Lee N. & D. Brown 1992. Quality control in environmental assessment. *Project Appraisal* **7**(1), 41–5.

Lee, N. & R. Colley 1992. *Reviewing the quality of environmental statements*, Occasional Paper 24, EIA Centre, Department of Planning and Landscape, University of Manchester.

McDonic, G. 1992. The first four years. In *Environmental assessment and audit: a user's guide*, 2–3. Gloucester: Ambit.

Milne, R. 1991. Statements striving for better standards. *Planning* **944**, 4–5.

Petts J. & P. Hills 1982. *Environmental assessment in the UK.* Institute of Planning Studies, University of Nottingham.

Robinson, J. 1991. *Environmental statement review: the need for a formal review procedure.* Bachelor's dissertation, School of Planning, Oxford Polytechnic.

Sheate, W. 1992. Broadening the base. In *Environmental assessment and audit: a user's guide*, 8. Gloucester: Ambit.

Tarling J. P. 1991. *A comparison of environmental assessment procedures and experience in the UK and the Netherlands*, MSc dissertation, University of Stirling.

Therivel, R. 1990, 1991. *Directory of environmental statements*. School of Planning, Oxford Polytechnic.

Wood, C. M. 1991. Environmental impact assessment in the United Kingdom. Paper presented at the ACSP-AESOP Joint International Planning Congress, Oxford Polytechnic, July 1991.

Wood, C. M. & C. Jones 1991. The impact of environmental assessment on local planning authorities. *Journal of Environmental Planning and Management* **35**(2), 115–27.

CHAPTER 9
Environmental impact assessment and new settlements

9.1 Introduction

This chapter examines the application of the UK Town and Country Planning (Assessment of Environmental Effects) Regulations to proposals for new settlements in the countryside. New settlements include a variety of activities and land-uses and provide some of the most comprehensive projects, akin to development plans, for the new procedures. The chapter aims to illustrate some of the issues arising from the implementation of the Regulations, including ambiguity over the need for EIAs for certain projects, the appropriate timing of the submission of the EIS, and the important rôle of the planning inquiry in the EIA process. In addition, two detailed case studies of specific new settlement proposals are examined, to highlight certain features of current good EIA practice.

The focus of the chapter is on proposals for *free-standing new settlements,* i.e. those whose boundaries are not immediately adjacent to an existing built-up area. However, substantial *village expansion* schemes, although not free-standing, are classified as "new settlements" where the scheme involves at least a doubling of the size of the existing settlement. Proposals for *extensions to existing towns* are not included, although EIA has been carried out in such cases.

9.2 The nature of new settlements

The idea of new settlements is not a new concept. Its development can be traced back to the garden city movement of Ebenezer Howard, followed by the series of new towns designated in the post-war years (Ward 1992). In the 1960s and 1970s, a small number of privately funded new settlements, such as New Ash

Green, Bar Hill and South Woodham Ferrers, were also developed. More recently, a series of proposals for "new country towns" was initiated by Consortium Developments Ltd in the mid-1980s. Other developers took up the concept in response to the opportunities presented by the round of structure plan reviews in the late 1980s and early 1990s. The number of new settlement schemes currently being promoted remains considerable, despite the lukewarm support for the concept in the DoE's revised Planning Policy Guidance Note 3 (PPG3) published in March 1992 (DoE 1992a) and a lengthening list of appeal refusals in recent years.

Between the implementation of EC Directive 85/337 in July 1988 and the end of 1992, planning applications were submitted for at least 31 free-standing new settlement schemes in England and Wales. By the end of 1992, only eight of these had yet to be determined. In addition, planning applications were expected to be submitted for at least a further 15 schemes being promoted at that time through the development plan process. Two-thirds of the schemes for which planning permission has been sought are located in East Anglia and the East Midlands, with most of the remainder in southeast England and the West Midlands (see Table 9.1). Just over half of these schemes are located in two counties: Cambridgeshire has seen the most applications (12), reflecting the A10 and A45 new settlement policies in the replacement structure plan of 1989; Leicestershire has also seen a large number of applications (4), although in this case no lead has been provided by the county's structure plan policies. The schemes currently awaiting progress with development plan work, before applications are submitted, are located mainly in southeast England (especially Hampshire, with three schemes), in East Anglia and in the area around York (three schemes).

Table 9. 1 Location of free-standing new settlement proposals submitted July 1988 to December 1992.

Region	Applications submitted	Applications expected
East Anglia	14	2
East Midlands	7	1
South East	5	7
West Midlands	3	1
Yorkshire & Humberside	1	3
South West	1	0
North	0	1
TOTAL	31	15

Sources: Journal of Planning and Environment Law, Therivel (1991), planning press, personal communications with local planning authority officers. The figure shows the position at the end of 1992.

Note: Only planning applications submitted since the implementation of the EC Directive in July 1988 are included. Expected applications are those for schemes currently awaiting the outcome of development plan work or the outcome of other neighbouring applications. Only free-standing schemes are included.

The applications submitted since July 1988 range in size from 200 dwellings to over 4000, although the average size is just under 1800. Very few proposals have been submitted during this period for new settlements of more than 3000 dwellings, with half of the schemes proposing 1500 or fewer (see Table 9.2). Almost all the schemes include village centres with shopping and community facilities, and most also include elements of commercial and/or industrial development.

Table 9.2 Size of new settlement proposals submitted between July 1988 and December 1992.

No. of dwellings proposed	Applications submitted	Applications expected	All proposals
Up to 500	5	2	7
501–1000	2	3	5
1001–1500	9	2	11
1501–2000	1	1	2
2001–2500	3	1	4
2501–3000	9	0	9
Over 3000	1	2	3
Not known	1	4	5
TOTAL	31	15	46
Average no. of dwellings	1794	1700	1769

Source: As for Table 9.1.

9.3 The need for EIA for new settlements

Guidance on need for EIA

Free-standing new settlement schemes are not specifically identified in either Schedule 1 or 2 of the EIA regulations. This has led to some confusion over the need for formal EIA for such schemes. An early ruling by the DoE that one of the new settlement proposals near Cambridge (for up to 3,000 houses and a business park) was neither a Schedule 1 nor 2 project, and that EIA was therefore not required, only added to this confusion.[1] In a further twist, the developers in this case eventually submitted an EIS for the scheme voluntarily, despite the Secretary of State's ruling.

It could reasonably be argued that new settlements are embraced within the term "urban development projects" (Sched 2.10b). Given that most proposals include some commercial, industrial or retail development, they may also contain an element of "industrial estate development" (Sched 2.10a). Granted that this *is* the case, which schemes are likely to require formal EIA? The DoE's indicative criteria and thresholds unfortunately provide little guidance on this matter. Most of the criteria for urban development and industrial estate projects appear to have been designed primarily for projects located within existing urbanized areas, rather than in free-standing locations – for example, close proximity to a significant number of dwellings is seen as a possible factor in determining the need for EIA, which is clearly of no use in the case of free-standing schemes. A floorspace threshold of 10,000 m^2 (for retail and commercial development) and a site area threshold of 20 ha (for industrial estate development) are suggested, but it is not made clear whether these are applicable to both free-standing and urban-area proposals. Rather confusingly, a higher floorspace threshold of 20,000 m^2 is suggested for out-of-town retail developments. No guidance is provided for housing schemes, whether free-standing or not. As one of the DoE's own commissioned research studies recently concluded:

> Several indicative criteria and thresholds in . . . the circular are ambiguous, especially those relating to "urban development" schemes and "redevelopment" projects, there being an absence of criteria against which to determine whether or not new settlements in the countryside should be subject to EA. (DoE 1991)

Clearly, therefore, local planning authorities have been left with a large amount of discretion in determining the need for EIA for such schemes, with little guidance on how this discretion might be exercised.

Interpretation of need in practice

In practice, EISs have been submitted in many cases for new settlement proposals. Table 9.3 provides further details on the 31 applications submitted since July 1988. EISs were submitted in two-thirds of these cases, either voluntarily or in response to a request by the local planning authority. The remaining third of cases were not subject to formal EIA and were treated as normal planning applications.

One possible reason for local planning authorities not requesting EIA in these cases is that the size of the new settlement was too small to justify the use of the formal EIA procedures. The number of dwellings proposed is an obvious indicator of the size of a new settlement, although the overall site area and the scale of any commercial, industrial or retail development proposed will also be important. The average number of dwellings proposed in schemes subject to EIA was 2100, which is well above the average of 1200 dwellings in schemes not subject to EIA. Indeed, half of the projects not subject to EIA were small schemes of 500 dwellings or fewer (see Table 9.3). Nevertheless, there are examples of larger schemes (of more than 2000 dwellings) that escaped the need for EIA, as well as much smaller schemes (of fewer than 1000 dwellings) that *were* subject to formal EIA. Factors other than the size of the proposal are clearly at work in determining the need for EIA.

Table 9.3 Size of new settlement proposals for which planning applications were submitted July 1988 to December 1992.

No. of dwellings proposed	EIS submitted	EIS not submitted	Total
Up to 500	1	5	6
501–1000	3	1	4
1001–1500	7	1	8
1501–2000	1	0	1
2001–2500	2	1	3
2501–3000	7	2	9
Over 3000	1	0	1
Not known	1	0	1
TOTAL	23	10	33
Average no. of dwellings	2112	1186	1823

Source: As for Table 9.1.
Note: Includes two voluntary EISs submitted prior to planning application.

A second possible reason for the local planning authority not requiring EIA is that the proposed site was already allocated for residential development in the relevant local plan. To date, this appears to have been the case for very few proposals. However, an example is provided by a proposal for up to 1150 houses on a redundant hospital site near Newark in Nottinghamshire. This application did not result in a request for EIA by the local planning authority, apparently because 700 dwellings were already allocated on the site in the deposit version of the Newark Area Local Plan. By contrast, an earlier application within the same district, for 1000 houses on a site not allocated in a local plan, *did* result in a request for EIA by the local planning authority. As discussed below, this type of situation may become more common in future as more new settlement proposals are pursued through the local plan process.

A study commissioned by the DoE (1991) suggests other possible reasons for EIA not being requested: (a) the developer may have already provided a substantial amount of supporting information with the planning application; (b) the local planning authority may intend to refuse the application and regard any subsequent appeal inquiry as being likely to deal adequately with the environmental implications of the development; (c) if the application is for outline planning permission, the design of the proposal may not be far enough advanced to allow EIA to be carried out; and (d) there may have been no formal consideration by the local planning authority of the need for EIA. It is not known whether any of these factors were applicable in any of the cases shown in Table 9.4.

Table 9.4 Free-standing new settlement schemes for which planning applications were submitted July 1988 to December 1992.

Name of scheme and location	Local planning authority	No. of dwellings	EIS submitted?	Planning decision
Hare Park A45 east of Cambridge	East Cambridgeshire DC	3000	no	1
Allington A45 east of Cambridge	East Cambridgeshire DC	3000	yes	1
Kennett North east of Newmarket	East Cambridgeshire DC	1500	yes	5
Westmere A10 north of Cambridge	East Cambridgeshire DC	1500	yes	1
Waterfenton A10 north of Cambridge	South Cambridgeshire DC*	1500	yes	1
Scotland Park A45 west of Cambridge	South Cambridgeshire DC	3000	yes	1
Highfields A45 west of Cambridge	South Cambridgeshire DC	3000	yes	1
Great Common Farm A45 west of Cambridge	South Cambridgeshire DC	3000	yes	2
Bourn Airfield A45 west of Cambridge	South Cambridgeshire DC	3000	yes	1
Swansley Wood A45 west of Cambridge	South Cambridgeshire DC	3000	yes	2
Belham Hill A45 west of Cambridge	South Cambridgeshire DC	3000	yes	1
Crow Green A45 west of Cambridge	South Cambridgeshire DC	3000	yes	5
Mangreen South of A47 Norwich Southern Bypass	South Norfolk DC	1500	yes	5
Leziate East of Kings Lynn	Kings Lynn & West Norfolk BC	450	no	6
Hilton Near Burnaston Toyota plant, A516 west of Derby	South Derbyshire DC	1120	yes	3
Kettleby Magna Great Dalby airfield, south of Melton Mowbray	Melton BC	1200	yes	4
Six Hills Village Alongside A46, between Melton Mowbray & Loughborough	Melton BC	1400	yes	5
Stretton Magna East of Leicester, between A6 and A47	Harborough DC	2400	yes	5

Name of scheme and location	Local planning authority	No. of dwellings	EIS submitted?	Planning decision
Wymeswold airfield East of Loughborough	Charnwood DC	2200	no	1
Bilsthorpe Village Expansion Between Mansfield & Newark	Newark & Sherwood DC	990	yes	6
Balderton hospital South east of Newark	Newark & Sherwood DC	1150	no	6
Marston Park Marston Moretaine, between Bedford & Milton Keynes	Mid Bedfordshire DC	800	no	1
Upper Donnington North of Newbury	Newbury BC	300	no	1
Chiltern Acres Near Stoke Mandeville, south of Aylesbury	Aylesbury Vale DC and Wycombe DC	400	no	1
Northwick Village Canvey Island	Castle Point DC	4300	yes	1
Otterham Quay East of Gillingham	Gillingham BC	200	no	1
Strensham upon Avon M5 between Cheltenham & Worcester	Wychavon DC	not known	yes	1
Brockhill Near Redditch	Redditch BC	1300	yes	8
Aston Prior Near Shifnal, east of Telford	Bridgnorth DC	360	no	1
Acaster Malbis South of A64 York southern bypass	Selby DC	2250	yes	7
Poundbury Near Dorchester	West Dorset DC	2500–3000	no	3

Source: As for Figure 9.1.

Note: Planning decisions are coded as follows:

1 = Appeal or called-in application dismissed by Secretary of State after public inquiry.
2 = As with 1, but application resubmitted in modified form and yet to be determined.
3 = Outline planning permission granted by local planning authority.
4 = Grant of outline planning permission by local planning authority is imminent, subject to completion of Section 106 agreement.
5 = Application or appeal withdrawn prior to determination.
6 = Decision by local planning authority awaited – dependent on progress with local plan.
7 = Holding direction issued by Secretary of State, preventing local planning authority determining application until completion of structure plan and Green Belt review.
8 = Appeal outcome awaited.

*Two schemes were submitted: one for 1500, one for 3000. Planning decision = 1 in both cases

9.4 EIA and the development plan process

Of the 31 planning applications submitted between July 1988 and December 1992 for new settlement proposals, 18 had already been determined at the time of writing, and a further five were withdrawn before determination. Of those determined, the vast majority (16) were either called in by the Secretary of State or taken to appeal following refusal by the local planning authority. The rôle of EIA and the EIS in resultant public inquiries has been discussed in §6.5. All of the appeals and called-in applications for the new settlements shown in Table 9.3 were dismissed by the Secretary of State. At the time of writing, only two schemes have been granted outline permission by the local planning authority: at Hilton in Derbyshire and at Poundbury in Dorset. Outline permission for a third scheme, at Kettleby Magna in Leicestershire, was imminent at this time.

The lack of success at appeal and the revised PPG3 indicate that proposals are likely to be successful only when promoted through the development plan process. It is therefore not surprising that, of the 15 schemes for which planning applications have yet to be submitted, all but one are being steered through either the local plan or structure plan process. Five of the sites involved have already been allocated in the draft or deposit versions of the relevant local plan, and applications can be expected once these plans are adopted (assuming that the allocations are retained). The furthest advanced of these proposals are at Cransley Lodge in Northamptonshire, Leybourne Grange in Kent and Red Lodge in Suffolk. In the first of these cases, Kettering Borough Council adopted a novel approach by inviting developers to submit proposals for a new village in the borough during the local plan consultation period. Cransley Lodge emerged as the borough council's preferred scheme, ahead of five other proposals, and was incorporated as an allocation in the deposit version of the local plan. The Local Plan Inquiry was held during 1992 and – at the time of writing – the inspector's report was being awaited. Another four schemes are being pursued as representations on the relevant draft local plan, on sites not currently allocated for housing. Two schemes in Hampshire are being promoted in representations to the Structure Plan Alteration Examination in Public, and three schemes have emerged in response to a draft study by the local planning authorities in the greater York area that identified the need for up to two new villages.[2]

The promotion of new settlement proposals through the local plan process now appears to be the dominant approach being adopted by developers, in line with the advice in PPG3. This raises issues about the rôle of EIA and the appropriate timing of the submission of the EIS in such cases. If a proposal has been incorporated as an allocation in the local plan after extensive consultation and the submission of considerable information by the developer, the local planning authority may decide that requesting an EIS with any subsequent application would be superfluous and might prejudice relations with the developer. There is therefore a possibility that such proposals might escape the need for formal EIA. In the case of the three furthest advanced proposals listed above, the local planning authorities concerned have indicated informally that EISs *are* likely to be required for the Cransley Lodge and Red Lodge schemes, notwithstanding the inclusion of the proposals in the relevant local plan. The attitude of the local planning authority concerned with the third scheme, at Leybourne Grange, is not known.

It could be argued that, for schemes promoted through the local plan process, formal EIA should be carried out at a stage earlier than the submission of a planning application. The resulting EIS could then be taken into account by the local planning authority and made available for public comment, either during the consultation period on the draft plan or at the Local Plan Inquiry. An alternative approach would be for the local planning authority to carry out an EIA of the local plan itself, which would include an assessment of the environmental impacts of any major residential allocations in the plan (see Ch. 13). To date, there is no evidence that local planning authorities have requested EISs from developers during the local plan consultation process and *prior to* the submission of a planning application. However, there *are* examples of developers voluntarily submitting EISs in such circumstances (see Table 9.5). One example is provided by a village expansion scheme at Lighthorne Heath, adjacent to the M40 in Warwickshire. The site involved had already been allocated for housing in the draft Stratford-on-Avon District Local Plan and the proposal had been worked up in considerable detail. The developers agreed to submit a voluntary EIS as part of their representations on the draft plan, "in view of the complex nature of the project and its possible impact on the area". The required scope of the EIA was determined largely by the district council. Statutory and other consultees were contacted during the preparation of the EIS, in accordance with the EIA regulations. However, as no planning application had been submitted, the requirements relating to publicity in the regulations were not regarded as applicable.

The submission of such voluntary EISs, before the planning application and outside the formal EIA regulations, could be expected to become more common as an increasing number of proposals are pursued through the local plan process. Whether a further or updated EIS would be submitted or requested at the time of the planning application in such cases is not certain. If not, then the requirements relating to publicity of the EIS and consultation with statutory and other bodies could potentially be bypassed.

9.5 Introduction to the case studies

The remainder of the chapter presents detailed case studies of EIA for two new settlement proposals. The aim is not to provide a comprehensive review of the entire EIA process, but rather to highlight certain features of current good practice. The first case study, in §9.6, examines the use of EIA for a substantial village expansion scheme in Nottinghamshire. Interesting features include the scope of the EIA, the important rôle of pre-application consultation, the approach to

Table 9.5 Free-standing new settlement schemes for which planning applications were expected or possible at the end of 1992.

Name of scheme and location	Local Planning Authority	No. of dwellings	EIS submission	Current status
Dunstall St. Michael	Kings Lynn & West Norfolk BC	1800	5	3
Expansion of Red Lodge A11 northeast of Newmarket	Forest Heath DC	1500	3	1
Bittesby Parva	Harborough DC	3400	5	6
Little Easton East of Stansted Airport	Uttlesford DC	2500	2	3
Dibden Bay Between Marchwood & Hythe	New Forest DC	1300	5	4
Micheldever Station North of Winchester, west of M3	Winchester CC	5000	4	4
Portway Grove, Palestine Between Salisbury and Andover	Test Valley BC	not known	3	3
Leybourne Grange Hospital site West of junction 4, M20	Tonbridge & Malling BC	700	5	1
Cransley Lodge West of Kettering and of A43	Kettering BC	350	3	1
Wootton – Quinton South of Northampton, north of M1	South Northants DC	650	5	2
Expansion of Lighthorne Heath Adjacent M40, east of Stratford upon Avon	Stratford upon Avon DC	500	1	2
Askham Richard Southwest of York, north of A64	Selby DC	not known	5	5
Hagg Wood Southwest of York, south of A64	Selby DC	not known	5	5
Ridsdale/Clock Farm Adjacent A64 York Southern Bypass	Selby DC	not known	5	5
Seven Wells Village Northwest of Newcastle upon Tyne	Castle Morpeth BC	800–1000	1	3

Source: As for Figure 9.1

Note: EIS submission is coded as follows:

1 = EIS already submitted voluntarily prior to planning application, as part of developer's representations on the local plan.

2 = Developer considering submitting EIS voluntarily prior to planning application, as part of representations on local plan.

3 = Local planning authority will/is likely to request an EIS at the time of the planning application.

4 = Developer intends to submit an EIS at the time of the planning application.

5 = Position on submission of EIS not known.

Note: Current status is coded as follows:

1 = Proposal is identified as an allocation in deposit version of the local plan; planning application is imminent.

2 = Proposal is identified as an allocation in draft version of local plan.

3 = Proposal is being pursued by the developer in representations on the draft local plan.

4 = Proposal is being pursued by the developer in representations to the Hampshire Structure Plan Alteration Examination in Public.

5 = Proposal is awaiting completion of the North Yorkshire Structure Plan Alterations and York Green Belt alterations.

6 = Other proposals not currently being pursued through the development plan process.

EIS submission and current status codes relate to the position at the end of 1992.

prediction and the assessment of significance, and the treatment of mitigation and monitoring. The second case study, in §9.7, concerns one of the many new settlement proposals in the Cambridge area. The way in which the scope of the EIA was determined and the methods used to assess the importance and significance of impacts are of particular interest.

9.6 A case study of EIA for an expanded village: Bilsthorpe, Nottinghamshire

Introduction

This first case study concerns the proposal to expand the existing village of Bilsthorpe in Nottinghamshire. The scheme was proposed by Nottinghamshire-based environmental consultants David Tyldesley and Associates. An outline planning application was submitted in December 1990, with funding provided by the owner of the site. At the time of writing, the application had yet to be determined.

The village of Bilsthorpe is located in Newark and Sherwood District, approximately seven miles east of Mansfield. The existing settlement lies immediately to the north and east of the application site, and the A614 trunk road forms the site's western boundary. The site comprises approximately 125 ha of land, almost entirely given over to arable farming (Fig. 9.1). The existing village of Bilsthorpe has a population of about 3,100, with a housing stock of about 1,100. No significant housing development has taken place since 1970, mainly because of the limited capacity of Bilsthorpe sewage works. Since the early 1970s, the village's population has declined slightly, with an increasing proportion in the older age groups. Community facilities, including the primary school, tend to be concentrated in the northern half of the village. Bilsthorpe colliery is the dominant employer, accounting for 60% of jobs. There are currently more jobs in Bilsthorpe than economically active residents, although there is substantial commuting both into and out of the village. The future of the colliery was thrown into doubt by British Coal's announcement in October 1992 of the proposed closure of up to 31 pits (including Bilsthorpe).

The proposed scheme, application and planning context

The proposal envisages the construction of just under 1,000 houses, almost doubling the size of the existing settlement. A total of 10 ha on the western edge of the site would be allocated for industrial development, potentially creating a total of 500 jobs. The total site area would be 125 ha, and 50 ha of this would be established as three new areas of woodland, alongside the A614, along a ridge line leading into the centre of the site and on a prominent slope in the northern part of the site. The planting and subsequent management of these areas would attempt

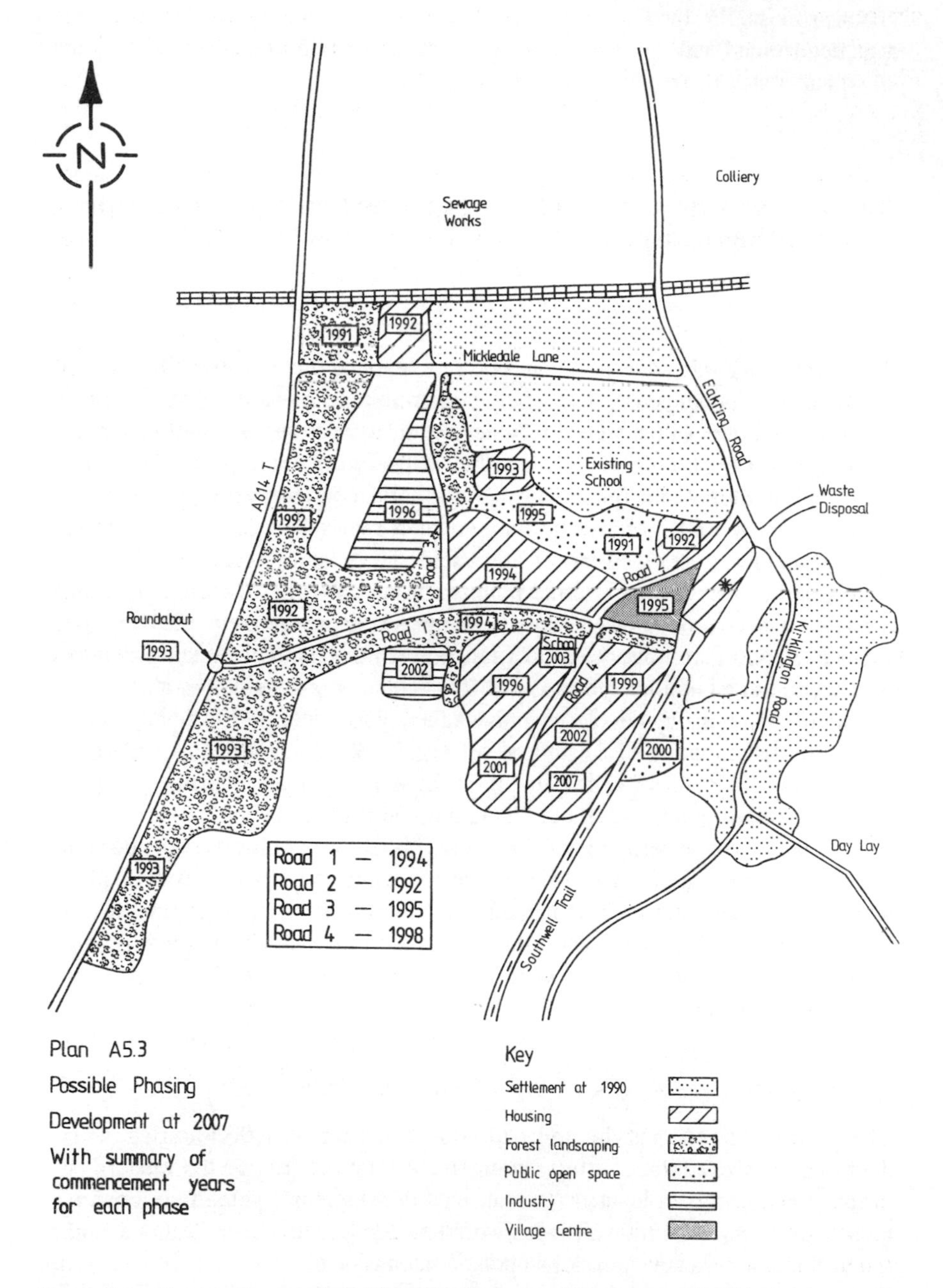

Figure 9.1 Proposed expanded village at Bilsthorpe. (*Source:* David Tyldesley & Associates 1990)

to recreate the natural habitats associated with Sherwood Forest. A new junction with the A614 is proposed, with closure of the existing junction. A new village centre would be provided, trebling the amount of retail floorspace and enabling the provision of services that the village cannot currently support. Sites are reserved for a new primary school and a community/sports centre. Development would be phased over a 15-year period, resulting in an annual average building rate of about 65 houses.

An outline planning application for the proposal was lodged with Newark and Sherwood District Council in December 1990. The Nottinghamshire Structure Plan Review Examination in Public had been held earlier in the year. The review had identified Bilsthorpe as one of several villages suitable for "limited" residential and employment development, although no definition of what was meant by "limited development" had been provided. At the time of the planning application, the Bilsthorpe area was not covered by a local plan. However, the district council was in the process of preparing three separate local plans. One of these covered the western part of the district, which includes the Bilsthorpe area. Public consultation on the draft version of this Western Area Local Plan did not take place until *after* the submission of the planning application, during the spring of 1991. The Western Area Local Plan will not proceed to the Local Plan Inquiry stage, but will be incorporated into a draft district-wide local plan.

At the end of July 1990, the local planning authority, Newark and Sherwood District Council, requested that an EIA of the proposal be undertaken. Accordingly, an EIS was prepared by David Tyldesey and Associates (1990) and submitted with the outline planning application in December 1990.

The scope of the EIS

The effects considered in the EIS were determined partly by reference to the "specified information" in Schedule 3 of the EIA regulations. However, certain effects included in the specified information were not considered relevant for this proposal and were therefore not included in the EIS – these included the effects on air quality and climate. Appendix 4 of the DoE publication, *Environmental assessment: a guide to the procedures*, contains a more detailed checklist of matters that may need to be considered in an EIS, although it is recognized that not all of these will be relevant to every project. The EIS identifies which of these matters were regarded as irrelevant or insignificant, and provides a brief justification of this conclusion. For example:

- The EIS does not address the effects of waste disposal; the effects of pollutants on water courses; and the effects on hydrology outside the site boundaries. It argues that such effects would be adequately handled by the normal requirements of the National Rivers Authority (NRA) and Severn Trent Water, which would incorporate any necessary mitigation measures.
- Consideration of the possible production processes or operational features of industrial land-users, and their potential effects, are not addressed. The EIS

argues that the use of the proposed industrial sites should be limited to occupiers in Classes B1, B2 and B8, and that this could be controlled by a condition attached to any planning consent.

- The EIS considers that the construction phase would not involve any unusual or unacceptable methods of construction. It assumes that mitigation of the effects of construction would not need to be considered, because of the physical separation of the site from residential areas and the speed and type of construction activities anticipated. Despite this, certain effects (e.g. the effects of construction traffic on existing residents before the completion of the new internal road network) *are* addressed in the EIS.
- Discussion of the main alternative sites considered is also not seen as relevant, given that the fundamental objectives of the project relate to the needs of Bilsthorpe.

Some effects not specifically identified in the EIA regulations, but considered to be relevant to the EIS, *are* included. These include employment, community and social effects, and the extent to which the proposals conform with, or contribute to the achievement of, statutory development plan policies and objectives. The effects considered in the EIS are listed below:

- community and social effects, including effects on commuting flows, employment opportunities and community and recreational facilities;
- effects on highways and traffic, and on rights of way;
- effects on existing water courses and on other infrastructure (e.g. sewers);
- employment effects;
- effects on landscape and visual amenity, including the impact on the general landscape setting and character of the area; the impact on views into the site from roads, public rights of way and residential properties; and the impact on views within and out of the site;
- effects on flora and fauna, including the loss, modification, reduction or extension of existing habitats on site; the creation of new habitats on site; disturbance to or loss of species or animals; and indirect or cumulative effects on habitats outside the site boundaries;
- changes in land-use, including agricultural land loss;
- effects on the cultural heritage;
- effects on buildings and other material assets;
- the extent to which the proposals conform with statutory development plan policies and objectives, and/or contribute to the achievement of certain policies;
- the interaction between the development and the British Coal proposal to construct a 150 MW coal fired power station in Bilsthorpe.

Consultations

During the preparation of the EIS, all statutory consultees and a number of other bodies were contacted, in order to obtain information about the site and its environs and to elicit initial comments on the proposals. The responses received from consultees are included in an appendix to the EIS. This allows the main issues raised at this stage to be easily identified, and shows how these comments were incorporated into later versions of the proposal and into the EIS. Examples of the type of issues arising out of the consultation process are summarized below.

The *Nature Conservancy Council* requested that the EIS should address "not only the impacts of the proposal on existing nature conservation resources and the measures to be taken to *minimize* these impacts, but also the steps to be taken to *create* further areas for nature conservation". In particular, they were keen to see proposals in the EIS for the creation of patches of the characteristic habitat of the Sherwood Forest, such as heathland, acid grassland and open oak and birch woodland. The establishment of 50 ha of such habitat is a key element of the proposal and is discussed in some detail in the EIS.

The *Department of Transport* indicated that the proposal to create a new road junction on the A614 would be likely to meet with a direction of refusal on highway safety grounds. The Department requested that this view be included in the EIS. However, the proposal was retained by the developers in the outline planning application. They argued in the EIS that the new junction would replace the existing one, rather than being additional, and that visibility for road users would be improved compared with the existing situation.

A watercourse flows through the site and has existing flooding problems. The *National Rivers Authority* indicated that it would object if the development exacerbated these problems or caused new ones. The EIS mentions the possible need to provide a surface-water storage facility, in order to prevent such impacts. The need for such a facility would be determined by the outcome of further, more detailed studies. However, such studies were not carried out as part of the EIA process and are not reported in the EIS.

Severn Trent Water Ltd pointed out that the existing public sewers and Bilsthorpe sewage works would be inadequate to cope with the development. They would look to the developer to finance the provision of all off-site works required to service the development. Major off-site mains reinforcement would also be required to enable a water supply to be made available. The EIS states that a new trunk sewer would be laid and the existing sewage treatment works extended. It suggests that any adverse effects of these measures would be mitigated by conditions attached to any discharge consent granted by the NRA.

Records held at the *Nottingham Natural History Museum* were consulted during the preparation of the EIS. This source of information revealed two possible impacts not previously anticipated during the drawing up of the project proposals. The first of these was the possible effect on the Southwell Trail, a disused railway line passing through the site. The trail was found to be a Biological Grade 2 alert site, of district-wide importance. A section of the trail would be in close proximity to residential and amenity elements of the proposal, and was therefore at risk of damage through the effects of increased public use. The second potential impact emerged from an examination of records of protected species of birds, animals and plants on the site and in its environs. This revealed that part of the site may be contained within the breeding territory of the barn owl. As a result of these findings, additional mitigation measures were incorporated into the EIS. These included: (a) preventing construction works within 50 m of the Southwell Trail during the breeding season; (b) ensuring that the layout and design of residential areas minimized the potential abuse of the trail; (c) monitoring the effects of the development on the habitats of the trail; and (d) support for survey and protection measures for the barn owl.

Local residents were also consulted. In October 1990, before the submission of the planning application, a public exhibition outlining the proposals was held at Bilsthorpe village hall. The developers also publicized the proposals in the local press at this stage. A total of 240 people visited the exhibition, with 130 completing questionnaires about the proposals. A majority (62%) were in favour of the scheme, a quarter were against, with the remainder (13%) undecided. The developers claim that "this level of public support for a large-scale greenfield development is entirely unprecedented, but is based on a common perception of Bilsthorpe's problems". As a result of the comments received, alterations were made to the project proposals. The most notable of these were the provision of additional leisure and community facilities, and the allocation of a larger area for the new village centre, to allow the provision of further facilities at a later date.

Prediction and assessment of significance

The EIS adopts a novel approach to impact prediction. For each predicted impact, an indication is given of the confidence in/probability of the prediction. This is followed by a qualitative assessment of the significance of the impact. Where appropriate, there is also an indication of the quantitative scale or magnitude of the impact (e.g. the number of dwellings affected; the percentage increase in traffic flows). Although this basic approach has been followed in other statements, the Bilsthorpe EIS develops it further by using a series of standard terms relating to the probability of predictions and the qualitative assessment of effects. Each term is identified on a numerical scale of 1–7 (Table 9.6). The use of this scale is intended to provide a consistent meaning to each term, relative to other terms, every time it appears in the text of the EIS.

Clearly, this approach has advantages and disadvantages. The main weakness is the large amount of subjective interpretation potentially involved. The classification of certain impacts as "unimportant", "minor" or "significant" may be affected by value judgements about the relative importance of different types of impacts (e.g. economic and social versus landscape or ecological impacts). There may also be an understandable tendency for those preparing the EIS to view beneficial effects of the development as of more significance than adverse effects. The approach would seem to be less relevant in the case of impacts that involve detailed technical calculations or models (e.g. the effects on air or water quality of specific emissions or discharges). Nevertheless, the approach has distinct advantages.

Table 9.6 Bilsthorpe EIS – approaches to confidence/probability of predictions, and qualitative assessment.

Confidence/probability of predictions:

7 Absolute certainty
6 Near certainty/very high probability
5 High probability – to be expected
4 Likelihood/normal anticipation – to be anticipated
3 Seriously anticipated possibility
2 Possibility
1 Remote possibility

Qualitative assessments of effects:

7 Total/consuming/eliminating
6 Profound/considerable/substantial
5 Material/important
4 Discernible/noticeable/significant
3 Marginal/slight/minor
2 Unimportant/inconsequential/indiscernible
1 Irrelevant/no effect

Positive effects are followed by a + sign and negative effects by a – sign

A number of examples illustrate the technique:

a) "The habitat would certainly be eliminated (7, 7–)" (i.e. total confidence of total impact).
b) "There is a possibility that the habitat could be substantially damaged (2, 6–)" (i.e. relatively low probability, but a relatively high level of impact if it occurred).
c) "It would be expected that slight damage may occur to the habitat (5, 3–)" (i.e. relatively high probability, but a relatively low level of impact).

Source: Adapted from David Tyldesley & Associates 1990.

It represents a useful means of describing the significance and likelihood of potential impacts in a consistent way throughout each section of the EIS, for a wide variety of different impacts.

Predicted impacts regarded as "at least significant" (i.e. 4–7 on the numerical scale) are easily identified in the EIS. They include a total of 12 adverse effects and 29 beneficial effects. The significant adverse effects identified consist mainly of the effects of increased traffic on existing residents and the adverse visual impacts of the development. Other significant effects include the irreversible loss of agricultural land, a potential increase in the flooding problems of an existing watercourse and a likely conflict with two policies in the approved structure plan. Half of the significant beneficial effects identified consist of community, social, employment and recreational benefits; these include the creation of about 600 jobs, reduced dependency on the colliery, a reduction in the proportion of out-commuters, improved shops and community facilities, and greater recreational opportunities, including the use of open space and rights of way. The other significant benefits of the development mainly consist of the landscape and ecological benefits of the establishment of the new woodland/landscaping areas.

Modifications to the development introduced during the EIA process

The developers state that "environmental objectives have been a fundamental element of the scheme from its commencement. Environmental assessment has been a continuous process throughout the preparation of the project . . . The proposed layout of the development and landscaped areas has therefore been modified in minor ways, many times". The EIS lists the most important changes introduced on environmental grounds during the preparation of a succession of nine different proposals plans. Examples of such changes, along with their main justification, are given below:

- extension of the landscaped area in the west of the site (undertaken three times), to refine the landscape and visual effects, and to enhance the Sherwood Forest regeneration and other local planning authority objectives;
- a reduction in the amount of residential development proposed, from 49 ha to 40 ha, to achieve a better balance between the numbers of economically active residents and job opportunities, and to minimize the residential and visual amenity effects of the new development on the existing village;
- relocation of one of the two industrial sites (undertaken twice), to improve its relationship with the proposed new road access and landscaped areas;
- introduction of a wedge of public open space between the historic part of the existing village and the new residential development, to minimize the residential and visual amenity effects of the new development on the existing village;
- introduction of improved leisure and community facilities in the proposal, and an extension of the area for the new village centre, to offset underprovision, widely expressed by the local community during the public exhibition of the proposals;
- modification of the western boundary of the development, to avoid potential effects on archaeological resources identified during the preparation of the EIS.

Approach to mitigation measures and to environmental monitoring

Many *mitigation* measures are proposed in the EIS, most of which were adopted during the preparation of the various plans for the proposals, rather than *after* the undertaking of the EIA. In other words, mitigation measures were incorporated into the design and layout of the project, rather than being "added on" after the EIA. For those mitigation measures that *did* result from the findings of the EIA, the EIS suggests that mitigation should generally be considered "for any negative effects which are at least anticipated and at least significant (i.e. 4–7)". A possible danger with this approach is that mitigation measures may not be considered for impacts with a relatively low level of probability, but a high level of significance.

The EIS suggests that the proposed mitigation measures could be included either in conditions imposed on any planning permission or in an agreement between the local planning authority and the developer under Section 106 of the Town and Country Planning Act 1990 and Section 33 of the Local Government (Miscellaneous Provisions) Act 1982. The EIS provides a suggested list of conditions to control the development, and a list of matters that could be subject to agreements with the local planning authority.

The EIS draws attention to the need for continued *monitoring* of baseline environmental information, in order that (a) the proposals can be refined to take account of changing circumstances and (b) the predictions on environmental effects can be confirmed. Specific monitoring proposals are put forward in the EIS, and it is suggested that these could be incorporated in conditions or agreements associated with any planning permission. The responsibility for, and funding of, each aspect of monitoring are discussed in the EIS. Two examples of the monitoring measures proposed are given below:

- The establishment of the proposed woodland and heathland areas would need careful monitoring. Different techniques for the re-establishment of these habitats are recommended and the monitoring of their outcomes is seen as essential. This should include (a) their landscape/visual effects and (b) changes in and development of habitats. The latter would necessitate regular surveying of plant and animal communities in all new areas. Relevant organizations would be invited to contribute their expertise as regards possible techniques for re-establishing these habitats and the development of an appropriate monitoring methodology.
- The EIS recommends that an annual monitoring report be produced on behalf of the developers, identifying the amount and type of new development and reporting on relevant planning considerations. The monitoring report would examine the following specific issues: demographic changes, the origin of new residents, the workplaces of residents, house type and tenure, job creation and availability, the number of economically active residents, transport and communication networks, and the progress made towards the fulfilment of objectives and targets. The report would be made available to the local planning authorities, the parish council, the village trust (if established), British Coal, the applicant and any other bodies suggested by the local planning authorities.

Current status of the proposal

The outline application for the expansion of Bilsthorpe, although submitted in December 1990, had yet to be determined by Newark and Sherwood District Council at the time of writing (in early 1993). Determination of the application was delayed while consultation took place on the draft Western Area Local Plan, and comments were invited on the Bilsthorpe proposals as part of this process. Meanwhile, a draft district-wide local plan is currently being prepared to replace the three area local plans within the district. Since the submission of the planning application and the EIS, further investigations have been carried out into the traffic and highways implications of the development. Subconsultants were commissioned to undertake further survey work and make new traffic forecasts. This work was undertaken in an attempt to resolve outstanding differences between the developers and Nottinghamshire County Council Highways Department. At the time of writing, these differences had been resolved.

9.7 A case study of EIA for a new settlement: Great Common Farm, Cambridgeshire

Introduction

The second case study concerns the proposal to construct a new settlement west of Cambridge on land at Great Common Farm and Bourn Airfield. The scheme was originated by the University of Manchester, with backing from Stanhope Properties plc. An outline planning application was submitted in April 1989. The scheme was one of eight new settlement proposals along the A45 corridor, considered at a joint public inquiry held between February and July 1990. All eight schemes were rejected by the Secretary of State in March 1992, although the inquiry inspector had recommended the granting of outline planning permission in the case of the Great Common Farm application.

The application site is located about six miles west of Cambridge, immediately to the south of the A45 (Fig. 9.2). The existing small villages of Caldecote and Highfields lie to the east of the site, and open farmland forms the western and southern boundaries of the site. An unclassified road bisects the site, linking the village of Bourn with the A45. The application site, which totals just over 400 ha, consists of two principal landholdings, Great Common Farm (owned by the University of Manchester) and the disused Bourn Airfield.

The proposed scheme, application and planning context

The proposal for the 400 ha site envisages the construction of 3000 houses, a town centre, a business park and industrial area, recreational space (including a golf course) and major new areas of woodland. A business park covering 35 ha would be constructed on the disused airfield, providing just over 1 million ft^2 of office and research space. The existing industrial uses on the site would be retained, and an adjoining 10 ha of land would be allocated for new industrial development. About 50 ha of new woodland would be planted on the periphery of the site, mainly between the development and the village of Highfields. The existing road junction with the A45 would be substantially modified, with the dualling of a 1.25 mile stretch of the A45 fronting the site and the construction of a new grade-separated interchange. Access to the new settlement from this interchange would

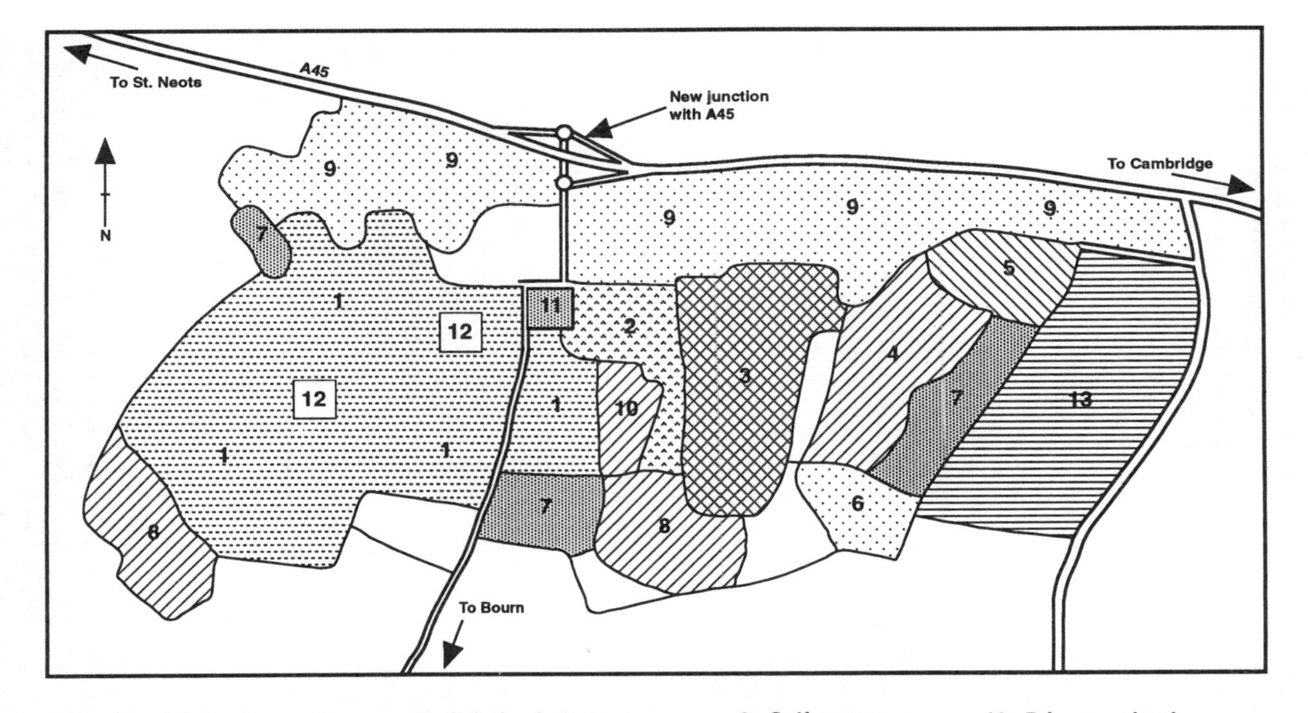

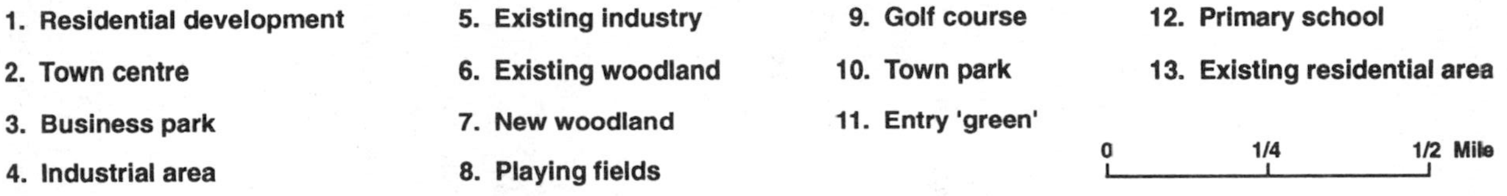

Figure 9.2 **Proposed new settlement at Great Common Farm, west of Cambridge.** (*Source:* Adapted from Land Use Consultants 1989)

be via a new dual link road. Development would be phased over a 15-year period, resulting in an annual average building rate of about 200 houses.

The impetus for the scheme was provided by the review in the late 1980s of the Cambridgeshire Structure Plan. The review had identified the need for two new settlements in the county, along the A10 corridor to the north of Cambridge and along the A45 corridor to either the west or east of Cambridge. Policy 20/2 of the approved structure plan outlined requirements that would need to be satisfied by any new settlement proposals along the A45 corridor:

> Provision will be made for a new settlement on the A45 corridor west or east of Cambridge which will:
>
> - be close enough to Cambridge to make a significant contribution to its development needs, but located outside the green belt;
> - complement the existing settlement pattern and not prejudice the extension planned for Papworth Everard;
> - make use as far as possible of land which is under-used or of little environmental value and minimise the loss of high quality agricultural land;
> - minimise infrastructure costs and flood risks;
> - provide the opportunity for business park development;
> - be capable of accommodating about 3,000 dwellings with some reserve capacity for future expansion, 2,000 of which should be available before 2001;
> - provide safe and easy access to the A45 trunk road.

The Secretary of State approved the structure plan review with modifications on 21 March 1989. Two weeks later, an outline planning application was submitted to South Cambridgeshire District Council to create a new settlement at Great Common Farm. Land Use Consultants was commissioned in May 1989 to undertake an EIA of the proposals, and a preliminary EIS was submitted to the district council in July 1989. The final version of the EIS (Land Use Consultants 1989) was submitted in December 1989, eight months after the submission of the outline planning application. Meanwhile, Ove Arup and Partners (consulting engineers) were commissioned to prepare a detailed report on the traffic and highways impacts of the development. This report was also submitted in December 1989, but its findings were not included in the EIS.

The scope of the EIS

All of the applications for new settlement proposals along the A45 corridor were called in by the Secretary of State for consideration at a joint public inquiry. Given this context, the scope of the EIS appears to have been largely determined by reference to the issues thought likely to be relevant to the Secretary of State's consideration of the various proposals. These issues were determined by advice from the DoE and DoT, and later by the Secretary of State's statement under Rule

6 of the Town and Country Planning (Inquiries Procedures) Rules 1988, i.e. the call-in letter. The Rule 6 statement identified the following issues as relevant to the consideration of the applications:

1 *The relationship of the developments proposed to development plan policies, namely*:
(a) the policies in the Approved Replacement Cambridgeshire Structure Plan (see above)
(b) the green belt local plan, and any subsequent modifications proposed by the County Council
(c) the policies proposed in the South Cambridgeshire Draft Local Plan.

2 *The scale and nature of each development, including its expected final size beyond the plan period.*

3 *The appropriateness of each site, in terms of*:
(a) its physical capacity
(b) the landscape
(c) nature conservation interests, particularly SSSIs
(d) architectural and historic interest, including listed buildings and conservation areas
(e) loss of agricultural land and the re-use of derelict land
(f) susceptibility to flooding and the adequacy of any proposed flood-prevention works.

4 *The effects of each development on*:
(a) the highway network and other transport facilities
(b) the character of the existing towns and villages in the area
(c) the housing and labour markets of the area, particularly the provision of low-cost homes
(d) the amenities of the area, including other services and facilities proposed or affected by the development.

The scope of the EIS was also influenced by consideration of the "specified information" in Schedule 3 of the EIA regulations, assessment of the environmental issues identified at earlier stages in the project proposals, and an examination of the characteristics of each of the other Cambridge new settlement proposals. The resulting list of impacts considered in the EIS is outlined below:

- effects on microclimate and air pollution;
- effects on geology and soils;
- effects on surface water and groundwater; effects of foul drainage;
- ecological impacts, including effects on existing habitats and creation of new habitats;
- effects on land-use, including effects on agriculture; existing industry and

infrastructure on site; existing users of Bourn Airfield; and recreation;
- landscape and visual impacts;
- effects on human beings, including (a) effects on the existing population (effects on local properties; coalescence of settlements; generation of additional traffic; noise; and socio-economic effects) and (b) the characteristics of the new community (population and community structure; employment creation; physical form; architecture; community facilities; and accessibility);
- effects on the cultural heritage;
- effects on material assets.

Methods used to identify, predict and assess impacts

The EIS is arranged so that the treatment of all environmental issues is structured in a similar way. The examination of each topic is divided into sections:

(a) A factual description of the *existing situation*, based on the findings of surveys conducted on site as well as desk studies.

(b) A description of the *potential impacts* of the development and the size of the area potentially affected. The relative importance of these impacts is defined as follows:

- Impacts are of national or regional importance if the effects would be sufficiently important to be relevant at the level of national policies or regional strategies. For example, direct impacts on an SSSI or National Nature Reserve would be of national importance, since such sites are part of a national register of protected nature conservation sites. The employment impact of the business park element of the new settlement might be regarded as of regional importance, if it is likely to affect strategies to direct employment development to particular parts of the region.
- Impacts are of county-wide importance if they would affect strategic decisions at a county level, i.e. the scale of impact would be such that it could undermine (or support) structure plan policies, such as those outlined in Policy 20/2 above.
- Impacts are of district-wide importance if they would be relevant within the context of a local plan. For example, this might include the long-term impact of increased traffic on local roads or impacts on existing local industry.
- Impacts are of local importance if they would be largely contained within the site itself. Examples might include the impact of construction traffic, effects on local properties, or increased noise levels.

The assessment of "relative importance" is based on the professional judgement of the relevant specialists involved in the preparation of the EIS. The inclusion of potential impacts in the EIS does not mean that they are necessarily expected to occur. Rather, they are included to indicate the issues that have been considered during the design process and investigated in the subsequent prediction of impacts.

(c) A description of the *predicted impacts* of the development, for both the construction and operational stages. In most cases, the predicted impacts cover a smaller range of issues and are of less significance than the potential impacts.

This is because of the incorporation of mitigation measures during the design process to eliminate or minimize those features likely to cause significant adverse effects. The predictions are informed views based on the professional judgement of the relevant specialists, rather than statements of fact.

(d) The scope for *mitigation* or amelioration of predicted impacts.

(e) A summary of the *residual predicted impacts* and a judgement about their *significance*. Impacts are classified as having "major", "some", "minor" or "no" significance. The assessment of the level of significance of particular impacts is again based on professional judgement. As with the Bilsthorpe EIS, the temptation to regard beneficial impacts as of greater significance than adverse impacts is a potential weakness of such an approach.

Although the EIS clearly sets out the method of determining the relative importance of potential impacts, there is obviously scope for subjective interpretation. Indeed, inconsistencies are apparent in the assignment of levels of importance to particular impacts in the EIS. For example, the need for safe and easy access to the A45 trunk road is specifically identified as a requirement of the new settlement in Structure Plan Policy 20/2. For this reason, this issue is regarded as of county-wide importance. However, certain other matters clearly identified in Policy 20/2 are classified as of only district-wide importance. These include the possible coalescence of communities, the loss of agricultural land and the re-use of derelict land.

One reason for these apparent inconsistencies stems from the specific circumstances of the Cambridge new settlement proposals – namely, the explicit support in the approved structure plan for a new settlement along the A45 corridor and the holding of a joint public inquiry to consider all eight applications together. The key issue at the inquiry was therefore *where* to locate the new settlement, rather than *whether* a new settlement was needed. In these circumstances, impacts that would arise wherever the new settlement was located, or that do not differ significantly between the alternative sites, might be seen as less critical in the final weighting of the competing proposals. Such impacts may therefore be regarded as of less importance than would otherwise be the case. This reasoning certainly appears to have been used in deciding the relative importance of certain impacts in the EIS.

The current status of the proposal

The planning application for the Great Common Farm new settlement was called in by the Secretary of State and considered at a joint public inquiry during 1990. The proposal did not receive support from any of the local authorities involved in the inquiry. However, the inquiry inspector recommended that the proposal should be granted outline permission; all seven of the other applications were recommended for refusal. In March 1992, after a lengthy delay, the Secretary of State rejected all eight applications, including that at Great Common Farm (DOE 1992b). The size of the business-park element of the scheme appears to have been the decisive factor in the Secretary of State's decision. Since this decision, revised applications have been submitted on two of the original sites, including Great Common Farm. The revised application was accompanied by an updated EIS; at the time of writing, it had yet to be determined.

9.8 Summary

This chapter has examined the application of the Town and Country Planning (Assessment of Environmental Effects) Regulations to proposals for new settlements in the countryside. There is ambiguity over the need for EIA for such projects, with little guidance provided by the DoE's indicative criteria and thresholds. In practice, a significant minority of schemes have escaped the need for formal EIA. Predictably, the size of the new settlement is an important influence on the need for EIA, but other factors are also at work.

Most new settlement proposals are eventually considered at a public inquiry, raising the issue of the extent to which EIA and consideration of the EIS are incorporated into such inquiries. There is a need for further research in this important area. The promotion of new settlement schemes through the local plan process raises questions about the appropriate timing of the submission of the EIS. It could be argued that, for such projects, formal EIA should be reported on at an earlier stage than the submission of the planning application. Indeed, some developers appear to be adopting this approach voluntarily, although there are concerns about the extent of publicity and consultation in such cases.

An examination of two case studies of specific new settlement proposals reveals some features of good EIA practice. Of particular interest are the scoping process, the rôle of consultation, the treatment of mitigation and monitoring, and the approach to prediction and the assessment of impacts.

References

David Tyldesley & Associates 1990. *Bilsthorpe 2000+: Environmental Impact Statement* (For C. A. Strawson Farming Ltd).

David Tyldesley & Associates 1991. *Bilsthorpe 2000+: a sustainable future*.

DoE (Department of the Environment) 1991. *Monitoring environmental assessment and planning*. Report by the EIA Centre, Department of Planning and Landscape, University of Manchester. London: HMSO.

DoE1992a. *Planning Policy Guidance Note 3: housing*. London: HMSO.

DoE 1992b. *Secretary of State's decision letter and Inspector's report: Cambridgeshire new settlements, A45 corridor*. London: DoE

Land Use Consultants 1989. *Great Common Farm, proposed new settlement and business park, environmental statement*.

Therivel, R. 1991. *Directory of environmental statements 1989–1990*. Impacts Assessment Unit, School of Planning, Oxford Polytechnic.

Ward, S. (ed.) 1992. *The garden city: past, present and future*. London: Spon.

Notes

1. The direction, dated 26 June 1989, is listed in the *Journal of Planning and Environment Law*, 1989, p. 856.
2. A fourth scheme near York, at Acaster Malbis, has already been the subject of a planning application.

CHAPTER 10
Environmental impact assessment of projects not subject to planning control

10.1 Introduction

This chapter examines the application of EIA to projects not covered by town and country planning legislation. Two case-study sectors are used to illustrate some of the key issues involved in the EIA of such projects. §10.2 and §10.3 consider the rôle of EIA in the assessment of trunk road schemes proposed by the Department of Transport (DoT). The planning of new trunk road schemes in the UK has been the sole responsibility of the DoT, at least until the recent advent of privately financed toll-road proposals. This has aided development of a consistent approach to the assessment of such proposals, including the consideration of alternatives. The recent critique of existing practice by the Standing Advisory Committee on Trunk Road Assessment (SACTRA) is outlined in §10.2 and a case study of a motorway proposal is examined in §10.3 in an attempt to uncover some of the strengths and weaknesses of the current approach to EIA. §10.4 then discusses the application of EIA to projects in the electricity supply industry. As with trunk roads, this sector had an established record in the evaluation of environmental impacts long before the implementation of EC Directive 85/337. However, the recent privatization of the industry means that many developers have become involved in the planning and EIA process in the sector, for a wide range of projects. This raises questions about the consistency of approach towards EIA within the industry. Other recent developments, such as the EC Directive on Large Combustion Plants and the introduction of integrated pollution control, also raise important issues of relevance to EIA.

10.2 EIA of trunk road proposals

The planning of trunk road proposals

The planning and decision-making process for new trunk road schemes in the UK is well established and it involves a number of key stages. Some knowledge of these is useful in appreciating how and when EIA is incorporated into the planning of such schemes. The main stages involved, and the way in which environmental matters are taken into account and presented, are outlined below (SACTRA 1992).

1. *The formal identification of a need or problem* by the DoT Regional Office. This may spring from environmental considerations. For example, village bypass schemes are often motivated by a desire to reduce the effects of heavy through-traffic on existing residents.
2. *A route identification study* is carried out in cases where the proposal involves a major route or corridor, such as upgrading a road over a long distance or improving a network of roads. This will typically include only a broad-brush consideration of environmental matters. Detailed assessment of impacts is difficult at this stage, since the precise line of any proposed routes may not be known. Once a route has been identified, it is divided into individual schemes for further study.
3. *Scheme identification studies* are carried out for individual parts of an overall route or for smaller-scale proposals that are not part of a major route or corridor. The study will examine the alternative options available for solving the problems identified, including do-nothing or do-minimum options. Again, the study will typically identify sensitive areas and potential effects, rather than contain a detailed environmental appraisal. Initial findings will be presented on the nature of the existing environment (e.g. landscape quality). The likely visual impacts of the alternative options under consideration may also be addressed.
4. *Consideration of the scheme by DoT headquarters and ministers, and entry of the scheme into the Roads Programme*. A list of schemes that have been admitted into the Roads Programme is issued by the DoT once every two years, arranged in order of priority. The announcement that a scheme has been admitted into the Roads Programme includes a statement of the significant environmental effects identified at that stage.
5. *Public consultation on alternative scheme options*. This consists of the local exhibition and presentation of different scheme options – these will typically include alternative lines for the same route, as well as do-nothing or do-minimum options. The first detailed EIA of schemes is undertaken during the period leading up to public consultation. The environmental effects of the alternative options are summarized and presented in a formal tabular framework for consideration during this stage.
6. *Publication of the Secretary of State's preferred route or scheme*. The Secretary of State's choice of route will be influenced by the results of the public consultation.
7. *Publication of Draft Line Orders, Side Road Orders and Compulsory Purchase*

Orders. The statutory EIS required by the EC Directive is published with the Draft Line Orders for the scheme. A period is allowed for objections or comments to be submitted on the Draft Orders and on the EIS.
8. *Consideration of Draft Statutory Orders at a Public Inquiry* in almost all cases. The EIS, and any comments on it by statutory consultees, will be considered at the Public Inquiry, along with other supporting information. The Inquiry Inspector submits a report and recommendations to the Secretaries of State for Transport and the Environment.
9. *Decision by Secretaries of State for Transport and the Environment*. The scheme will be approved, with or without modifications, or rejected. The decision will have regard to the environmental impact of the scheme under consideration.

The environmental appraisal of trunk road schemes prior to the implementation of the EC Directive

Since the 1960s, new trunk road schemes proposed by the DoT have been subject to formal appraisal procedures. Until the mid-1970s, these procedures tended to concentrate on the assessment of the economic or financial implications of schemes, often using a cost–benefit analysis (CBA), with little formal appraisal of physical environmental or social impacts (SACTRA 1992, NCC 1990). In CBA, the relative costs and benefits of different scheme options are typically compared with a do-nothing or do-minimum option. The variables in a CBA of trunk road proposals include the value of travelling time (saved or lost as a result of the scheme), the value attached to injuries and deaths in road accidents (avoided or made more likely by the scheme), vehicle operating costs, the construction costs of the scheme, and road maintenance costs. The stream of future costs and benefits flowing from each option is discounted, and expressed as a "net present value". Comparison of these net present values allows the scheme providing the largest net economic benefit (or smallest net cost) to be identified (see Ch. 5).

By the mid-1970s, such methods had become well established within the DoT. However, the development of methods for assessing the various "non-economic" (physical environmental and social) impacts of trunk road proposals had not proceeded at the same pace. Reflecting these concerns, an internal DoT report in 1976 (the Jefferson Report) recommended the inclusion of physical environmental and social factors in road appraisal, although not in CBA. In the same year, the DoT appointed the Standing Advisory Committee on Trunk Road Assessment (SACTRA) to comment on its appraisal methods and traffic-forecasting techniques. The committee concluded that there was an imbalance in the current methods of assessment and that the various non-economic effects of road schemes should be evaluated and brought formally into the assessment process. However, it was thought that most of these effects could not satisfactorily be translated into monetary units and could not therefore be incorporated into an expanded form of CBA. Instead, the committee recommended that all predicted effects, whether quantified or not, should be presented together in a special tabular format or "framework". This would mean that all types of effects – economic and non-economic – would be given equal prominence in the appraisal and decision-making process. This tabular framework was developed in more detail in the subsequent SACTRA Report of 1979.

In response to these developments, the DoT issued guidance on the EIA of trunk road schemes in its *Manual of environmental appraisal* (MEA), first published in 1983. This provided advice on two main issues:

1. *How information on environmental and other effects should be presented.* Guidance on how to draw up the tabular framework recommended by the SACTRA Report was provided. The framework was designed to summarize the effects of alternative scheme options on six specific groups:

   Group 1 – effects on travellers
   Group 2 – effects on occupiers of property
   Group 3 – effects on users of facilities
   Group 4 – effects on policies for conserving and enhancing the area
   Group 5 – effects on policies for development and transport
   Group 6 – financial effects.

   The effects of each scheme option were to be summarized in a separate column within the framework, so that the different schemes could be easily compared with each other. The framework was to include all effects on the six groups, including those from the CBA. This framework has since become the standard means used by the DoT to present information on the effects of road proposals, both at public consultation and at public inquiries.

2. *Appropriate techniques to be used in assessing the environmental effects of scheme proposals.* The MEA recommended that 11 specific types of effects should be examined, although not all of these would be relevant in all cases:

   1. traffic noise
   2. visual impact
   3. air pollution
   4. community severance
   5. effects on agriculture
   6. effects on heritage and conservation areas
   7. ecological impact
   8. disruption due to construction
   9. effects on pedestrians and cyclists
   10. view from the road
   11. driver stress.

   The environmental effects identified as a result of this assessment would be summarized and presented in the tabular framework, by allocating them to the appropriate groups in the framework. For example, severance might affect vehicle users or cyclists (group 1), residents (group 2) and bridle-way or footpath users (group 3). The framework presents as many impacts as possible in terms of monetary costs and benefits, such as the savings in time and vehicle-operating costs for travellers. As many as possible of the remaining impacts are presented in quantitative but non-monetary form, for instance in terms of the number of houses

Table 10.1 Part of a UK Department of Transport trunk road proposal appraisal framework.

GROUP 2: OCCUPIERS. Sub-Group: Residential

Effect	Units	Proposed scheme	Do nothing	Comments
Properties demolished	Number of properties	1	0	Cost of acquisition & demolition is included in Group 6 (financial effects)
Noise increase	Number of houses experiencing an increase of:			Noise changes take into account proposed mitigation measures.
	More than 16 dB(A) L10	9	0	The changes are the difference between the forecast for 2012
	11–15 dB(A)	22	0	and the existing levels in 1997 prior to the opening of the road.
	6–10 dB(A)	8	0	The units are dB(A) L10 for 18 hours, 6am to midnight.
Noise decrease	Number of houses experiencing a decrease of:			
	More than 16 dB(A) L10	59	0	Properties are along the existing A556 trunk road. The
	11–15 dB(A)	0	0	changes are as described above for noise increases.
	6–10 dB(A)	7	0	
	3–5 dB(A)	7	0	
Visual obstruction	Number of properties within 300 m of centreline subject to:			
	High	6	0	
	Moderate	3	0	
	Slight	7	0	
Visual intrusion	High	8 (15)	0	Numbers take account of proposed landscaping measures.
	Medium	18 (30)	0	Figures in brackets are without landscaping
	Low	31 (12)	0	
Reduction of existing severance		Substantial relief to properties in Mere and Bucklow Hill fronting A556	None	

Source: DOT/Allot & Lomax (1992).

affected by specified noise increases or levels of visual intrusion. Remaining issues that cannot be quantified, such as the effects on policies for conserving an area, are simply described in the table. In each case, the impacts of a proposed route are presented assuming both high and low growth in traffic. Table 10.1 shows a section of a road proposal appraisal framework.

The implementation of EC Directive 85/337

The existing methods and procedures of environmental appraisal have heavily influenced the way in which EC Directive 85/337 has been interpreted and implemented by the DoT. For trunk road schemes, the Directive was implemented by amendments to the Highways Act 1980 (Section 105). The DoT subsequently issued guidance, in the form of a Departmental Standard, on how the Directive and the provisions of the amended Act were to be followed in practice (DoT 1989). The guidance addresses three issues:

1. *Which trunk road schemes will require a formal EIS?* The construction of new motorways and of "Special Roads" under the Highways Act 1980 are Schedule 1 projects and therefore always require formal EIA. All other trunk road schemes are Schedule 2 projects, for which EIA is discretionary. Chapter 3 discusses the precise circumstances in which EIA is required for such projects. In practice, a not-inconsiderable number of trunk road proposals appear to have escaped the need for formal EIA under the EC Directive. A Parliamentary Written Answer in December 1990 revealed that, at that time, 49 trunk road schemes for which line orders had been confirmed since July 1988 had not been subject to the requirements of the EC Directive (CPRE 1991b). These schemes included some major bypasses.
2. *When should an EIS be published?* An EIS should be submitted with the Draft Line Orders for the Secretary of State's preferred scheme. It is then considered at the Public Inquiry, along with other supporting material.
3. *What is the appropriate format and content of an EIS?* The Departmental Standard advises that the EIS should always include:
   - A description of the published scheme and its site. This should contain a detailed description of the line of the published scheme and should indicate the proposed standard of the road, the heights of embankments and depths of cuttings. Roads that are expected to experience substantial changes in traffic flow as a result of the scheme should be identified. A description of the general area through which the route runs is to be provided, including any areas protected by statute or by the policies of a national or local authority.
   - An outline description of the measures proposed to mitigate adverse environmental effects. It is suggested that these may include route diversions, cuttings or tunnels, noise barriers or bunds and landscaping measures.
   - Sufficient data to identify and assess the main effects that the scheme

is likely to have on the environment. The tabular framework derived from the MEA should be used to present the information on environmental effects in the EIS. However, some effects not included in the MEA are listed in Annex III (Clause 3) of the EC Directive, namely the effects on soil, water and climate. The DoT states that these effects should be included in the EIS, if they are likely to be significant. However, it advises that individual schemes will not have a significant effect on climate and, in most cases, are unlikely to have significant effects on soil or water. The implication is therefore that EISs will rarely need to address such matters (CPRE 1991a). Conversely, those effects included in the MEA, but not specifically listed in Annex III of the Directive, are apparently to be included in the EIS. These include traffic noise, severance, effects on agriculture, disruption caused by construction, effects on pedestrians and cyclists, view from the road, and driver stress.

Annex III (Clause 4) of the Directive lists other issues that should be addressed in the EIS: impacts due to the existence of the project (e.g. consequential development pressures), impacts caused by the use of natural resources (e.g. aggregates for road construction), impacts from the emission of pollutants, the creation of nuisances and the elimination of waste (e.g. the long-term contribution of pollutants to greenhouse emissions), and a description of the forecasting methods used to assess the effects on the environment. However, the Departmental Standard states that none of these issues needs to be addressed in EISs for trunk road proposals (CPRE 1991a, DoT 1989).

- A non-technical summary. This should include a plan of the local road network showing the line of the published scheme and of the main alternatives considered. In cases where the alternative options presented at public consultation had significantly different environmental effects from those of the published scheme, the EIS should include two further elements (the following two points).
- A summary description of the main alternatives presented at public consultation. Their routes should be shown on a plan, identifying any adjoining areas of particular environmental sensitivity. *However, the EIS does not need to include details of the environmental effects of these alternatives.*
- The reasons for the choice of the published scheme. These may include economic, operational and environmental considerations.

The SACTRA Report: assessing the environmental impact of road schemes

In September 1989, the Government asked SACTRA to (a) review the methods currently used for assessing the environmental impacts of trunk road schemes; and (b) re-examine the interrelationship between "economic" and "environmental" effects in the decision-making process. SACTRA's report, and the Government's response, were published in March 1992 (SACTRA 1992, DoT 1992). SACTRA's work raised specific issues about the way in which the EC Directive has been implemented in practice, as well as more general concerns about the treatment of environmental matters in trunk road appraisal and decision-making. Some of these issues had already been publicized by the Council for the Protection of Rural England in submissions to the EC before the publication of the SACTRA Report (CPRE 1991a,b). The key issues raised by SACTRA are outlined below.

(a) *Is formal EIA being carried out and reported on at the appropriate stage in the decision-making process? What are the implications of this for the treatment of alternatives in the EIA process?* The lengthy planning and decision-making process for new trunk road proposals has been outlined earlier in the chapter. The statutory EIS is submitted towards the end of this process, with the publication of Draft Line Orders for the Secretary of State's preferred route or scheme. Formal EISs are not required at the earlier public consultation stage, for any of the alternative options under consideration. SACTRA compared this procedure with that for projects covered by the town and country planning legislation. In such cases, the EIS must be submitted at the time of the planning application. SACTRA felt that there was a need for the formal presentation of the results of EIA at a much earlier stage than is currently the case. Accordingly, it recommended that the following reports should be submitted for all schemes:

- An environmental assessment report, which would be submitted on the entry of a scheme into the Roads Programme. Although this would not be made publicly available, its findings would be summarized in the later reports.
- A more detailed environmental assessment report, which would be submitted at the public consultation stage on the alternative scheme options. This report would be made available for public scrutiny during the public consultation period. Detailed EIA would therefore be carried out and the results formally reported, for all alternatives considered, before the choice of the preferred scheme.
- The statutory EIS required by the EC Directive, which would continue to be submitted on publication of Draft Line Orders.

These reports would deal successively with national and wider environmental issues, corridor and regional effects, and detailed, local environmental effects arising out of the alternative scheme options.

(b) *The format of EISs – is information on environmental impacts being presented in the appropriate way in EISs?* The DoT has indicated that the appropriate way to present information on environmental effects in the EIS is to use the framework derived from the MEA. All that is required to produce the EIS is therefore to add text describing the published scheme and its site, the proposed mitigation measures and a non-technical summary. As part of their work, SACTRA examined EISs prepared for trunk road proposals since the implementation of the EC Directive. In most cases, the tabular format of the framework (using the six groups specified in the MEA) had been used as the centrepiece of the EIS.

No modifications were made to the framework for the purposes of the EIS, so that the financial and economic data were included alongside the summary of the environmental effects. In most cases, no accompanying text was included in the EIS outlining the environmental effects in more detail, or discussing effects not included in the framework. SACTRA thought that the practice of using the framework as the means of presenting information on environmental effects in the EIS should be abandoned, and that the EIA, however it was presented, should not be tied to the six groups identified in the MEA. SACTRA recommended an expanded format for future EISs, which should include the following sections:

- Description of the site, including its main environmental features.
- Description of the scheme and its environmental effects, including a statement of the scheme's main objectives, a full description of the scheme, a statement of its effects and how these relate to the environmental features of the site.
- Mitigation of environmental effects, including descriptions of measures to mitigate, avoid, offset or repair environmental impacts. This should include measures to replace assets lost as a result of the scheme.
- Data and methodology, including a full presentation of the environmental data. This should not be in summary form, since this is the function of the non-technical summary.
- Non-technical summary.

(c) *Is mitigation being addressed satisfactorily in EIA and EISs, in accordance with the requirements of the EC Directive?* Annex III of the EC Directive requires EISs to include "a description of the measures envisaged to prevent, reduce and where possible offset any significant adverse effects on the environment". SACTRA argued that the DoT appeared to have executed this requirement somewhat narrowly, in advising that EISs should include a description of those "measures intended primarily to reduce an adverse effect on the environment" (DoT 1989). SACTRA clearly had in mind the possibility of relocating or recreating environmental assets suggested by the term "offset". It recommended a rephrasing of the relevant guidance, to include explicitly the possibility of preventing or offsetting significant effects, as well as simply reducing them.

In cases where road proposals threaten assets of environmental quality (such as SSSIs), SACTRA recommended that a formal series of reviews should be undertaken as part of the EIA process. These reviews would address the questions:

- Can the damage be prevented altogether?
- If not, can it be reduced?
- If neither, can it be offset? (i.e. can something else be done which creates an environmental benefit to set against the damage?)

It was recommended that the findings of these reviews should be reported in the various environmental assessment reports and the EIS.

(d) *Should assessment be led by explicit environmental objectives?* SACTRA was concerned that the EIA system should not exist in a policy vacuum, but should derive from explicit environmental policy objectives. EIA should examine the extent to which these objectives are likely to be met by alternative schemes, including the do-nothing or do-minimum options. It was recommended that the performance of schemes should be evaluated against both national environmental objectives and constraints (e.g. a commitment to reduce CO_2 emissions or to protect SSSIS); and regional and local environmental objectives (e.g. those of local planning authorities). To be of any use, such an evaluation would need to take place during the early stages of the decision-making process, before the admission of schemes into the Roads Programme. SACTRA argued that "the results of this early appraisal should be available at the public consultation and public inquiry stages, so that the public may be satisfied that these matters have been fully taken into account, and to allow the results to be subjected to public scrutiny" (SACTRA 1992).

(e) *The comprehensiveness of current appraisal methods – are all environmental impacts being addressed, both at the local and wider strategic levels?* SACTRA questioned the comprehensiveness of current appraisal methods at two levels:

- The assessment of local environmental impacts. It was felt that the six groups in the framework placed too much emphasis on effects that directly affected people (i.e. travellers, occupiers, users) rather than the environment for its own sake. Effects on the natural environment are described in the framework only to the extent that they affect people (e.g. Group 3, users of the countryside) or policies for conserving and enhancing the area (Group 4). SACTRA recommended that the list of impacts to be considered should be lengthened, to include the effects of land take, loss of open space, water pollution, vibration and public health.
- The assessment of wider environmental and strategic impacts. SACTRA was concerned about the way in which major road improvements or "routes" are divided into a series of separate "schemes". In such cases, EIA is carried out separately for each component scheme, but not for the overall route and its regional or cumulative impact. It was felt that EIA on a scheme-by-scheme basis could not adequately deal with strategic concerns or wider environmental issues (such as air pollution). To capture such impacts, SACTRA therefore concluded that an earlier, strategic level of assessment was needed. (The issue of strategic environmental assessment is discussed in more detail in Chapter 13.)

(f) *Should environmental effects be quantified and expressed in monetary terms?* SACTRA argued that cautious steps should be taken towards expressing some environmental effects in monetary terms. Further research into this area was recommended.

(g) *Is there a need for revised guidance on environmental appraisal?* The recommendations outlined above clearly have implications for the existing MEA. SACTRA therefore recommended the publication of a revised version, incorporating a number of additional features:

- additions to the checklist of impacts to be included in EIA
- revised guidance on how to measure and quantify these impacts, and judge their significance
- guidance on the treatment of mitigation measures
- guidance on how to assess alternative schemes against national, regional and local environmental policies
- guidance on the preparation and presentation of the EIS and the recommended "environmental assessment reports".

The government's response to the SACTRA Report

The DoT expressed general support for the recommendations in the SACTRA Report and promised to implement them. However, SACTRA's recommendations are unlikely to result in a reappraisal of existing schemes currently at various stages in the decision-making process. The DoT has indicated that "decisions already made . . . in the development of an individual scheme will not be reviewed, although as [schemes] progress towards later decision-making, [they] will be subjected to updated assessment techniques, depending upon the time at which those new techniques are formally adopted" (Wood 1992). Implementation of one of SACTRA's recommendations – the need to update the MEA – was already in hand before its report was submitted. The revised MEA is due to be published during 1993 and the DoT has indicated that it will take account of SACTRA's recommendations, with a wider range of impacts brought into the assessment process (DoT 1992).

The only major area of disagreement between the DoT and SACTRA concerned the continued use of the framework in EISs. The DoT intends to continue to use the framework as the means of presenting information on environmental impacts in the EIS, although the need for some modifications to it is accepted (Wood 1992, DoT 1992). However, an examination of EISs submitted since the publication of the SACTRA Report indicates that the importance of the framework as a means of presenting information appears to have declined considerably compared with earlier EISs. This development is discussed in more detail in the next section.

10.3 A case study of EIA: the proposed M6–M56 motorway link road

Introduction

This section presents a case study of EIA for a new motorway proposal in the northwest of England. The EIS for this scheme was submitted in October 1992, some six months after the publication of the SACTRA Report. The detailed environmental studies upon which the EIS is based were largely carried out during 1991 and 1992, during and immediately after SACTRA's deliberations. It is therefore of interest to examine how this recent example of EIA deals with some of the issues raised by SACTRA. Of particular interest are the scope of the EIA, the way in which information is presented in the EIS, and the treatment of mitigation, alternatives and indirect or consequential impacts.

The proposal

The proposed scheme involves the construction of a new motorway link between junction 19 of the M6 and junction 7 of the M56, between Knutsford and Altrincham in Cheshire (see Fig. 10.1). The existing link between the M6 and M56 is provided by the A556 trunk road, which is mainly of four-lane single carriageway construction. This route has a high volume of traffic (up to 20% of which consists of heavy goods vehicles), peak-hour congestion and a poor accident record. The scheme would be a three-lane dual-carriageway construction for most of its 6.5 mile length. The existing A556 would lose its trunk road status.

The land surrounding the existing A556 is rural and predominantly agricultural. There are three Areas of Special Landscape Value in the vicinity, at Tatton Park to the east, the Bollin Valley to the north and Tabley Park to the southwest. Ecologically designated sites include Rostherne Mere near the existing junction with the M56, which is a Wetland of International Importance, an SSSI and a National Nature Reserve. There are also six Sites of Biological Importance, as well as many patches of undesignated woodland, unimproved grassland, lane-side hedges and ponds of local importance; wildlife corridors connect these features.

The proposed scheme was presented for public consultation in November 1989 and confirmed as the Secretary of State's preferred route in December 1990. Draft Line and Side Road Orders were submitted in October 1992 and a public inquiry was held during 1993. Substantial modifications to the proposals have been made at various stages throughout this process.

Types of impacts addressed and format of the EIS

The range of impacts addressed in the EIS appears to have been based largely on those listed in the MEA. However, the effects on water and drainage – although not included in the MEA – are examined and a comprehensive series of mitigation measures proposed. The broad structure of the EIS is in accordance with the advice in the Departmental Standard, but the presentation of information is substantially different from that found in earlier EISs and criticized by SACTRA. Most notably, the tabular framework is not used as the main means of presenting the information on environmental effects. The framework is relegated to the status of an appendix, with the environmental impacts of the scheme being discussed in the main body of the text of the EIS. There is no attempt to structure this discussion around the six groups in the framework. A full, rather than summarized, presentation of most of the environmental effects is provided, as recommended by SACTRA. More detailed assessments prepared by various consultants are incorporated into three

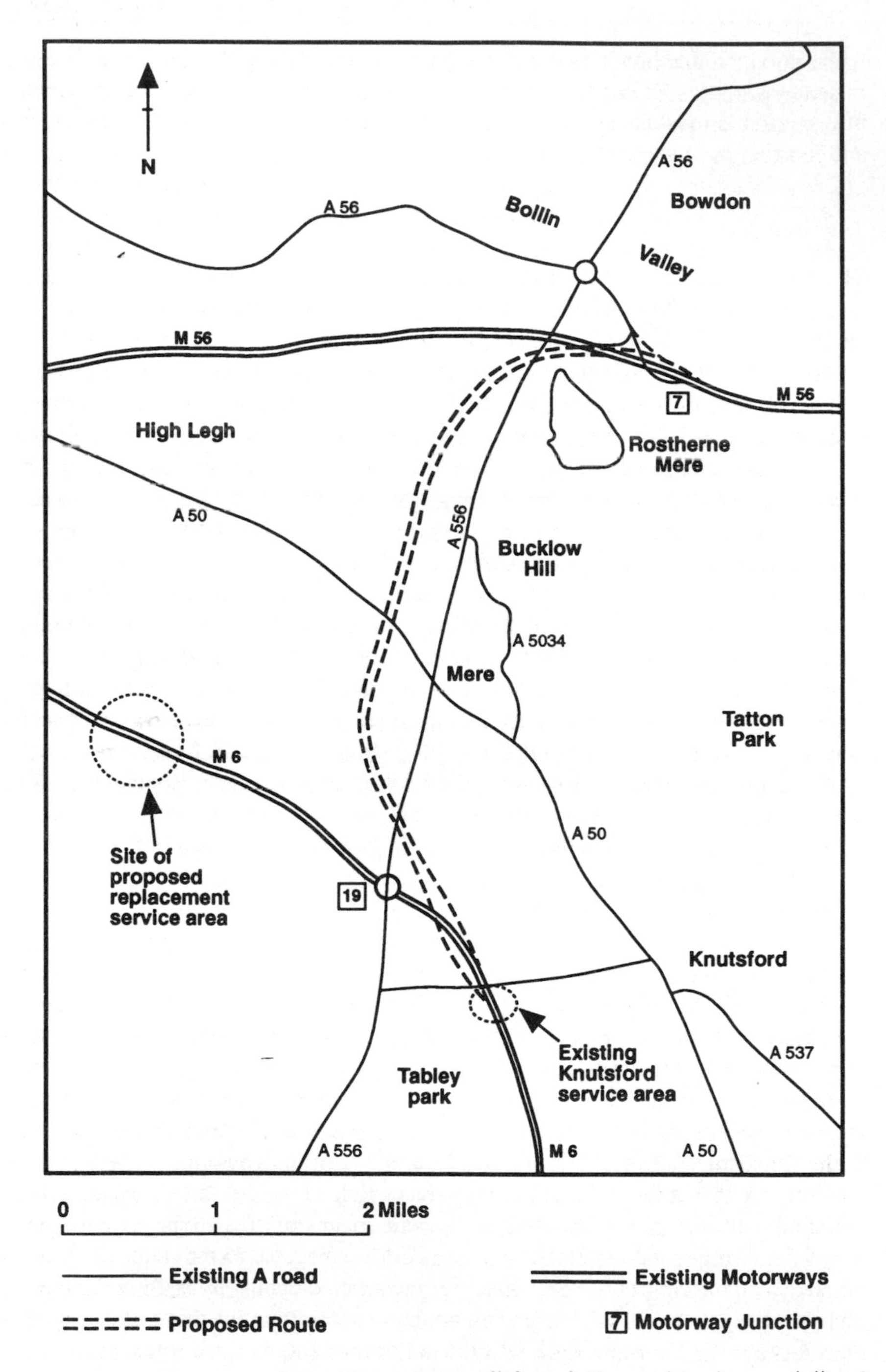

Figure 10.1 The proposed M6–M56 motorway link road. (*Source:* based on DoT/Allott & Lomax 1992)

additional, separately bound volumes, which form part of the EIS. These include an agricultural assessment, ecological survey and assessment, archaeological assessment, air quality report, and road traffic and construction noise report. However, detailed technical reports in support of the conclusions in the EIS are not provided for the effects on water and drainage, landscape, severance or effects attributable to construction. At the Public Inquiry into the scheme, the DoT presented a detailed technical report covering the effects on water and drainage.

The treatment of mitigation

The SACTRA Report emphasized the need explicitly to include the possibility of preventing or offsetting adverse environmental impacts, as well as simply reducing them. The mitigation measures proposed in the EIS are mainly designed to prevent or reduce adverse impacts. However, some of the measures suggested could be regarded as means of offsetting adverse effects. These mainly concern the ecological impacts of the proposal. It is conceded that sites for habitat creation will largely be limited to the motorway verges and the areas enclosed by the motorway access roads. The creation of scrub/woodland, species-rich grassland, and shallow pools or ponds with extensive, dense cover alongside (suitable for amphibians), is recommended. Existing ponds may also be improved. It is argued that "the creation of [such habitats] will enhance the wildlife-carrying capacity of the area, and may encourage birds such as the barn owl and hobby to breed". The EIS contains suggestions of possible mitigation measures, rather than a detailed outline of specific proposals. More specific measures will be devised during the detailed design stage.

The treatment of alternatives

Six alternative schemes were considered by the DoT before the public consultation stage. These included a do-minimum option (involving an on-line improvement of the existing A556) and five alternative off-line routes, including the preferred scheme. Compared with the preferred scheme, these alternative routes involved a more westerly line and/or a more westerly location for the new interchange with the M56. Those alternatives not considered included the do-nothing option and alternative forms of transport, such as public transport or park-and-ride. The EIS states that, as a major strategic route for motorway traffic between the Midlands and Manchester, "only a new road was judged to be appropriate or effective in coping with the forecast growth in demand for traffic movement".

Five of the six alternatives were rejected by the DoT prior to public consultation. Only one option was therefore presented at the public consultation stage in November 1989, and this was subsequently confirmed as the Secretary of State's preferred route in December 1990. Modifications to the preferred route were made prior to the submission of Draft Line Orders and the EIS in October 1992. The EIS was therefore submitted almost two years after the announcement of the preferred route and three years after the public consultation stage. Clearly, the key decisions about the line of the route had been taken within the DOT long before the EIS appeared, and indeed before the public consultation stage. It is therefore not surprising that the treatment of alternatives in the EIS is far from satisfactory.

The guidance in Departmental Standard HD 18/88 suggests that, since no options other than the preferred route were presented at public consultation, the EIS does not need to address the issue of alternatives at all. Notwithstanding this, a brief description of all six original options and the reasons for the choice of the preferred route are included. However, the comparison of the environmental impacts of the various options occupies only one page of the EIS, with most of the discussion centring on the relative economic, traffic and safety implications of the different schemes.

The EIS states that "an assessment of the five Do Something options revealed that there was little to choose between them in environmental terms", with the exception of a more serious impact arising out of the more westerly M56 interchange locations, which would be sited in a conspicuous location. This conclusion may or may not be true, but the EIS does not contain the detailed information on the effects of each option to support such a statement. Indeed, the more serious impact of the westerly M56 interchange locations was questioned by English Nature in its comments on a preliminary version of the EIS:

> The choice of the most easterly option [for the interchange] has been based on traffic, operational, safety and economic grounds. There appears to have been *no* consideration of the considerably greater impact of the chosen location on Rostherne Mere [a National Nature Reserve and SSSI].

English Nature argued that this impact (visual intrusion, increased air pollution and noise levels) had not been adequately addressed in the draft EIS. The final version of the EIS, although including English Nature's comments in an appendix, made no response to its criticisms.

Subsequent to public consultation and the announcement of the preferred route, significant alterations were made to the proposed route which, the DOT argued, reinforced its selection. However, the EIS does not identify these alterations, nor does it justify them, either in environmental or any other terms. Further modifications to the proposals were made after the submission of the EIS. The Department's interpretation of the EC Directive, as contained in Section 105 of the amended Highways Act 1980 and the Departmental Standard, clearly limits the assessment in the EIS of alternatives already rejected earlier in the planning and decision-making process.

The consideration of certain indirect and consequential impacts

The EIS states that the proposed scheme would have a substantial effect on the existing M6 motorway service area at Knutsford. The existing north-facing slip roads onto the M6 would be stopped up, so that the service area would no longer serve M6 through-traffic. The service area would remain, but would only serve traffic using the new route. The DOT indicated in the EIS that the capacity of the existing service area is inadequate to cope with predicted increases in M6 traffic. It may therefore need to be extended or replaced at some point in the future, even if the proposed scheme does not go ahead. Nevertheless, the proposals turned this future possibility into a more immediate probability. The need for a replacement M6 service area in the vicinity, its possible site and its environmental effects, are not addressed in the EIS. However, at an earlier stage in the scheme proposals, a site had been identified by the DOT for a replacement service area east of Arley Hall, some two miles west of Junction 19, and a notice of proposed development had been submitted. The DOT requested that this site be included in the ecological survey and impact assessment commissioned in 1991, but it subsequently abandoned this element of the proposals before the EIS was submitted. The EIS does not therefore identify or discuss this or any other service area site.

Following the publication of Draft Line Orders and the submission of the EIS in October 1992, the scheme proposals were subjected to further modifications. These revised proposals involved the existing Knutsford service area remaining open to M6 through-traffic. However, given the predicted growth in M6 traffic, the need for extension or replacement at some point in the future was still implied. Meanwhile, in March 1993, a private-sector company submitted an outline planning application to Macclesfield Borough Council for the extension of the existing service area. This will be considered as a separate planning application under the town and country planning legislation. The borough council reluctantly decided not to request an EIS with the application, since advice from the DOE indicated that the project did not fall within any of the relevant indicative criteria or thresholds for Schedule 2 projects. The application to extend the existing service area cannot be seen as a direct consequence of the proposed motorway scheme. However, the two proposals are clearly linked, in that the slip roads from the M6 to the new scheme would form a boundary to future expansion of the service area (beyond that proposed). This is important, given that any expansion of the existing service area would involve encroachment into the green belt. Whether motorway scheme EISs should or could discuss the need to provide or replace service areas – and the environmental impacts of such provision – is open to debate. What is clear is that the DOT's interpretation of Directive 85/337 does not require the treatment of this or any other type of consequential development in the EIS.

Availability of the EIS

The Department of the Environment has indicated that "the developer should make a reasonable number of copies of the [environmental impact] statement available for sale to the public. A reasonable charge reflecting printing and distribution costs may be made" (DOE 1989). The EIS for the M6–M56 link proposal is priced by the DOT at £120 for Volume I (the main EIS) and at £277 for all four volumes (including the detailed technical reports and assessments). Such prices appear far from "reasonable"!

Conclusions

The case study indicates that some of SACTRA's recommendations appear to have been incorporated into the DOT's most recent EISs. The most notable changes are the consideration of impacts not included in the MEA (e.g. water and drainage), the abandonment of the framework as the means of presenting information on environmental effects, and the inclusion of detailed technical reports and assessments in the EIS. The quality of the DOT's EISs has undoubtedly improved substantially compared with the earlier examples criticized by SACTRA. However, there are continuing concerns about the Department's interpretation of Directive 85/337 with respect to trunk road proposals, as contained in Section 105 of the amended Highways Act 1980 and in Departmental Standard HD 18/88. The arrival of the EIS at a time when many of the key decisions about the scheme have already been made and the limited treatment of alternatives, indirect and consequential impacts, are well illustrated by the case study.

10.4 EIA and the electricity supply industry

Introduction

The consideration of the environmental impact of project proposals in the UK electricity supply industry pre-dated the introduction of EC Directive 85/337 by several years. Evaluation of certain environmental impacts, especially those on air quality, can be traced back to the early years of the Central Electricity Generating Board (CEGB) in the late 1950s and early 1960s. At this time, a series of large 2000 MW coal-fired power stations were being planned. Baseline studies of SO_2 concentrations and dust-fall in the vicinity of the proposed sites were carried out prior to operation, and monitoring was continued throughout the operational life of these new stations. Such information was used to test stack-plume rise and pollutant dispersion calculations, and it assisted in the development of present-day dispersion models (Manning 1991). Since its formation in 1957, the CEGB had also been obliged under the Electricity Act 1957 to take account of the effects of its projects on "local amenity" – namely the natural beauty of the countryside; flora, fauna and geological or physiographical features of special interest; and buildings and other objects of architectural or historic interest.

Over the years, a considerable body of expertise was built up within the CEGB on the environmental implications of its projects. Environmental research was undertaken within the Board into a variety of issues, such as air pollutants, effects on the ecology of rivers and estuaries, and the restoration of ash dumps to agricultural use (CEGB 1979, Howells & Gammon 1980, Sheail 1991). The CEGB also commissioned research from outside bodies, an example being studies of the socio-economic impacts of its proposals carried out by Oxford Polytechnic during the 1980s (Glasson 1984, Glasson et al. 1987).

By the early 1980s, after initial misgivings, the CEGB had become a firm supporter of the draft EC Directive on EIA. Sheail suggests that a pivotal event in helping to shift attitudes may have been the Sizewell B public inquiry, the length of which made "a more formal environmental assessment procedure [begin] to appear more attractive" (Sheail 1991). Indeed, the Board was one of several major utilities, including the coal, oil and gas industries, to submit EISs with its proposals before the implementation of the Directive. Statements were prepared in 1987 and early 1988 for a PWR power station at Hinkley Point, three coal-fired stations at Fawley, Kingsnorth and West Burton, and a flue gas desulphurization plant at the existing Drax station. By the time the Directive came into force, substantial experience had been gained within the industry in producing EISs for a range of different projects. An excellent historical review of the treatment of environmental matters within the CEGB is provided by Sheail (1991).

Projects subject to EIA under EC Directive 85/337

For projects within the electricity supply industry (ESI), EC Directive 85/337 was implemented by means of the Electricity and Pipeline Works (Assessment of Environmental Effects) Regulations 1989. Revised regulations were issued in early 1990 as a result of the privatization of the ESI under the Electricity Act 1989. In addition, certain projects in the ESI are covered by the Town and Country Planning (Assessment of Environmental Effects) Regulations. The need for EIA for different types of projects is summarized below.

Projects subject to the Secretary of State's consent[1] (and deemed planning consent): Electricity and Pipeline Works (AEE) Regulations:

- the construction or extension of a nuclear power station (Schedule 1)
- the construction or extension of a non-nuclear generating station with a heat output of at least 50 MW (Schedule 1 if 300 MW or more; Schedule 2 otherwise)
- the construction or diversion of an oil or gas pipeline of at least 10 miles in length (Schedule 2)
- the placement of an overhead transmission line (other than a service line) of at least 10 miles in length (Schedule 2).

For Schedule 2 projects, EIA is required "where the Secretary of State takes the view that the project would be likely to have significant environmental effects".

Projects not subject to the Secretary of State's consent: Town and Country Planning (AEE) Regulations:

- the construction or extension of a non-nuclear thermal power station, an installation for the production of electricity, steam and hot water (e.g. combined heat and power) or an installation for hydroelectric energy production, with a heat output of less than 50 MW (Schedule 2)

- the construction or diversion of an oil or gas pipeline of less than 10 miles in length (Schedule 2)
- the placement of an overhead transmission line of less than 10 miles in length (Schedule 2).

There is some ambiguity in the regulations over the need for EIA for wind power proposals. Such projects will almost always have a heat output below the 50 MW threshold and therefore do not require the Secretary of State's consent. They are therefore covered by the Town and Country Planning (AEE) regulations, but are not specifically identified in either Schedule 1 or 2. In practice, EISs for wind-turbine developments have been requested by local planning authorities and the DoE under Schedule 2–3(a), i.e. "an installation for the production of electricity, steam and hot water". However, the CPRE has correctly argued that this category would appear to refer to combined heat and power schemes rather than wind turbines, which do not produce steam or hot water. The CPRE is concerned that, given the lack of any explicit inclusion of wind-power developments in Schedule 2, a developer challenging the need for EIA may well be successful (CPRE 1991b). The DoE's consultation paper on the operation of the existing EIA system (see §12.3) proposes to extend Schedule 2 to cover wind farms.

Overhead transmission lines under 10 miles in length do not require the Secretary of State's consent and therefore again fall under the Town and Country Planning (AEE) regulations. However, such projects are defined as permitted development under the General Development Order 1988 and do not therefore require planning permission. Consequently, there is "no effective consent procedure to which an EIA requirement can be tied . . . EIAs are therefore not generally required for [such developments]" (CPRE 1991b).

Current issues in EIA in the electricity supply industry

Divided consent and EIA procedures for individual project components

Most of the new power station proposals submitted since the implementation of the EC Directive on EIA have been gas-fired developments. Although existing power station sites have been selected in a number of cases, these projects have usually necessitated the construction of new gas pipelines to the sites, as well as additional transmission connections off site. Each of these project components is often carried out by different companies and is the subject of separate consent and EIA procedures. A good example is provided by the proposal to construct a 1725 MW gas-fired combined heat and power (CHP) station at the existing ICI chemical complex at Wilton, Teesside, whose EIS states:

> The overall project involves five major components:
>
> 1. A new natural gas pipeline from the North Sea . . . to a landfall in the Teesside area to be built and run by others.

> 2. A gas reception and processing facility in the Teesside area to be built and run by others.
> 3. [The] combined heat and power (CHP) plant . . . [at ICI Wilton].
> 4. A [gas] pipeline from the processing facility to the CHP facility to be built and run by others.
> 5. National Grid system upgrades by the National Grid Company will be necessary [ie. transmission lines and substation improvements].
>
> Each component will require a separate environmental assessment, with individual planning applications being submitted to the relevant authority as appropriate. (Cremer & Warner 1990)

The application for the CHP generating station was approved by the Secretary of State for Energy in November 1990. It later transpired that the project would necessitate the construction of extensive 400 kV overhead lines through open countryside (CPRE 1991b). This led the CPRE to lodge a formal complaint with the European Commission about the Secretary of State's decision. The CPRE argued that consent had been granted "without the implications of overhead transmission lines being properly addressed as part of the project EIA . . . The need for transmission lines from a power station is entirely dependent on the existence of the project and therefore should be considered as part of the EIA". The implication was that the Secretary of State had granted consent "without having [all] the necessary information available on which to base his decision" (CPRE 1991b).

This example is by no means unique and clearly highlights a problem with the current procedures. The separate consent procedures for different components of the same "project" divide the responsibility for EIA among separate developers. EIA may well be carried out and reported on at different times for each component, depending on the timing of each consent application. The result is that all the environmental effects of the "project" are not assessed and presented together, for consideration by the Secretary of State.

The consideration of adjacent concurrent developments

A related issue concerns the extent to which EIA of projects in the ESI takes account of adjacent development proposals. A review by Therivel et al. (1992) suggests that such cumulative impacts are rarely addressed. An example is provided by the proposals for two gas-fired power stations at Killingholme in Humberside. The CEGB's successor companies, National Power and PowerGen, submitted applications within a few months of each other for almost identical developments on adjacent sites. However, neither EIS examined the interaction between the two proposals or their likely cumulative impacts. Other examples are provided by applications for wind-farms on adjacent sites, often involving the same developer. The consideration of cumulative impacts in such cases is usually inadequate, if not non-existent.

Manning (1991) suggests that such examples raise a number of issues. Although the combined impacts of the different proposals "might be acceptable even if all the schemes went ahead, . . . this would have to be demonstrated by an integrated assessment. Otherwise, a form of "first come, first served" approach would arise by default" (Manning 1991). But who should carry out such an integrated assessment? If developers are required to take account of other related projects in the locality, problems can be anticipated. Access to the necessary information about competing companies' proposals may be problematic, and developers will surely wish to present the impacts of their own projects in a light more favourable than those of their competitors. An alternative would be for the local authority concerned to carry out or commission an integrated assessment of the various proposals itself. Given that projects emerge at different times and may be subject to modification, when should such an assessment be carried out? In addition, if the assessment were to draw on information from each developer, "there [could] be scope for chaos if each interested party uses different assessment methods" (Manning 1991). An example would be the use of different pollutant dispersion models by each developer to assess the impact of atmospheric emissions.

It should be pointed out that, for certain projects, integrated assessments have been undertaken. An example concerns two applications for wind-farms on adjacent sites in Powys. The applications were submitted within a month of each other by the same developer. In this case, the local planning authority and the developer agreed that a single EIS should be submitted covering both sites.[2]

The treatment of alternatives

Ideally, EIA of individual projects in the ESI should include an assessment of alternative sites, fuels and technologies, or refer the reader to a higher tier of EIA where such an assessment has taken place. The final choice between these various alternatives should result from this EIA process. This does not mean that the least environmentally damaging options will necessarily be selected. However, it does mean that environmental considerations will be weighed with the relevant technical, commercial and other factors in arriving at the final decision. In practice, the consideration of alternatives in EISs for projects in the ESI tends to be rather limited. Therivel et al. (1992) report the results of a review of almost 60 energy sector EISs submitted since the implementation of the EC Directive. The authors found that, for power station projects, "consideration of alternatives was patchy, with better consideration of different locations than different processes". EISs for gas-fired power stations were particularly weak in this respect. For wind-power developments, "specific alternative sites were scarcely considered after general siting [criteria had been] discussed; . . . alternative processes were not discussed at all". By contrast, EISs for pipelines and transmission lines were more satisfactory, discussing both alternative routes and different methods of achieving the same results.

The need for EIA of plans and programmes in the electricity supply industry

The limited treatment of cumulative impacts and alternatives in EIA at the project level suggests the desirability of an earlier, more strategic, level of assessment.

The need for EIA of plans and programmes in the electricity supply industry has been recognized by a number of bodies in recent years and would appear to have become more pressing in the current privatized regime. For example, the CPRE (1990) has argued for an EIA to be conducted into the national plan to implement the 1988 EC Large Combustion Plants (LCP) Directive.

The LCP Directive requires Member States to draw up programmes for the progressive reduction of emissions of SO_2 and NO_x from large combustion plant. The National Plan drawn up by the UK Government to comply with the Directive requires phased reductions in SO_2 emissions of 20, 40 and 60% by 1993, 1998 and 2003 compared with the 1980 level. Emissions of NO_x are to be reduced by 15 and 30% by 1993 and 1998. As part of the National Plan, Her Majesty's Inspectorate of Pollution (HMIP) has set limits for total annual emissions by National Power and PowerGen. Fór example, National Power must reduce its total SO_2 emissions from 1,600 kilotonnes in 1991 to 660 kilotonnes in 2003. Within this overall framework, the generators have been left to decide how they will achieve these reductions. Possible responses by the ESI to meet HMIP's targets include the following:

- the use of low-sulphur coal in existing coal-fired power stations
- the installation of flue gas desulphurization (FGD) equipment to existing larger coal-fired power stations
- a shift away from coal towards other fuels, especially gas, which involves minimal SO_2 emissions – this would involve the construction of new gas-fired power stations.

Other possible responses might include demand-side measures, such as improved energy efficiency. However, such measures are not the responsibility of the privatized electricity generators. The CPRE (1990) has argued that an EIA of all the alternative ways of reducing emissions, including demand-side measures, should be conducted by the government. Such an assessment should identify the options likely to cause the least environmental damage before decisions on fuel mix and abatement technologies are made by the generators. Project-level EISs cannot perform such a rôle, since they are unable to influence the priority to be given to each option in the overall programme to reduce emissions.

An issue of particular concern to the CPRE (1990) is the priority to be accorded to the installation of FGD plant as a means of reducing emissions and the choice between the alternative FGD methods available. It comments: "CPRE is . . . concerned about the potential impact of retro-fitting FGD equipment to power stations when the wider environmental effects of such a programme have not been adequately assessed beforehand". Two main FGD methods are available: the limestone–gypsum method and the regenerative process. These methods have different environmental implications, and one method may be more suitable at certain power station sites than at others. The limestone–gypsum method: (a) requires large quantities of quarried limestone; (b) produces large quantities of gypsum (some of which will find a commercial market as plasterboard, with the remainder requiring disposal); and (c) involves the need to transport these materials to and from the site. The regenerative method requires different materials and produces different waste products (with a potential commercial use in the chemicals industry). To date, the generators have consistently favoured the limestone–gypsum method (Sheail 1991). CPRE complains that the EISs for FGD plant at Ratcliffe (Nottinghamshire) and Ferrybridge (North Yorkshire) do not even mention the alternative regenerative process. It claims that these project EISs have been prepared without a comparative assessment of the environmental impacts of the two methods, either overall or at the specific power station sites concerned or, if such an assessment has been carried out, it is not reported on in the EIS (CPRE 1990).

The potential for the application of EIA to other policies, plans and programmes in the ESI is discussed in more detail in Therivel et al. (1992). Chapter 13 includes further discussion of strategic environmental assessment.

The types of impacts to be addressed in the EIS

Local versus regional and global impacts Most EISs for power station projects concentrate largely on the local impacts of the development. For example, the consideration of atmospheric emissions will focus on the implications for SO_2 and NO_X concentrations in the vicinity of the site, with much more limited discussion, if any, of the far-afield effects of these emissions. As Manning states, "it is unrealistic to expect something as complex as a global warming analysis to be applied in the context of individual planning applications for combustion sources" (Manning 1991). He goes on to argue for a clearer distinction to be made in the EIA process between local and wider regional and global issues. EIA at the project level should continue to focus on an assessment of local impacts, with issues such as visual intrusion, noise and dust probably deserving more attention than at present. Manning notes that the wider regional and global impacts of projects are increasingly being controlled by national and international regulation, such as the EC LCP Directive. He suggests that EIA at the level of policies, plans and programmes is the appropriate context in which to deal with such concerns. However, in the period before the implementation of such a strategic level of EIA in the ESI, "the extent to which [project] EIAs should address wider issues is a matter for interpretation" (Manning 1991).

Socio-economic versus physical environmental impacts EISs for projects in the electricity-supply industry generally include consideration of socio-economic impacts as well as the more conventional physical environmental impacts. This aspect of current practice is to be welcomed. Formal prediction and monitoring of such impacts dates back to the late 1970s, when the CEGB commissioned research into the local social and economic effects of its construction projects and its existing operational stations. These studies have continued to the present day. Although socio-economic impacts can be negative as well as positive, developers have clearly welcomed the opportunity to include the employment and wider economic benefits of their schemes within the format of the EIS.

Integrated pollution control

All power stations are required by the pollution control inspectorate, HMIP, to limit their emissions to air, water and soil. Under the new arrangements for Integrated Pollution Control in the Environmental Protection Act 1990, generators must obtain a formal authorization from HMIP to operate all new and existing power stations. The information submitted in authorization applications to HMIP is made available to the public and certain statutory consultees, and a period for consultation is allowed. HMIP then determines permitted emission levels and issues an authorization, subject to certain terms and conditions (Harris 1992). New power station developments (above 50 MW) therefore require both the consent of the Secretary of State under Section 36 of the Electricity Act 1989 (which includes deemed planning consent) and an authorization from HMIP to operate.

These developments mean that much of the information required in the EIS is also required by HMIP with the authorization application. Manning (1991) poses several questions about how the two requirements should be integrated: "Should one application [precede] the other? What standards of consistency should be employed? Should the grant of an authorization [by HMIP] be taken as deemed indication that some of the impacts are judged acceptable and if so, is this only in the national context or in the local context too?" (Manning 1991). In particular, the DoE's guidance on the requirements of the best available technique not entailing excessive costs (BATNEEC), and HMIP's judgement concerning the best practicable environmental option (BPEO) – both of which are requirements of integrated pollution control – may affect the alternatives discussed in the EIS.

The treatment of uncertainty

The issue of uncertainty raises problems in EIA. Uncertainty caused by the continual refinement and modification of project proposals during and after the preparation of the EIS is not unique to projects in the ESI. However, several factors suggest the likelihood of particular problems with such projects. First, power station design and layout may be dependent on the choice of the main contractor used to construct the station, a choice that is unlikely to have been made at the time of the EIS submission. For example, National Power's EIS for a gas-fired station at Didcot makes the following statement: "contractors' plant and station designs are known to differ significantly . . . and the final plant configuration and layout will depend on the choice of main contractor following competitive tendering" (National Power 1990). The result is that the EIS identifies three different cooling options for the new station: (a) natural draught cooling towers, 114 m high; (b) mechanical draught cooling towers, 18–22 m high; or (c) air-cooled condensers, about 30 m high. A fourth option was presented after the submission of the EIS. Clearly, each of these options would have very different visual impacts (including the visibility of plumes as well as the towers).

A second reason for uncertainty is that certain design details will be subject to approval or modification by HMIP. Again, the Didcot EIS states that: "exhaust gases from [each] boiler would be discharged to the atmosphere via a chimney, the height of which . . . is subject to approval by HMIP, but is expected to be about 65 metres above ground level". There would be six of these chimneys, although again this would be subject to approval by HMIP. Any modifications to stack height or the number of stacks would have implications for the assessment of both air quality and visual impacts.

Several approaches are available to developers in dealing with such uncertainties in their EISs. The first would be simply to assess the effects of the most likely option; a second would be to assess the option likely to give rise to the most significant impacts (the worst case); and a third would be to assess the effects of all realistic options in the EIS. The second and third approaches would appear to be the most satisfactory way of acknowledging uncertainty in the EIS.

Local authority review

Although most power station proposals are subject to the Secretary of State's consent, local authorities will of course wish to examine the physical environmental and socio-economic consequences of such projects. Any objection by a local authority will cause a public inquiry to be held into the proposals. However, the complexity of certain impacts arising out of power station projects is likely to give rise to difficulties for local authorities in their consideration of such schemes. Local authorities may not possess the in-house expertise to conduct a thorough review of all aspects of an EIS. They may therefore decide to use outside consultants to review project proposals and EISs in such cases.

An example is provided by the decision of Oxfordshire County Council to commission environmental consultants to review National Power's EIS for its proposed gas-fired power station at Didcot (Environmental Resources Ltd 1991). The consultants were asked to review specific sections of the EIS, namely those dealing with atmospheric emissions, noise and vibration, water use and (more briefly) socio-economic issues. More detailed studies on noise and air-quality impacts, involving new survey work and the use of different pollutant dispersion models, were also commissioned. Impacts not included in the external review were those on flora, fauna and transport (which were reviewed in-house by the county council) and landscape and visual effects (which were reviewed by consultants commissioned by the district council). Other issues, strictly outside the remit of the EIS, were also addressed in the review. These included the need for the project, possible alternative fuels, the effects of FGD installation at the existing coal-fired station at Didcot, and the scope for a combined heat and power scheme on the site. The review was carried out within a five-week period, and its results were used by the county council during the public inquiry into the scheme proposals.

Environmental research and EIA

As was noted earlier, the former CEGB built up a considerable body of expertise on the environmental implications of its projects during the 32 years of its existence. With the privatization of the ESI in 1989, the CEGB was divided into three successor companies: National Power, PowerGen and Nuclear Electric, and a National Grid company. In addition, the privatized regional electricity companies and other private-sector companies became involved in many proposals, mainly for new gas-fired power stations.

In this new climate, the issue of which body should undertake environmental research and the development of assessment methodologies is open to debate. There is clearly a danger of wasteful duplication of effort. At the same time, concerns have been expressed by National Power and PowerGen about the public access to information required by the Environmental Protection Act 1990. Under the new legislation, registers of potentially polluting operations will be available for public inspection. "An operator will be required to allow almost all the details of his operation to be placed on the register" (Harris 1992). This will include the details of the operator's authorization application to HMIP. Operators can seek to keep certain commercially confidential information off the register. For instance in November 1992 National Power persuaded the Secretary of State that information about the future fuel consumption at 18 of its power stations was not directly relevant to the determination of the application for authorization, but could benefit National Power's competitors and fuel suppliers (*Simmons and Simmons Environmental Newsletter* 15). However, "failure to persuade HMIP that information ought to be withheld from the . . . register could result in millions of pounds of research and development funds being . . . thrown away. Competitors will be able to 'catch up' at the cost of paying HMIP's . . . reasonable photocopying charges" (Harris 1992).

In these circumstances, the willingness of the privatized generators to fund research into environmental monitoring and the development of EIA methodologies may become more limited. Although National Power and PowerGen continue to support a joint programme of environmental research, the Assessment Effects Manager of National Power has commented: "We are in a time of transition concerning how the necessary studies are funded and executed. It is no longer appropriate for electricity companies to carry the costs of the research needed for policy formulation but, as yet, compensatory provisions have not been made by the government" (Manning 1991).

10.5 Summary

The formal environmental appraisal of trunk road schemes had been in operation well before the implementation of EC Directive 85/337. Indeed, the existing methods and procedures of environmental appraisal have heavily influenced the way in which the Directive has been interpreted by the DoT. A recent report by the Standing Advisory Committee on Trunk Road Assessment (SACTRA) raised a number of specific concerns about the implementation of the Directive. A number of SACTRA's recommendations, particularly those concerning the presentation of information and the range of impacts to be addressed, appear to have been incorporated into the DoT's most recent EISs. The quality of trunk road EISs has undoubtedly improved substantially compared with the earlier examples criticized by SACTRA. Despite this, there are continuing concerns about the adequacy of the wider EIA process for trunk road proposals, including the arrival of the EIS late in the decision-making process and the limited treatment of alternatives, and indirect and cumulative impacts.

As with trunk road proposals, the assessment of environmental impacts in the electricity-supply industry pre-dated the introduction of EC Directive 85/337 by several years. Over the years, a considerable body of expertise was built up within the former CEGB on the environmental implications of its projects. Much of this experience has been inherited by the new privatized generating companies. Current weaknesses in the EIA of projects in the sector include the separate consent and EIA procedures for linked project components and the limited treatment of adjacent developments and alternatives. Flowing from these concerns, it has been argued that scope exists for a more strategic level of assessment in the electricity supply sector. This echoes the views expressed by SACTRA in relation to trunk road schemes. Other issues include the extent to which EIA should concentrate on localized impacts; the treatment of uncertainty in the EIA process; the overlap between the EIA and pollution authorization procedures; the requirements of local authority review; and the funding of future environmental and EIA research in the industry.

References

CEGB (Central Electricity Generating Board) 1979. *Electricity supply and the environment*. London: HMSO.

CPRE1990. *Response to the government consultation paper on implementation of the Large Combustion Plants Directive: the proposed statutory plan for reductions of emissions of sulphur dioxide and oxides of nitrogen from existing large combustion plants in the United Kingdom*. London: Council for the Protection of Rural England.

CPRE 1991a. *A complaint by the CPRE to the European Commission, UK implementation of environmental assessment for roads*. London: Council for the Protection of Rural England.

CPRE (Council for the Protection of Rural England) 1991b. *The environmental assessment Directive – five years on, submission to the European Commission's five-year review of Directive 85/337/EEC on environmental assessment*. London: CPRE.

Cremer & Warner Consulting Engineers and Scientists 1990. *Environmental assessment of proposed combined heat and power generating facility: technical report*.

DoE (Department of the Environment) 1989. *Environmental assessment: a guide to the procedures*. London: HMSO.

DoT (Department of Transport) 1983. *Manual of environmental appraisal*. London: HMSO.

DoT 1989. *Departmental standard HD 18/88, environmental assessment under EC Directive 85/337*. London: HMSO.

DoT 1992. *Assessing the environmental impact of road schemes, response by the Department of Transport to the report by the Standing Advisory Committee on Trunk Road Assessment.*London: HMSO.

DoT/Allott and Lomax Consulting Engineers 1992. *A556 (M) improvement (M6–M56) environmental statement, vols I and II.*

Environmental Resources Ltd 1991. *Environmental review of the proposed combined-cycle gas turbine power station at Didcot: final report.* London: ERL.

Glasson, J. 1984. Local impacts of power station developments. In *Energy policy and land use planning*, D. Cope, P. Hills, P. James (eds), 123–46. Oxford: Pergamon.

Glasson, J., M. J. Elson, D. Van der Wee, B. Barrett 1987. *Socio-economic impact assessment of the proposed Hinkley Point 'C' power station: final report.* School of Planning, Oxford Polytechnic.

Harris, R. 1992. Integrated pollution control in practice. *Journal of Planning and Environmental Law*, 611–23

Howells, G. D. & K. M. Gammon 1980. Rôle of research in meeting environmental assessment needs for power station siting. Paper submitted to British Ecological Society Conference on Ecological Aspects of Environmental Impact Assessment, University of Essex, April 1980.

Manning, M. 1991. The rôle of environmental assessment in power system planning. Paper submitted to conference on Advances in Environmental Assessment, London, October 1991.

National Power 1990. *Environmental statement: Didcot B proposed combined-cycle gas turbine power station.* London: National Power.

NCC (Nature Conservancy Council) 1990. *The treatment of nature conservation in the appraisal of trunk roads, submission to the Standing Advisory Committee on Trunk Road Assessment.*

SACTRA (Standing Advisory Committee on Trunk Road Assessment) 1979. *Trunk road proposals – a comprehensive framework for appraisal.* London: HMSO.

SACTRA 1992. *Assessing the environmental impact of road schemes.* London: HMSO.

Sheail, J. 1991. *Power in trust: the environmental history of the Central Electricity Generating Board.* Oxford: Oxford University Press.

Therivel, R., E. Wilson, S. Thompson, D. Heaney, D. Pritchard 1992. *Strategic environmental assessment.* London: RSPB/Earthscan.

Wood. D. A. 1992. Assessing the environmental impact of road schemes: the SACTRA report. Paper presented to Planning and Transport Research and Computation (PTRC) XXth Summer Annual Meeting on European Transport, Highways and Planning, University of Manchester Institute of Science and Technology, September 1992.

Notes

1. In England, the consent of the Secretary of State for Energy was required prior to the incorporation of energy matters into the Department of Trade and Industry. The consent of the Secretary of State for Trade and Industry is now required.
2. The applications were submitted in 1991 by a joint venture company led by Ecogen Ltd, for windfarms on sites at Penrhyddlan and Llidiartywaun in Montgomeryshire District, Powys.

CHAPTER 11
Comparative practice

11.1 Introduction

This chapter considers five different systems of EIA from Europe, North America, Australia and Asia, to illustrate the range of existing EIA systems and to act as comparisons with the UK and EC systems discussed earlier. They span the range from "model" systems generally praised for their comprehensiveness and effectiveness (e.g. the Netherlands, Canada) to considerably more problematic and less well developed systems.[1] For four of the countries, a case study is presented to highlight some of the successes and failures of the systems in practice.

The Netherlands was chosen as an example because the country is well known for its progressive and well developed environmental policies which are based on the principle of sustainable development. The Dutch EIA system incorporates a particularly high level of public consultation, and uses an independent EIA Commission to scope each EIA and subsequently review its adequacy. The case study of a demonstration integrated gasification combined-cycle power station shows the broad range of alternatives addressed and the efficiency of the system.

Canada is also known for its progressive environmental policies. Its federal EIA system has good procedures for public participation and review, and its particular strength lies in its emphasis on monitoring of a project's actual impacts after construction. Canada's provinces have separate EIA systems for projects under their jurisdiction. The case study of British Columbia reveals some of the strengths and weaknesses of one provincial EIS system, and proposals for and perspectives on reform.

In *Australia* the responsibility for EIA is also shared between the national and state governments. However, a high level of government discretion and low level of public participation render this system less powerful than those of Canada and the Netherlands. The case study of the third runway at Sydney Airport shows that environmental considerations may well be marginalized in the economic debates around a project.

Japan has no national legal requirement for the preparation of EIAs, in contrast to the other systems discussed here, but instead it has EIA guidelines for the national ministries to follow. At the local level, some EIA regulations and (mostly) guidelines have also been established. Generally this system seems to ensure that the most environmentally harmful proposals are avoided. The case study of the Trans-Tokyo Bay Highway shows how the different national and local EIA procedures interact.

Of the five countries studied here, *China* has environmental policies that are most closely restricted by the need to harmonize with plans for economic development. The Chinese system of EIA makes extensive use of mathematical modelling, but incorporates no public participation and does not consider alternatives to projects.

For each system of EIA, first the legislative framework is set out, then the procedures are explained, and a case study is given. Emphasis is placed on best practice (e.g. review in the Netherlands and monitoring in Canada).

11.2 The Netherlands

The Netherlands, with its small area of densely populated and highly industrialized land, has developed a worldwide reputation for powerful and progressive environmental legislation. The National Environmental Policy Plan (NEPP) of 1989 and an update (NEPP-plus) of 1990 (Ministry of Housing, Physical Planning and Environment 1989, 1990) established a national environmental strategy based on the concept of sustainable development. These plans in turn were based on an earlier report, *Concern for tomorrow*, published by the National Institute of Public Health and Environmental Protection, which specified targets (e.g. for emission controls) that would need to be met for the Netherlands to achieve sustainable development.

The Netherlands, on the basis of NEPP and NEPP-plus, is implementing far-reaching changes in a wide range of areas. For instance, waste-disposal policies aim to reduce waste disposal on land from 55% to 10% of all wastes produced by 2000; targets have been set for air pollution emissions for 2000, broken down by sector; and transport and land-use policies are being jointly developed to minimize the need for car travel. The cost of these measures is expected to be 0.9–2.6% of Dutch GNP in 2010.

Legislative framework

EIA in the Netherlands, as in the UK, is required by EC Directive 85/337. This is implemented through the Environmental Protection (General Provisions) Act of

June 1986; the EIA Decree of May 1987, which designates activities subject to EIA; and the Notification of Intent EIA Decree of July 1987, which designates the contents and requirements of the notification of intent. EIA is required for a so-called "positive list" of projects that are considered to have a significant impact on the environment. This list is based on Annexes I and II of Directive 85/337, with further additions, and with the exemption of projects that are expected to have no serious harmful environmental consequences.[2] In addition, EIA is currently required for sectoral plans on waste management, drinking water supply, energy and electricity supply, and some land-use plans. EIA procedures are also being developed for other policies, plans and programmes; this is discussed further in §13.4.

Procedures

Figure 11.1 summarizes the Netherlands' EIA procedures. Once a (public or private) developer decides to carry out an activity included in the "positive list", it informs the competent authority, a "notification of intent" is published, and the Minister of Environment notifies the EIA Commission.[3] Once the Commission is notified, the chairman sets up a panel of five to eight experts, which carries out a scoping exercise to assess the range and magnitude of the impacts and alternatives that the EIS should cover. Within two months, it must present an advisory note of project-specific guidelines to the competent authority. The competent authority in turn produces formal EIS guidelines for the action, specifying alternatives and key environmental impacts that the EIS must address. The first scoping stage must take place within three months.

The developer is responsible for preparing the EIS, which must include:

- a statement of the purpose and reason for the activity
- a description of the activity and "reasonable" alternatives (including that least harmful to the environment, and the do-nothing option)
- an overview of the specific decision(s) for which the EIS is being prepared, and of decisions already taken regarding that activity
- a description of the existing environment, and the future state of the environment if the activity is not carried out
- a description of the environmental impacts of the proposed activity and alternatives, and methods used in determining these impacts
- a comparison of the activity and each alternative
- gaps in knowledge, and
- a non-technical summary.

The developer submits the EIS to the competent authority, which has six weeks to determine whether it meets the criteria of the guidelines or whether any corrections or amendments are required. The findings of this inspection are made public. Once it has been accepted, the EIS is made publicly available for one month. During this time, a public hearing must be held, and the public may comment on the EIS. Bodies such as the Regional Inspector for Environmental Protection also provide advice at this time to the competent authority concerning the contents of

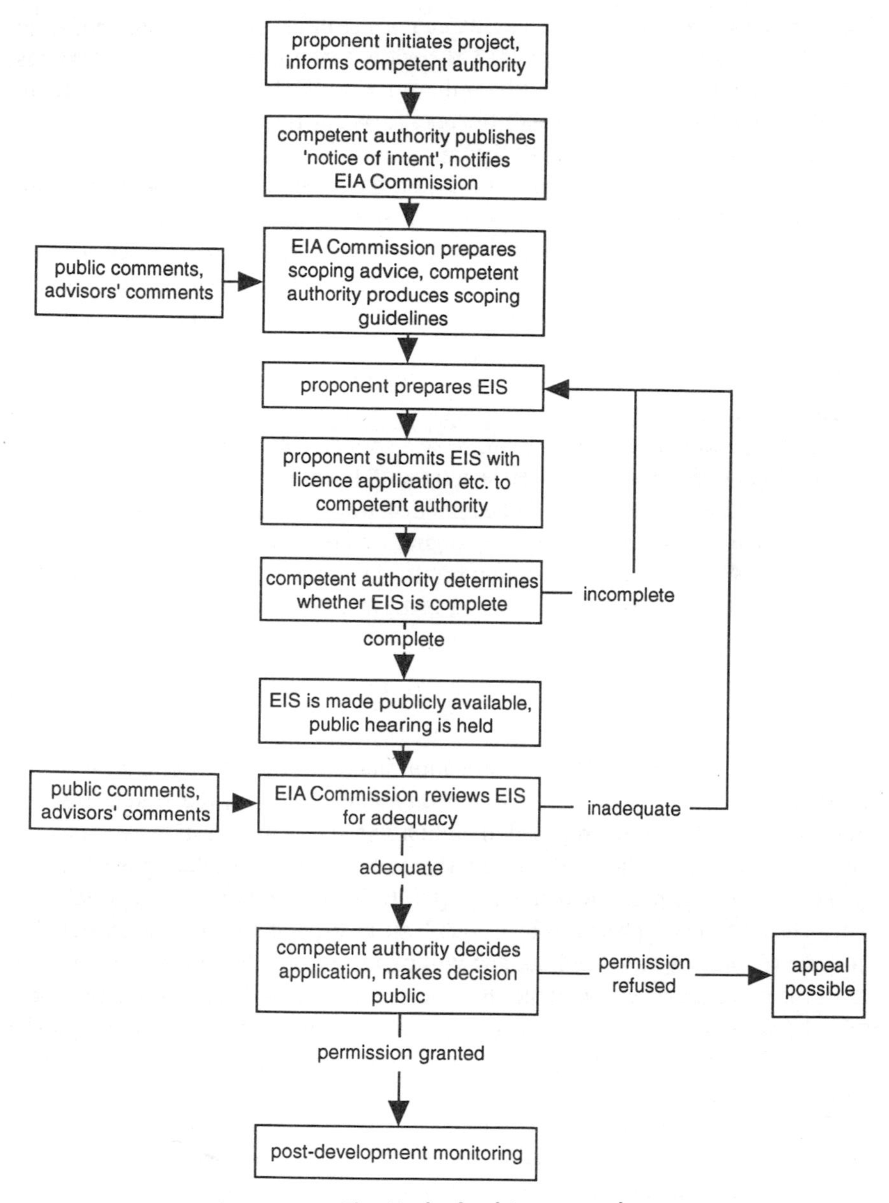

Figure 11.1 The Netherlands' EIA procedures.

the EIS. A record of the public review and other advice is then passed to the EIA Commission.

The Commission receives a copy of the EIS once it has been prepared, and checks the statement against current legislation and the EIS guidelines. It also considers the advice and public review.[4] The Commission's review is generally guided by two issues: whether the EIS can assist in decision-making and, if so, whether it is complete and accurate. The review concerns the adequacy of the EIS, not the environmental acceptability of the activity. Within two months of receiving the EIS, the Commission sends the results of this review to the competent authority, which makes the final decision.

The competent authority makes a decision based on the EIS, the advisors' comments, the Commission's review, and the results of the public hearing. It makes the results of this decision known, including how a balance was struck between environmental and other interests, and how alternatives were considered. The competent authority is subsequently required to monitor the project, based on information provided by the developer, and make the monitoring information publicly available. If actual impacts exceed those predicted, the competent authority must take measures to reduce or mitigate these impacts.

Forty EISs were prepared in 1988, 57 in 1989, and about 70 in 1990. By 1 July 1990, 159 EIA procedures had been begun for 191 activities (some EIAs cover more than one activity). The 191 activities include waste management (89), land-fills (37), chemical waste processing (23), incineration installations (15), plans (14), and roads (13). Further information on the Dutch EIA system can be found in Koning (1990), van Haeren (1991) and UNECE (1991).

Case study: the integrated gasification combined-cycle (IGCC) demonstration plant at Buggenum

In response to the policy for diversification and competitive pricing of electricity supplies set out in the Dutch Electricity Plan 1989–98, the Dutch utilities' parent organization SEP[5] proposed to build a 285 MW demonstration IGCC plant at Buggenum in the southeast of the country, near the existing Maas power station (Jones 1992). The plant, which is expected to begin operating in late 1993, will burn clean "syngas" produced by a coal gasifier located on site. It is expected to cost approximately NFL 1 billion, and will be the world's largest IGCC plant. The EIA process began when Demkolec BV, the company established by SEP to undertake the IGCC project, handed in a notice of intent to the Province of Limburg. The provincial authorities drew up EIA guidelines with advice from the EIA Commission and the public.

The EIA considered the proposed project's impacts on air quality, water quality and availability, noise, safety and landscape, as well as the effects of air contaminants. It evaluated alternatives to the main proposal:

No action alternative:
A – 3 units operating at Maas, reference year 1987
B – 1 unit operating at Maas, 1992

Proposed project:
C – 1 unit operating at Maas and the IGCC, 1993
D – Maas decommissioned, the IGCC only, 2000

Process alternatives:

E – improving desulphurization from 98% to 99.5%
F – a cooling tower for the entire IGCC capacity
G – sealed storage of coal gasification slag and fly ash
H – future action: improved reduction of NO_x emissions.

The EIS was carried out within 12 months, at about 0.0004% of the project cost. It was 292 pages long and it discussed the purpose of and need for the project, the proposed project and its alternatives, the decision-making process, the existing environment, the environmental effects of the proposed project and its alternatives, a comparison of the environmental effects of the project and its alternatives, and uncertainties in the EIA process. It summarized the most important impacts of the proposed project and its alternatives in a table, part of which is shown in Table 11.1. It concluded that situations C and D, with E and then H would give the preferred alternative from an environmental point of view.

The EIS was submitted with the planning application to the Province of Limburg in August 1989. It was reviewed by the Province for adequacy and was made public in early October. A hearing was held in late October and public objections/advice were received for one month after the announcement. The EIA Commission examined the EIA for one additional month, and one objection to the project was resolved in the High Court within weeks. The entire EIA process after submission of the EIS took three-and-a-half months, and the project was approved in April 1990.

Table 11.1 Part of table summarizing the most important environmental impacts of the demonstration IGCC project and its alternatives.

Scenario	Water	Landscape	Cost
A	Thermal discharge 690 MWt, little chemical influence	some impact	n/a
B	Thermal discharge 298 MWt, chemical influence <A	=A	n/a
C	Thermal discharge 590 MWt, additional NaOCl discharge	somewhat <A	n/a
D	Thermal discharge 285 MWt, better than C	less than A and C	n/a
E	=C/D	=C/D	NFL 1.1M/yr
F	Thermal discharge decreases depending on operation of cooling tower, no additional NaOCl discharge	greater impact	NFL 2.25M/yr
G	=C/D	=C/D	NFL 1.15M/yr
H	=C/D	=C/D	n/a

Adapted from Demkolec 1989.

11.3 Canada

Canada has also set up a powerful system of environmental legislation, but under conditions different from those in the Netherlands. Its wealth of natural resources, which were originally plundered indiscriminately by the giant "trusts" in coal, steel, oil and railroads; its lack of strong planning and land-use legislation; and the conflicting needs of its powerful provincial governments - all prompted the development of a mechanism by which widespread environmental harm could be prevented. Recently, Canada and the USA have co-operated in joint ventures on monitoring and protection of such assets as the Great Lakes (Ledgerwood et al. 1992).

Legislative framework

The responsibility for EIA in Canada is shared between the federal and the provincial governments. The federal Environmental Assessment and Review Process (EARP) establishes an overall framework and requires EIA for federal-level programmes and activities. It is administered by the Federal Environmental Assessment Review Office (FEARO), which operates at arm's length from the federal environmental agency, Environment Canada, and on behalf of the Minister of the Environment. The provinces have separate, and widely differing, EIA processes for projects under their own jurisdictions (Smith 1991).

At the national level, EIA procedures are laid out in the form of guidelines rather than in legislation. The EARP was created by Cabinet directives of 1973, and fine-tuned in 1977 (FEARO 1978). However, it was only in 1979, with the Government Organization Act, that the process was made mandatory. A revised guide which outlines the mechanics of the assessment process was published in 1979, and the Cabinet directives of 1973 were superseded in 1984 by an Order-in-Council (SOR 84/467). The EARP Guidelines were judged to be legally binding in 1989. Additional support for the EARP is provided by the Fisheries Act, the Clean Air Act, etc. (FEARO 1978).

In most of the provinces, a similar situation exists. For instance in New Brunswick the basis for EIA is set out by the Clean Environment Act, and specified in the guidelines *Environmental impact assessment in New Brunswick*. In Alberta, EIA is required by the Clean Air Act, Clean Water Act and Land Surface Conservation and Reclamation Act, and procedures are laid out in the Alberta Environmental Assessment System Guidelines. Only Ontario has specific legislation on EIA, the Environmental Assessment Act 1975 (Effer 1984). Smith (1991) discerns three clear quality divisions of provincial EIA systems. Those in the first division, including Saskatchewan, Newfoundland and Quebec, are characterized by a clear, inclusive definition of the environment, invoked by excellent institutional arrangements, on the basis of public participation in the assessment process. As adherence to these features softens, assessment provision declines in quality.

Federal procedures

Figure 11.2 summarizes Canada's federal EIA procedures. The federal EARP process is in two stages. An initial *self-assessment by the agency proposing the action* determines whether the action is likely to have a significant environmental impact. At this stage, agencies consider technical information, expert opinion, initial public reactions, and any other surveys and studies carried out in the time available. In this, the agency is helped by one of five Regional Screening and Co-ordinating Committees, and by Screening Guidelines prepared by the FEARO (FEARO 1978).[6] The purpose of this preliminary screening stage is to describe the proposal's likely

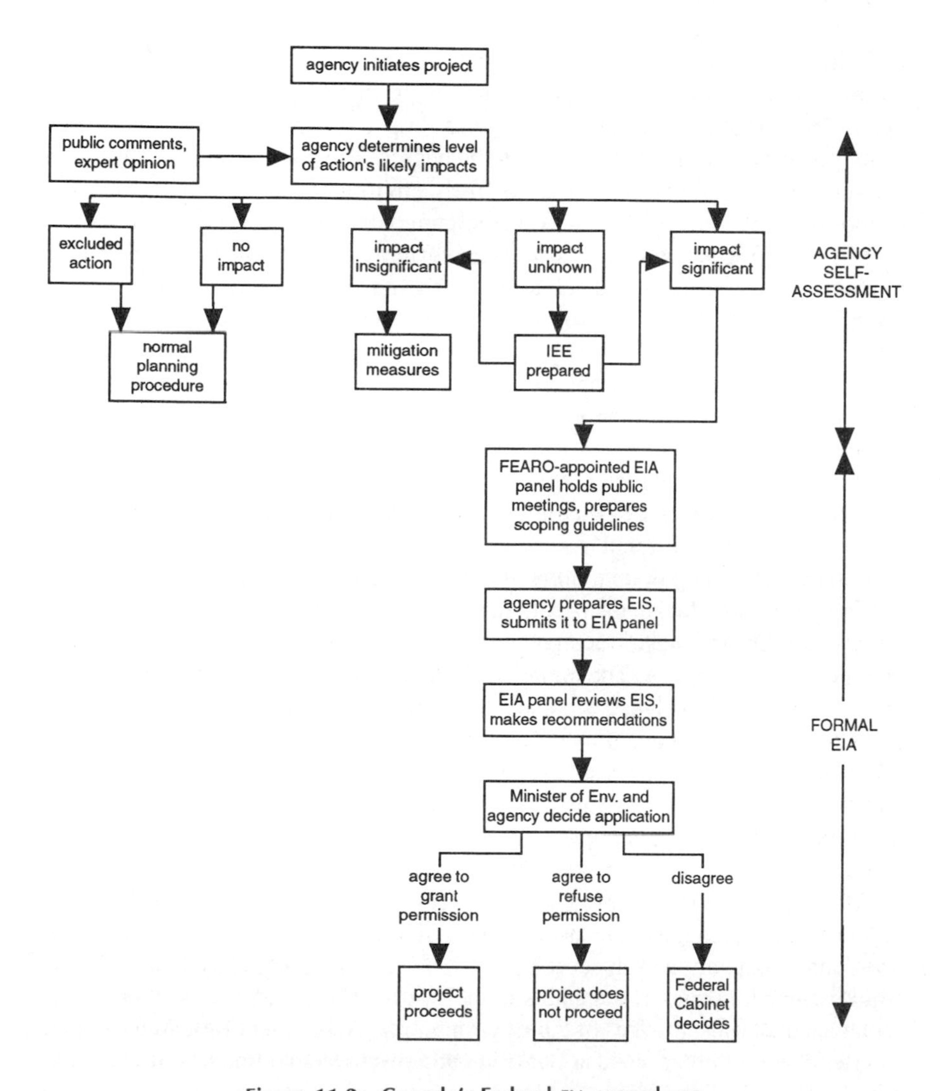

Figure 11.2 Canada's Federal EIA procedures.

environmental impacts and evaluate their significance. If an action is not on the list of excluded actions, the agency must determine whether:

- it has no environmental impacts, so it may proceed
- it has insignificant impacts, which can be mitigated without need for further assessment
- its impacts, or ways to mitigate them, are unknown, so a further Initial Environmental Evaluation (IEE) is needed to allow a definite appraisal to be made, or
- it is likely to have significant environmental impacts and may be of public concern, in which case the FEARO is called on to instigate a formal review.

If an IEE is required, the agency must prepare a document consisting of a project description, existing environment and resource use, potential impacts of the action and alternatives, and proposed mitigation measures (FEARO 1976). The action would again be subject to screening based on this document. If an action is determined to have significant environmental impacts, then the second stage of the EARP process, a *full formal, public EIA*, is begun. The purpose of this stage is to provide enough information about the likely environmental consequences of the proposal to allow the panel to make a recommendation concerning the proposal.

The FEARO appoints an independent Environmental Assessment Panel of four to eight experts, which holds public meetings and consequently prepares action-specific EIS guidelines. The project proponent then prepares an EIS, which generally discusses:

- project location and description
- description of the existing environment
- need for the project
- alternatives to the chosen option
- potential impacts of the chosen option and each alternative
- mitigation measures, and
- impacts remaining despite mitigation measures (FEARO 1979).

The EIS is submitted to the panel, and a public hearing may be held. The panel reviews the EIS and makes recommendations to the Minister of the Environment and the initiating agency. The initiating agency then decides "the extent to which the recommendations should become a requirement of the Government of Canada prior to authorizing the commencement of a proposal" (SOR 84/467), and the relevant minister is made responsible for dealing with the panel's recommendations.[7] The Minister of the Environment and the minister of the relevant agency determine if the project is to proceed; in case of disagreement between the ministers, the federal Cabinet makes a decision.

Although monitoring of project impacts is not yet mandatory in Canada, many monitoring programmes have taken place, and some major research projects have been undertaken on the subject (Sadler 1988). Environment Canada and FEARO commissioned a series of EIS audits in the 1980s. These audits were themselves reviewed under the auspices of the Environmental Assessment Research Council (CEARC).[8] A conference held at Banff in 1985 discussed the findings of these and other studies (Sadler 1987).

Since 1973, thousands of activities have been subject to initial assessment, hundreds have undergone IEE, and a few dozen have undergone full EIA. Approximately three full EIAs for federal projects are carried out annually (Wallace 1985). Canada's EIA system has been criticized because:

- its self-assessment procedures are felt to be open to bias
- its areas of application are limited, although amendments of 1984 expanded the EARP to include offshore projects, trans-boundary impacts, foreign aid projects, and Crown Corporations
- the make-up and activities of the Environmental Assessment Panels are controversial, and
- public participation in the process is limited and not guaranteed (Effer 1984, Bowden & Curtis 1988).

The possibility of establishing a uniform EIA system in Canada has been considered, but no common framework has yet been established. Difficulties include the fact that most jurisdictions have not defined the rôle of EIA within overall decision-making, the need to ensure that the EIA process does not constrain economic development, uncertainties about the need for a multi-tier EIA system, and lack of agreement on public consultation procedures (Effer 1984). Further information on Canada's EIA process can be found, e.g. in Bowden & Curtis (1988) and Smith (1991). Smith draws particular attention to a feature endemic to most EIA systems: change. As such, any review can only be a snapshot at one point in time.

Case study: reforming EIA in British Columbia

This case study is of procedures in one province of Canada and it provides a recent example of proposals to reform the EIA system (Province of British Columbia 1992). EIA procedures in British Columbia are orientated to projects and project-related issues and impacts, and are currently addressed through three formal review processes for energy projects, mine developments and major projects. The strengths of these processes are seen as including a "one window" government contact point for each sector, the use of inter-ministry committees, and comprehensive approaches to review. A particularly clear feature of EIA practice in British Columbia is the broad context attached to the term "environmental", which encompasses the biophysical, socio-economic, cultural, human health, safety factors and other related aspects of human activity. This is partly explained by the significance of some of the projects for economic development in remote areas of the province. By way of example, the Mount Milligan Ore Extraction Project, planned to extract 22 million tonnes per annum of gold and copper ore at a location north of Fort James, devotes two of its five substantial volumes to "community and regional socio-economic impact assessment", and "socio-economic assessment of native communities" (Continental Gold Corporation 1991). These socio-economic studies include social as well as economic assessment, in considerable and disaggregated detail. Some of the predictions

Table 11.2 Reforming EIA in British Columbia: some of the recommendations, and responses from consultation.

Recommendations	Example of reported consultation response
• A combination of category and threshold inclusion criteria should be used to determine which developments will be subject to EIA. There should be provision in the legislation for the minister(s) to require other projects to be subject to the EIA process where it is considered to be in the public interest.	• All projects, large and small, that have a serious significant effect on the environment must be given serious consideration for the EIA process (BC Wildlife Federation)
• Consistent with current practice, the legislation should state that it is the proponent's responsibility to identify and manage all direct environmental impacts associated with a project.	• The direct environmental impacts associated with a project are an integral part of project development and therefore the proponent's responsibility (Cominco Metals) • Making the project proponent responsible for the technical analysis on which the impact assessment is based is the most serious flaw in current impact assessment process. The impact assessment is not trusted by directly affected parties and members of the public. Thus, there is no commonly accepted factual foundation on which to base negotiations to achieve consensus (Irving Fox, University of BC)
• The legislation should specify that the proponent's responsibility for monitoring compliance be included in the project approval certificate and should enable the minister(s) to withdraw the certificate for breach of its conditions.	• The background paper indicates that the operator of a project should be responsible for monitoring impacts. This has the same weakness as assigning a fox to guard the chickens (Irving Fox, University of BC) • We agree that the proponent should have as much direct involvement and responsibility as possible in the assessment process. Therefore the onus for verifying and reporting compliance should remain with the proponent. (Noranda Inc.)
• Responsibility for chairing and directing the EIA process should be assigned to one of the following: • Min. of Environment, Lands and Parks • Min. of Environment, Lands and Parks and a lead agency; or • a neutral agency reporting to the Cabinet Committee on Sustainable Development.	• Wherever possible, we favour the use of a neutral agency to direct the EIA process – this would provide the best guarantee of independent and impartial review, consistent application of review requirements, and balanced consideration of a project's ecological, economic and social implications. (Alcan) • Environment, Lands and Parks - one only to avoid duplication of effort and conflicting priorities. (Canadian Earthcare Society)
• The legislation should outline public notification requirements, procedures for the release of documents, and public consultation. • The legislation should provide for the involvement of the public in issue identification and throughout the review process.	• The root of public frustration, animosity and civil unrest has been the exclusion of them from the resource development planning and management process. They will not be satisfied with legislation that merely states 'the minister may decide to include the public in review processes'. (J. Stelfox).
• The government is formulating a policy on direct participant funding which will guide its application to EIA. Advice on how to deal with this important policy issue would be appreciated.	• There should be no strings attached to participant assistance. (Sierra Club) • It is necessary to set limits on participant funding. (BC Hydro)

Source: Adapted from *Province of British Columbia 1992.*

bear some similarity to the monitoring information on Sizewell B in the UK (discussed in Ch. 7), for example:

> Some increases in impaired driving, assaults and other criminal activity in the community are possible because of the size of the construction labour force and the transient element that may be attracted to the area in the hope of gaining employment on the project.

However, there has been growing concern about deficiencies in British Columbia's EIA system, including definitions of categories of projects for review, limited public participation and lack of comparability between the different procedures. There is concern about potential bias in the procedures, with proponents of projects submitting their reports to the ministry responsible for promoting and regulating the industry. In addition, the province wishes to promote sustainability further, widen the scope of assessment to include cumulative impacts, integrate with federal legislation, and integrate Native American people's participation. The relevant ministries have adopted an interesting way forwards by producing a Legislation Discussion Paper containing 45 recommendations for change. This was mailed to over 6000 groups and individuals in the province; 600 written submissions were received, and many meetings, "open houses" and a workshop were held. Some recommendations and consultation responses are included in Table 11.2. This reveals some of the different and often conflicting perspectives – from government, private sector and the public – on the EIA process.

11.4 Australia

Like Canada, Australia also has a federal (Commonwealth) system with powerful individual states. Its environmental policies, including those on EIA, are generally not as powerful as those of Canada or the Netherlands.

Legislative framework

Australia's Environmental Protection (Impact of Proposals) Act 1974 requires EIA for actions that are carried out by the national government or require approval by the government (e.g. railways and airports, defence facilities, activities requiring export licences) and are likely to have a significant environmental impact. The Act was implemented by Administrative Procedures of 1975, and substantially amended in 1987 by the Environment Protection (Impact of Proposals) Administrative Procedures. EIA is also required by Australia's individual states:

- Australian Capital Territory's Environmental Assessment and Inquiries Act 1991
- New South Wales' Environmental Planning and Assessment Act 1979

- Northern Territory's Environmental Assessment Act 1982
- Queensland's State Development and Public Works Organization Act 1971
- South Australia's Development Act 1993
- Tasmania's Environmental Protection Act 1973
- Victoria's Environmental Effects Act 1978, and
- Western Australia's Environmental Protection Act 1986.

Of these, only New South Wales's EIA procedures are stronger than the national procedures. Where a proposal affects both state and national decisions, arrangements have been made to facilitate and streamline EIA procedures.

Commonwealth procedures

Figure 11.3 summarizes Australia's Commonwealth EIA procedures. The EIA process begins when a developer prepares a "notice of intent". This includes a description of the proposed action, the environment that would be affected, expected positive and negative impacts, alternatives to the action, and proposed environmental protection measures. This notice is submitted to the Department of Arts, Sport, the Environment, Tourism and Territories (DASETT), which determines the level of EIA needed:

- no further reports, provided that specified conditions are met
- a full EIS
- a simpler and less comprehensive Public Environment Report (PER), or
- examination by a Commission of Inquiry.

A PER is generally required when a proposal is expected to have only a few impacts or to be focused on a few specific issues, but where the issues still require consultation with the public. It briefly outlines the proposal, examines its environmental implications and describes safeguards needed to protect the environment. Where an EIS or a PER is required, the DASETT, in consultation with the developer and other organizations, prepares guidelines for the preparation of the document. A draft EIS or a PER is prepared, is announced in the *Commonwealth of Australia Gazette* and advertised in newspapers, and is made available for public review and comment (subject to commercial confidentiality) for at least 28 days. The Minister of the DASETT may also call for "round table" discussions between the DASETT, the developer and the public.

In the case of an EIS, comments by the public and relevant agencies are then sent to the developer, which revises the draft EIS; the developer prepares a final EIS and submits it to the DASETT for assessment, and it is again made publicly available. In the case of a PER, no further revision is needed. The DASETT then determines:

- whether the proposal meets the objectives of the Act
- whether further environmental information is needed, and/or
- any recommendations or conditions that should be associated with approval of the proposed action.

The results of this examination (for an EIS) or of the first-stage review (for a

PER) are presented to the Minister as an environmental assessment report. The assessment report may again be preceded by "round table" discussions. The Minister of the DASETT may make recommendations to the competent authority, which must be taken into account when the competent authority makes its decision.

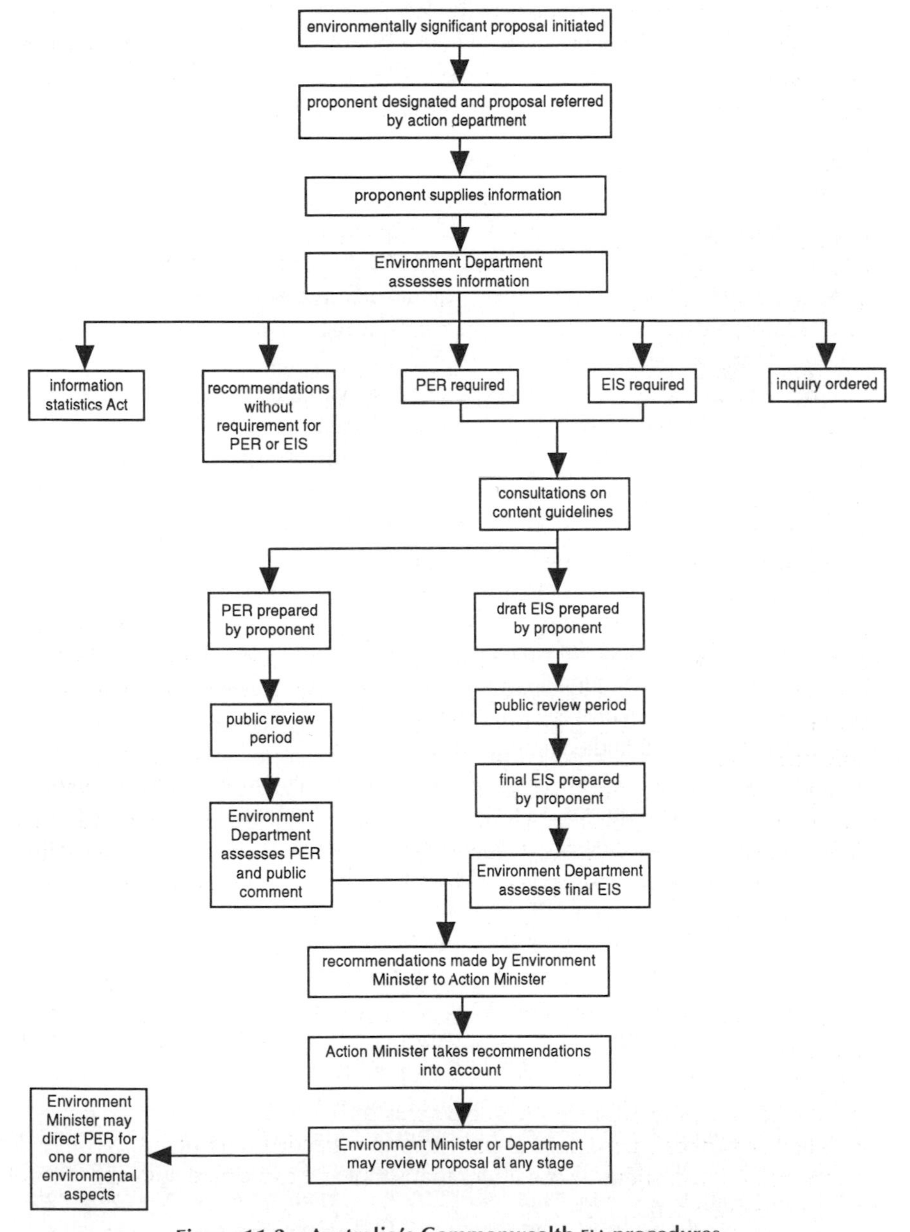

Figure 11.3 Australia's Commonwealth EIA procedures.

About 750 notices of intent for non-road projects are submitted annually in Australia; of these, about one-third require some form of environmental assessment. Of the latter, about 4% have required a full EIA (Gilpin 1992, Formby 1987). Between 1975 and the end of 1991, 131 Commonwealth-level EISs had been prepared (DASETT 1992). These concern mineral exploration (about 35%), transport development (25%), military developments (10%), communications, timber processing, and power generation and transmission. To date, only two public inquiries have been held into environmental matters: one concerning sand extraction at Fraser Island in 1975, and the Ranger Uranium Environmental Inquiry of 1977.

Australia's EIA procedures have been criticized for several reasons. Most of these are linked to government secrecy which allows a great deal of ministerial and administrative discretion regarding, for example, whether an EIS is needed, the scope of the EIS, whether the draft EIS is to be made publicly available, whether a public inquiry should be conducted, whether additional information is required from the proponent, and whether monitoring is needed. They also restrict the legal standing of interest groups (Formby 1987).

There has also been growing concern about the variation in EIA procedures, and their implementation, between States in the Commonwealth of Australia. The Australian and New Zealand Environment and Conservation Council (ANZECC) established a working group to pursue the issues. A *National approach* on EIA, and *Background paper*, were produced in 1991 (ANZECC 1991). They identified key areas of agreement between States on the objectives of EIA in Australia, and outlined the principles of EIA for the following groups: assessing authorities, proponents, the public and government. They also recommended a single national agreement for EIA between the Commonwealth, States and Territories, with schedules to accommodate individual legislative arrangements. The national approach included proposals on many issues of concern in Australian EIA, including the need to integrate ecological and economic considerations, to consider cumulative and long-term impacts, for public participation much earlier in the process, and for post-development auditing of projects. An Intergovernmental Agreement on the Environment was subsequently released, and a section of this agreement relates to EIA. At the time of writing (1993) one example of the implementation of the agreement is the production of Commonwealth-wide guidelines and criteria for determining the need for and the level of EIA in Australia.

In parallel with the ANZECC studies, a series of Commonwealth Government Working Groups produced a number of sector studies on ecologically sustainable development (ESD). Several of these studies addressed the role of EIA. For example, the Tourism Sector Working Group on ESD highlighted several recommendations on EIA procedures including: the clear definition of a triggering process for EIA that cannot be bypassed, the formal extension of the scope of EIA to include assessment of the social and cultural impacts of proposed developments, and the integration of the principles of ESD. *A national strategy for ecologically sustainable development* was produced in 1992 (Commonwealth of Australia

1992). Harvey (1992) discusses further the relationship between ESD and EIA in these recent Australian studies.

Case study: third runway at Sydney Airport

By the late 1980s, Sydney's Kingsford Smith Airport, Australia's main international gateway, was reaching its practical runway capacity (Gilpin 1992, FAC 1990). Two solutions to this problem had already been discussed for years: either supplementing the existing 4000 m north–south runway and 2500 m east–west runway with a 2400 m north–south runway ("the third runway"), or building a new airport 70 km away at Badgery's Creek. The Federal Airports Corporation commissioned a group of engineering consultants to prepare an EIS for the third runway; because the proponent was a Commonwealth authority, the national EIA procedures were used.

The EIS considered several alternatives to the main proposal of a third runway:
(a) no action
(b) development of the third runway and staged development at Badgery's Creek
(c) no third runway but directed development at Badgery's Creek, and
(d) active traffic management of the two existing runways and staged development at Badgery's Creek.

It also considered different spacings between the existing north–south runway and the proposed third runway ($<$760 m, 760–1525 m, $>$1525 m), and alternative ways of operating the three runways.

Public input into the EIA process was encouraged through the use of community access centres, telephone inquiry lines, newsletters, attitudinal research, and consultation with interest groups. A draft EIS was released in September 1990. This seemed to take a rather limited view of the environment: in its summary of the comparison of alternatives, it compared options (b), (c), and (d) above on the basis of capacity, cost, noise and other implications (operations, market trends, aviation sectors, timing and uncertainty). Copies of the draft EIS were put in public libraries and council offices, and were made available for purchase for A$25. Many submissions were received from members of the public, interest groups and local authorities during the ensuing three-month review period. Noise was a central issue. A final EIS was then prepared, reviewed by the DASETT, and considered by the national government. In 1991, the government decided to construct the third runway and also to start a staged development of a new international airport at Badgery's Creek. The third runway was expected to be operational in 1995.

11.5 Japan

In the 1970s, Japan responded to severe environmental degradation by adapting, developing and applying the newest technology for pollution control and energy efficiency. More recently, it has actively sought a more global rôle in environmental affairs, and is putting itself forward as a leader in resolving global environmental problems[9] (Barrett 1991). However, Japan's large ministries (for example, transport, construction and industry) tend to quash any environmental policies that are likely to harm the nation's economic development. Similarly, environmental policies that would require major social changes, such as reduced car use, are unpopular (Barrett & Therivel 1991).

Legislative framework

The establishment of federal EIA legislation in Japan was discussed with increasing intensity from the early 1970s until 1983. A bill concerning EIA was first proposed by the Environment Agency in 1976, was discussed with other agencies, and was presented to the Diet (parliament) in 1981. However, after the bill was discussed at Diet level for two years, the Diet was dissolved and the bill nullified. Instead, the Cabinet adopted non-mandatory federal EIA guidelines (*Implementation of environmental impact assessment*) in August 1984. Since then, the various ministries have established EIA guidelines for developments under their jurisdiction. Recently the Environment Agency has once again proposed an EIA bill, but this is being opposed by the other large ministries on the grounds that it may harm economic development and lead to lawsuits. About half of Japan's local authorities have established separate EIA regulations or guidelines, most of which are more stringent than the national guidelines.

National procedures

Figure 11.4 summarizes Japan's national EIA procedures. The Cabinet decision of 1984 requires the preparation of an EIA for certain listed projects. These include roads of four or more lanes more than 10 km long; dams with a water surface of more than 200 ha; airports with runways of 2500 m or longer; and industrial estates, residential developments, urban development projects and land readjustment projects of 100 ha or more.

For these listed projects, the developer prepares a draft EIS based on guidelines prepared by the responsible government ministry. Copies of the EIS are sent to the governor(s) and mayor(s) of the area the project will affect (as determined by the developer), and are made publicly available for one month. During that month, the developer also holds explanatory meetings for residents of the affected area "if possible". Residents can send written comments to the developer for one calendar month plus two weeks after publication of the draft EIS; the developer summarizes these and passes them to the governor(s) and mayor(s).

Within three months of receiving the residents' opinions, the governor(s), in consultation with the mayor(s), comments on the EIS, focusing on issues of pollution control and conservation of the natural environment. The developer then prepares a final EIS, which includes revised information from the draft EIS, a summary of the residents' opinions and the governor's opinion, and measures to respond to these opinions. This final EIS is publicly available for one month. If permission is needed from a government ministry, the final EIS is presented with

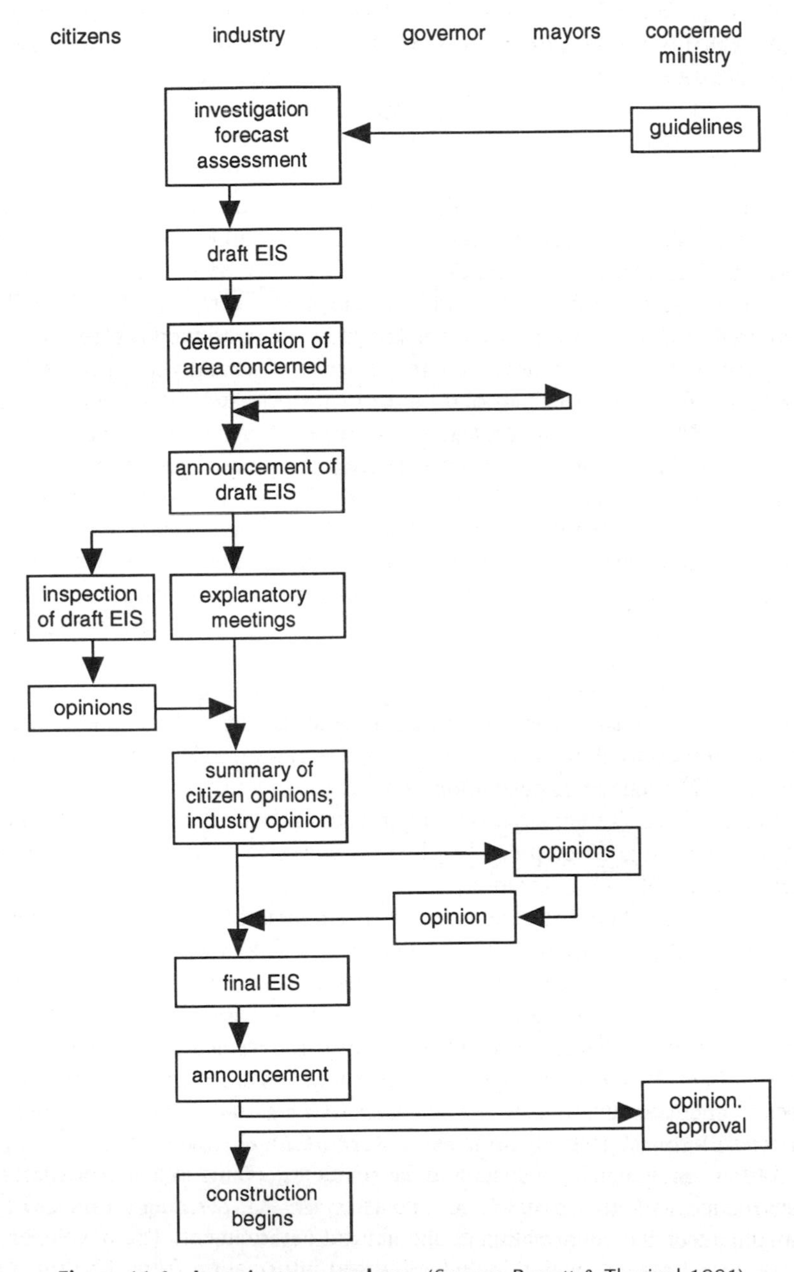

Figure 11.4 Japan's EIA procedures. (*Source:* Barrett & Therivel 1991)

the application; before it grants a licence, the ministry must ensure that the EIS properly considers pollution control and nature conservation.

Local authority procedures

Local authority EIA procedures in Japan are broadly similar to national procedures, but they generally apply to more projects, have a broader scope, and include more public participation. For instance, many local authorities require EIA for waste treatment plants, recreational projects, and water-supply projects, none of which is required at the national level. Some local authorities require the consideration of, for example, sociocultural impacts, climate, and the obstruction of sunlight (which is particularly important in Japan's densely populated urban areas). Again, these are not required by the national guidelines. Some local authorities require the developer to hold public hearings, and to collect and publicize monitoring information during the construction and operation of a project.

Approximately 70 federal-level EIAs are prepared annually in Japan. Of these, about half are for ports or harbours; most of the rest are for power stations. In addition, about 75 local authority EIAs are prepared annually. These vary more widely, including reclamation projects (approximately 25%), residential developments (18%), roads (13%), power stations (9%) and railways (7%).

Japan's EIA procedures have been criticized for several reasons. They are non-mandatory and therefore non-enforceable. Some development projects that are bound to have significant environmental impacts – such as crude oil refineries, integrated chemical installations and nuclear power stations – are not subject to EIA. Alternatives to the proposed project do not have to be considered. Public consultation requirements are not strong. Finally, the multiplicity of national government and local authority EIA procedures leads to duplication and confusion. EIA in Japan seems all too often to be used as a tool to justify development decisions and overcome local opposition.

Case study: the Trans-Tokyo Bay Highway

The Trans-Tokyo Bay Highway (TBH), when completed, will be a six-lane highway 15 km long (10 km bridge and 5 km tunnel), which will connect the cities of Kawasaki and Kisarazu by traversing the Tokyo Bay (Barrett & Therivel 1991). The TBH will connect with the Tokyo Bay Coastal Expressway to make a loop linking the greater Tokyo region, and it is expected to add further impetus to the development of that region's economy. Construction began in 1988, and it is estimated that it will take 10 years at a cost of ¥1.15 trillion (approximately £6 billion). The Japan Highway Public Corporation (JHPC) was responsible for planning the highway and preparing its EIA. The TBH had been discussed since the early 1960s. In 1986, the TBH Company was established as a private company to design, construct and operate the highway.

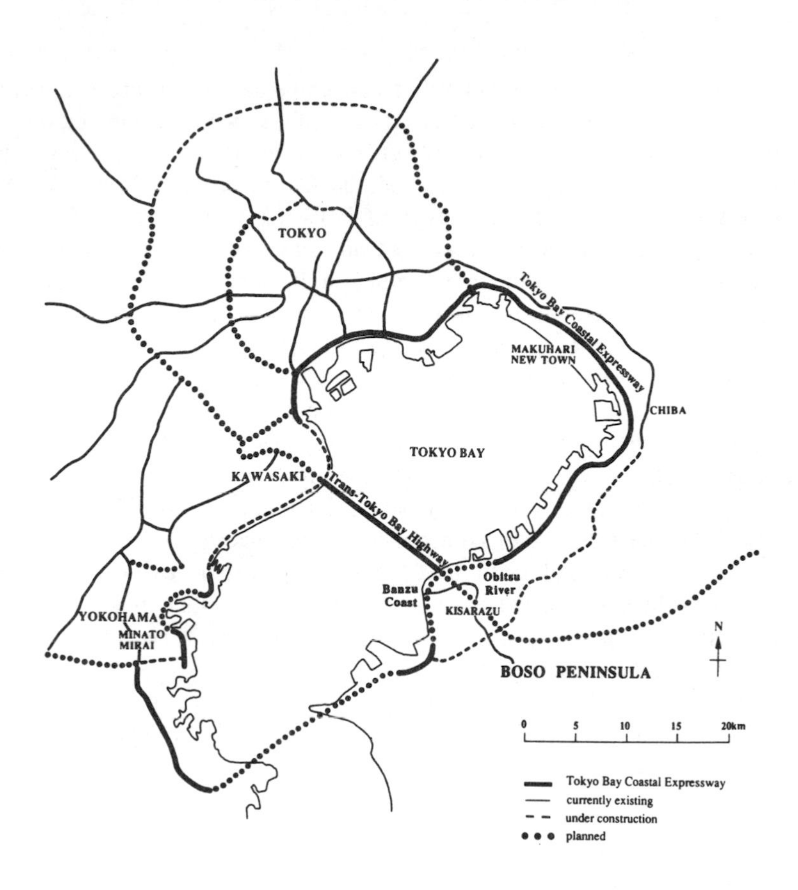

Table 11.3 Environmental impacts considered in the EIA for the Trans-Tokyo Bay Highway.

	Air pollution	Water pollution	Noise	Vibration	Ground config.	Land plants	Land animals	Marine life	Scenery
Construction									
Man-made island	X	X						X	
Tunnel									
Bridge	X	X	X	X				X	
Roads	X		X	X		X	X		
Operation									
Man-made island		X						X	X
Tunnel									X*
Bridge		X			X			X	X
Roads						X	X		
Traffic	X	X	X	X		X	X	X	

*air ventilation towers

Source: Barrett & Therivel (1991).

The project affected 11 local authorities, and could have been subject to seven different EIA guidelines or regulations. Some of these were eliminated on the grounds that the local authority was not sufficiently involved; in other cases the stronger EIA procedures were used. In the end, the EIA procedures of Kawasaki city, Chiba prefecture and the federal Ministry of Construction were used. The JHPC followed the (weaker) Ministry of Construction guidelines, and the local authorities followed their own procedures regarding public consultation and review.

Table 11.3 shows the environmental impacts considered in the EIA for the TBH. This type of simple matrix is commonly used in Japanese EIAs. No attempt was made to specify the key impacts within this framework, or to discuss the potential for shipping accidents or the ecosystem as a whole. Public participation in the EIA process was widespread. In total, 1770 opinion statements were received concerning the EIS, of which 1746 opposed the project. In addition, 154 people attended a public hearing at Kawasaki. Concern focused on future NO_2 levels, the loss of tidal flats, and the continued decline of water quality in the bay.

The draft EIS was completed in mid-1986, and was then reviewed over the course of a year by the 11 local authorities. Kawasaki city alone took six months to analyze the EIS in specialist committees. As a result of this review process, the JHPC agreed to implement expensive measures to contain emissions. The final EIS was 962 pages long, with 312 pages of supporting data. It was submitted to the Ministry of Construction in mid-1987. The ministry approved the development one month after receiving the EIS.

11.6 China

With its centrally planned economy, huge population, and relentless industrialization and urbanization, China is establishing environmental policies that attempt to balance the need for economic growth and environmental protection. With the proposals for and implementation of a range of major projects, and with China's increasing participation in international affairs, it is likely that existing EIA legislation will need to be strengthened in the near future.

Legislative framework

The main impetus for the introduction of EIA in China was the adoption of the Environmental Protection Law of the Peoples' Republic of China in September 1979, which states that:

> All enterprises and institutions shall pay adequate attention to the prevention of pollution and damage to the environment when selecting their sites, designing, constructing and planning production. In planning new construction, reconstruction and extension projects, a report on the potential

> environmental effects shall be submitted to the Environmental Protection Department and other relevant departments for examination and approval before designing can be started . . . (Article 6)

Over the years, guidelines for the implementation of this law have been prepared, including the *Management rules of environmental protection of basic projects* of 1981 and the *Management guidelines on environmental protection of construction projects* of 1986 (EPC 1986). The National Environmental Protection Agency carries out EIA for activities undertaken or organized by the central government. Some provinces have additional EIA regulations, and carry out EIA for enterprises under their own jurisdiction. Monitoring and enforcement is carried out by district agencies and neighbourhood committees.

Federal procedures

Figure 11.5 summarizes China's federal EIA procedures. EIA in China is required for a range of projects, including those in industry, transportation, water conservancy, forestry, commerce, health, culture and education, and tourism (EPC 1986). The EIA process begins when a developer asks a competent authority to determine whether a proposed action requires a full EIA or not. Large-scale projects with significant impacts, or smaller-scale projects in inappropriate locations, require such an EIA. Other smaller-scale projects, or projects with fairly small environmental impacts, require only the preparation of an environmental impact form that contains a description of the project and a brief statement of its environmental impacts. The competent authority personnel, sometimes with assistance from outside experts, conduct a preliminary study and then make a ruling. If an EIA is needed, its management is entrusted to a team of state-approved experts that works to a brief prepared by the competent authority. If the agency responsible for the EIA does not have the necessary expertise, some of the work may be contracted out to academic institutions (professional environmental consultants do not exist (yet) in China). A lead group will co-ordinate the EIA if more than one group is involved. In this scoping stage, those factors most likely to affect the environment are identified and given importance weightings.

The impacts identified at the scoping stage are then analyzed in greater detail and compared against relevant environmental quality standards. Baseline environmental assessments are carried out if the project is proposed for an area of low industrial activity where environmental standards are high. The impacts are then predicted, often using a systems approach and simulation techniques. They are evaluated for their impact on human health, ecosystems, and sometimes social systems. Mitigation measures may be proposed. An EIS is then produced, which, according to the *Management guidelines on environmental protection of construction projects* (EPC 1986), needs to include the following information:

- general legislative background
- description of the proposed development, including materials consumed and

produced
- baseline environmental conditions and the surrounding area
- short-term and long-term environmental impacts of the project
- proposals for monitoring
- cost–benefit analysis of the environmental impacts
- an assessment of impact significance and acceptability, and
- existing problems and proposals for addressing them.

After being evaluated and commented on by the competent authority, the EIS, or environmental impact form, is submitted to the environmental protection

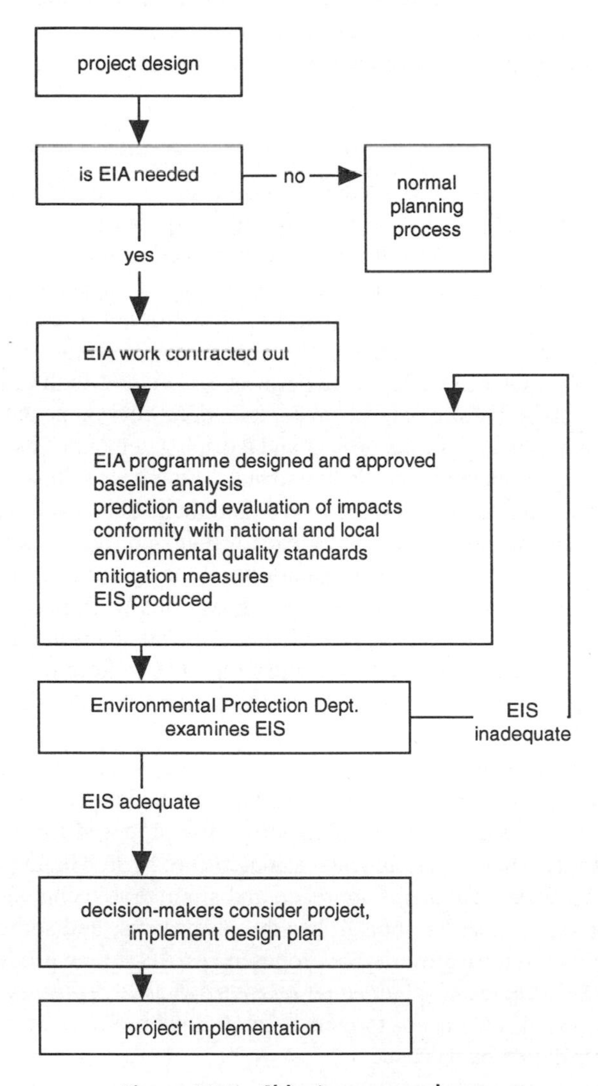

Figure 11.5 China's EIA procedures.

department of the province in which the project is proposed, and to the National Environmental Protection Agency. The environmental protection department checks the proposal against environmental standards, and then makes a decision within two months, considering the comments of the competent authority and outside experts. For controversial projects, or projects that cross provincial boundaries, the document is submitted to the higher environmental protection department for examination and approval. If the project is approved, various conditions for environmental protection may be included, such as monitoring and verification procedures. The competent authority must submit a report that states how the project will be carried out, and how any required environmental protection measures will be implemented. Once this has been approved by the provincial authorities (within one and a half months), a certificate of approval is issued.

Since 1979, more than 500 EIAs have been prepared, covering most of China's 30 provinces, autonomous regions and municipalities (Li Xingji 1987). These include smelters, chemical and petrochemical plants, nuclear power stations, light industries, water conservation projects, coal mines and airports (Wenger et al. 1990). However, it is possible that many projects in China are being constructed without adequate assessment of their environmental impacts, particularly collective and private-sector projects built in rural areas.

China's EIA system has no formal procedures for public consultation: the system is solely administrative, and makes no provisions for informing residents or for eliciting their views. All Chinese counties and cities have a form of ombudsman's office that receives complaints on all types of matters, including environmental issues. However, this is quite different from a formal system of public participation. Other criticisms of China's EIA system are that locations for projects are generally decided before EIA begins, that too much stress is often placed on mathematical modelling techniques rather than practical assessment methods, that the biotic component of EIA is often poorly assessed, and that many decision-makers seem to feel that they must put economic concerns before environmental issues (Wenger et al. 1990).

However, EIA in China does seem to have broadened the traditional form of decision-making, which considered only economic criteria. Some projects (e.g. a coal-gas project at Lanzhou) have been stopped because of EIA, and others have been substantially modified. Further information on China's system of EIA can be found in Ning et al. (1988) and Wenger et al. (1990).

11.7 Summary

EIA systems can be found in many countries, as was shown in Figure 2.2. The nature of the systems varies – influenced by a particular country's resource base, the nature of its institutional environment and the development actions concerned; all have their strengths and weaknesses. However, an overview of at least some of these systems, as set out in this chapter, can provide valuable comparative experience. Chapters 12 and 13 seek to draw on some of the ideas discussed here, and elsewhere, to identify possibilities for the future, focusing primarily on the UK system, but set in the wider European Community and global context.

References

ANZECC 1991. Environmental impact assessment: a national approach; background paper to the national approach. Canberra: Australian and New Zealand Environment and Conservation Council.

Barrett, B. F. D. 1991. Japan and the global environment: a case for leadership. *Japan Digest* (July 1991), 29–35.

Barrett, B. F. D. & R. Therivel 1991. *Environmental policy and impact assessment in Japan*. London: Routledge.

Bowden, M. & F. Curtis 1988. Federal EIA in Canada: EARP as an evolving process. *Environmental Impact Assessment Review* **8**, 97–106.

CEARC (Environmental Assessment Research Council) 1986. *Philosophy and themes for research*. Ottawa: Minister of Supply and Services.

Commonwealth of Australia 1992. *National strategy for ecologically sustainable development*. Canberra: Commonwealth of Australia.

Continental Gold Corporation 1991. *Mount Milligan project*. Placer Dome Inc.

DASETT (Department of the Arts, Sport, the Environment, Tourism and Territories) 1992. *List of proposals on which environmental impact statements have been directed under the administrative procedures – Commonwealth Environmental Protection (Impact of Proposals) Act 1974*. Canberra: DASETT.

Demkolec BV 1989. *EIS for the integrated gasification combined cycle (IGCC) demonstration plant at Buggenum*. Arnhem, The Netherlands: Demkolec BV.

Effer, W. R. 1984. Ontario Hydro and Canadian environmental impact assessment procedures. In *Planning and ecology*, R. J. Roberts & T. M. Roberts (eds), 113–28. London: Chapman & Hall.

EPC (Environmental Protection Commission – People's Republic of China) 1986. *Management guidelines on environmental protection of construction projects of the People's Republic of China* (26 March 1986).

FAC (Federal Airports Commission – Australia) 1990. *Summary draft EIS for the proposed third runway, Sydney (Kingsford Smith) Airport*.

FEARO (Federal Environmental Assessment Review Office) 1976. *Guidelines for preparing initial environmental evaluations*. Ottawa: FEARO.

FEARO 1978. *Guide for federal screening*. Ottawa: FEARO.

FEARO 1979. *Revised Guide to the Federal EARP*. Environmental Assessment Panel, Environment Canada.

Formby, J. 1987. The Australian government's experience with environmental impact assessment. *Environmental Impact Assessment Review* **7**(3), 207–26.

Gilpin, A. 1992. *Environmental impact assessment in Australia*. Unpublished report.

Harvey, N. 1992. The relationship between ecologically sustainable development and environmental impact assessment in Australia. *Environmental and Planning Law Jour-*

nal **9**(4).

Jones, T. 1992. *Environmental impact assessment for coal*. IEAcr/46. London: IEA Coal Research.

Koning, H. 1990. EIA in the Netherlands. *EIA Newsletter* 5, EIA Centre, University of Manchester.

Ledgerwood, G., E. Street, R. Therivel 1992. *The environmental audit and business strategy*. London: Pitman/Financial Times.

Li Xingji 1987. Discussing the methods and their selective principles adopted in the EIA. Paper presented at the International Environmental Impact Assessment Symposium, Beijing, October 1987.

Ministry of Housing, Physical Planning and Environment 1989. *To choose or to lose: national environmental policy plan*. The Hague: Government of the Netherlands.

Ministry of Housing, Physical Planning and Environment 1990. *National environmental policy plan plus*. The Hague: Government of the Netherlands.

Ning, D., H. Wang, J. Witney 1988. Environmental impact assessment in China: present practice and future developments. *Environmental Impact Assessment Review* **8**, 85–95.

Province of British Columbia 1992. *Reforming environmental assessment in British Columbia: (i) A legislation discussion paper; (ii) A report on the consultation process*. Victoria, BC: Ministry of Environment, Lands and Parks, British Columbia.

Sadler, B. 1987. *Audit and evaluation in environmental assessment and management: Canadian and international experience*. Proceedings of the Conference on Follow-up/ Audit of EIA Results, The Banff Centre, Banff, Alberta, 13–16 October 1987.

Sadler, B. 1988. The evaluation of assessment: post-EIS research and process development. In *Environmental impact assessment: theory and practice*, P.Wathern (ed.), 129–42 London: Unwin Hyman.

Smith, L. G. 1991. Canada's changing impact assessment provisions. *Environmental Impact Assessment Review* **11**, 5–9.

UNECE (United Nations Economic Commission for Europe) 1991. *Policies and systems of environmental impact assessment*. New York: United Nations.

van Haeren, J. 1991. EIA in the Netherlands. *EIA Newsletter* **6** (EIA Centre, University of Manchester).

Wallace, R. R. 1985. *Public input to government decision-making*. FEARO Occasional Paper 13. Calgary, Alberta: FEARO.

Wenger, R., H. Wang, X. Ma 1990. Environmental impact assessment in the People's Republic of China. *Environmental Management* **14**(4), 429–39.

Notes

1. Little publicly available information exists on EIA systems in Africa, so these are not considered, but they would generally come into the latter category.
2. By mid-1990, 21 exemptions had been granted, of which seven were for projects covered by Annexes I or II of the Directive. As a result, the EC informed the Dutch government in a letter of April 1990 that changes were needed. In May 1991, the Dutch government published "The report of the government on the working of the EIA regulations" which proposed changes to address the issues raised by the EC, and a bill has been proposed that would amend the Environmental Protection Act to conform to the Direc-

tive.

3. The EIA Commission is an independent body which carries out research on the EIA system, and which advises on the scope and adequacy of each EIA. The core of the Commission is composed of a chairman, who is appointed by the Council of Ministers; two vice-chairmen; and a full-time staff of about 25. In addition, about 200 members who are experts in EIA-related fields assist on a case-by-case basis.
4. The same group that carried out the scoping exercise usually also reviews the EIS.
5. Samenwerkende Electriciteits Produktiebedijuen.
6. The Screening Guidelines use a series of matrices to first identify the activities that are likely to have an environmental impact, and second focus on specific problem areas. Federal agencies are expected to establish their own screening guidelines, including exclusion lists. These are lists of specific actions considered to be environmentally benign and are excluded from the requirements for eia, for instance routine maintenance, interior renovation of buildings, and surveys and inventories.
7. This process, instituted in 1984, was felt to introduce more accountability for implementation into the system (Bowden & Curtis 1988).
8. The Environmental Assessment Research Council (CEARC) was established in 1984 to "advise on ways to improve the scientific, technical and procedural basis for environmental impact assessment" (CEARC 1986). The CEARC is composed of 12 members from the government, academic and private sector, who are appointed by the chairman of the FEARO for three-year terms. The CEARC promotes and reviews research related to improving the EIA system, in particular the integration of EIA with strategic planning, the incorporation of ecological and social sciences within EIA, the incorporation of social values in EIA evaluation, and the strengthening of policy and institutional frameworks related to EIA. More recent research has focused on EIA of government policies, the relation between EIA and economics, the rôle of EIA in sustainable development, and the possibility of "intervenor funding", namely funding of those participants who would be significantly affected by the proposal under review (Bowden & Curtis 1988).
9. Barrett (1991) notes: "One Japanese journalist recently argued that no country other than Japan could realistically adopt such a rôle. The United States and the Soviet Union are 'pollution superpowers' who have neglected the development of pollution control technology as they vied for military supremacy. Both countries must now devote a significant proportion of their declining economic power in order to overcome the environmental threats they are currently facing. European countries, moreover, are not particularly interested in new environmental technologies and the ecological disaster now facing their eastern partners will keep the Europeans pre-occupied for some time. Meanwhile, developing countries, already burdened with debt, cannot afford to develop new technologies to counter the growing number of environmental problems they will have to face. The only country with the ability to effectively react to global environmental problems, therefore, is Japan".

PART 4
Prospects

CHAPTER 12
Improving the effectiveness of project assessment

12.1 Introduction

Overall, the experience of EIA to date can be summed up as being like the proverbial curate's egg: good in parts. Current issues in the EIA process were briefly noted in §1.5: they include screening, scoping, EIA methodology, the relative rôle of the participants in the process, EIS quality, monitoring, and the extension of EIA to more strategic levels of decision-making. The various chapters on steps in the process have sought to identify best practice, and Chapter 8 provides an overview of the quantity and quality of UK practice from 1988 to 1993. Detailed case studies of good practice and comparative international experience provide further ideas for possible future developments. The lack of experience in EIA among the key actors in the process - from consultants, local authorities, central government, developers and affected parties - explains some of the current issues.

However, less than five years after the implementation of EC Directive 85/337, there is less scepticism from most quarters and a general acceptance of the value of EIA. There are still some fundamental shortcomings and there is considerable scope for improving quality, but practice and the underpinning knowledge and understanding are quickly developing; EIA is on a steep learning curve. The procedures, process and practice of EIA will undoubtedly evolve further, as evidenced by the comparative studies of other countries. The EC countries can learn from such experience and from their own experience since 1988.

This chapter focuses on prospects for project-based EIA, focusing primarily on the EC and the UK. The following section briefly considers the array of perspectives on change from the various key actors in the EIA process. This is followed by a consideration of possible developments in some important areas of the EIA process and in the nature of EISs. The chapter concludes with a discussion of the parallel and complementary development of environmental audits. The nature and types of audits are explained, and their important relationships to EIA are discussed. Chapter 13 closes the book by widening the scope of EIA from projects, to programmes, plans and policies.

12.2 Perspectives on change

The practice of EIA under the existing systems established in the EC Member States has improved rapidly (see Ch.8). This change can be expected to continue in the future, as the provisions of Directive 85/337 become better understood and used. However, the Directive itself is also likely to change, as is anticipated in its own Article 11: the Member States and the Commission of the European Communities (CEC) are required to exchange information on experience gained in applying the Directive and, on the basis of such information, the CEC must inform the European Parliament and Council on the application and effectiveness of the Directive, and may propose measures to further the application of the Directive in a sufficiently co-ordinated manner (CEC 1985). All of this was supposed to have been completed by July 1990, five years after the notification of the Directive. However, the final report (CEC 1993) was published only in April 1993, in part because of delays in the Directive's implementation. The report is discussed further in §12.3. Changes in EIA procedures, like the initial introduction of EIA regulations, can generate considerable conflict between levels of government: between federal and state levels, between national and local levels and, in the particular case of Western Europe, between the CEC and Member States. They also generate conflict between the other actors in the process: the developers, affected parties, and facilitators.

The *Commission of the European Communities* is generally seen as positive and proactive with regard to EIA. The CEC welcomes the introduction of common legislation as reflected in Directive 85/337, the provision of information on projects, and the general spread of good practice, but is concerned about the lack of compatibility of EIA systems across frontiers, the opaque processes employed, the limited access of the public, and lack of continuity in the process. The CEC is committed to reviewing and updating EIA procedures, which may involve formal changes in Articles and Annexes of the Directive. In contrast with the CEC, *Member States* can be seen to be more defensive and reactive. They are generally concerned to maintain "subsidiarity" with regard to activities involving the EC; this has been an issue with EIA, as reflected in the exchanges between the EC Commissioner for the Environment and the UK Prime Minister in 1991/92 (see §6.5). Governments are also sensitive to increasing controls on economic development in difficult economic times, although this may only be a short-term factor.

Within the *UK government*, the Department of the Environment (DoE) is concerned to tidy up ambiguities in the existing project-based procedures, and to improve guidance and informal procedures, but is wary of new regulations. The

DOE has commissioned research reports on an EIA good practice guide, and on the evaluation and review of environmental information. *Local government* in the UK is still coming to terms with EIA, but there is evidence that those authorities with considerable experience (e.g. Kent, Cheshire) learn fast, apply the regulations and guidance in "user friendly/customized" formats to aid developers and affected parties in their areas, and are pushing up the standards expected from project proponents. "The overwhelming response from planning officers interviewed was that EIA was a useful exercise and a positive contribution to informed decision-making in the planning process" (Wood & Jones 1992).

Pressure groups - exemplified in the UK by bodies such as the Council for the Protection of Rural England (CPRE), the Royal Society for the Protection of Birds (RSPB) and Friends of the Earth (FoE) - and those parties affected by development proposals, view project EIA as a very useful tool for increasing access to information on projects, and for advancing the protection of the physical environment in particular. They are keen to develop EIA processes and procedures; see, for example, the reports by CPRE (1991, 1992). Many *developers* are less enthusiastic about change in the regulations, but would welcome clarification on ambiguities - especially on whether an EIA is needed in the first place for their particular project. For *facilitators* (consultants, lawyers, etc.), EIA has been a welcome boon in economically depressed times; their interest in longer and wider procedures, involving more of their services, is clear.

Other actors in the process, such as the Institute of Environmental Assessment (IEA), the Association of Environmental Consultants, academics and some environmental consultancies, are carrying out ground-breaking studies into topics such as best-practice guidelines, the use of monetary valuation in EIA, and approaches to types of impact studies. At the time of writing, the IEA had recently produced guidelines on assessing traffic impacts (IEA 1993a), and was working on guidelines for ecological and landscape impact assessment. The production of over 300 EISs a year is generating a considerable body of expertise, innovative approaches, and comparative studies. EISs are also becoming increasingly reviewed and it is hoped that bad practice will be exposed and reduced. Training in EIA skills is also developing.

12.3 Possible developments in the EIA process

Legal basis

Implementation of Directive 85/337 has been slow, and is still not complete in some Member States. This can be attributed to a number of factors, including the resistance to greater public participation and to designated environmental authorities in Member States with no previous experience of this; the need to implement EIA at several tiers of government (regional as well as national) in some Member States; the need to pass several different regulations in Member States where more than one ministry has jurisdiction over relevant development actions; and difficulties in transposing some concepts (e.g. "significant environmental effects") into formal legislation (CEC 1993). It is likely that the outstanding regulations needed to implement the Directive fully will become implemented soon. There would also appear to be a good case for rationalization of regulations. In the UK the wide array of regulations can cause fragmentation of EIA activity for linked elements of a project, as is revealed in Chapter 10.

There are strong variations between the EIA procedures of various Member States. This can already be seen in the number of EISs prepared annually, ranging from fewer than 15 in Denmark to more than 5000 in France (see Appendix 2). Of course, there will always be variation in the EIA process and in EISs, according to the relative project and institutional context. Certainly from an EC perspective, there is probably no one best EIA approach, because the process must be sensitive to the procedures and policies of a particular country. However, many of the possible improvements in procedures and practices can apply widely. These are discussed next.

Screening

The report commissioned by the UK DoE (1991) on the operation of the existing EIA system proposed a clarification of screening criteria to allow the competent authority to determine the projects that are likely to have a significant environmental impact. In particular, the report comments and recommends:

- The interpretation of the indicative criteria in the Circular advising when EA may be required for Schedule 2 projects tends to vary. Several of the indicative criteria and thresholds in the Appendix to the Circular are ambiguous, especially those relating to 'urban development schemes' and 'redevelopment' projects. There is an absence of criteria against which to determine whether or not new settlements in the countryside should be subject to EA. The indicative criteria and thresholds in the Circular should be revised to reduce the ambiguity and omissions in these areas.
- . . . the Circular should be revised to recommend that local planning authorities introduce an internal check before the application is entered on the planning register, to determine whether a proposed development exceeds any of the criteria and thresholds which indicate that EA might be appropriate. (DoE 1991)

The CPRE (1991) has suggested that screening criteria should include the environmental sensitivity of the site (the quality and relative abundance of resources, the absorption and diffusion aptitude of the natural environment and regenerative capacity of the resources) and the characteristics of the main and associated development projects (pollution and nuisances, use of environmental resources, accidents and risks). In its five-year review of the UK's implementation of Directive 85/337, the CEC also noted that "[m]ore specific guidance should be issued to reduce any ambiguity in the interpretation of the indicative criteria and thresholds for Annex II projects; the application of these criteria and thresholds should be monitored, on a sample basis, to ensure satisfactory compliance" (CEC 1993).

The DoE responded to these concerns by issuing a consultation paper in September 1992 (DoE 1992). This contains, *inter alia*, proposals to extend the categories of project subject to planning control which may require EIA to include water treatment plants, wind-power generators, motorway and other service areas, coast protection works, golf courses and privately financed toll roads; by June 1993 this had not yet been enacted as legislation. The forthcoming DoE best-practice guidelines should help to further clarify thresholds for the remaining Schedule 2 projects.

As a result of its five-year review of Directive 85/337, the CEC may move some Annex II projects to Annex I, making EIA compulsory for them. The vexed question of exemptions to the EIA procedures, under Article 2.3 of the EC Directive, and the differential interpretation of this article by Member States, poses another and as yet unresolved screening issue for European EIA procedures.

Scoping

Scoping is generally poorly defined in UK EISs; it often occurs too late in the process, with minimum opportunity for consultation with relevant parties. The EC's five-year review for the UK notes that:

> [m]easures should be taken to ensure that the EA process starts sufficiently early and that its effectiveness during the early stages is strengthened by placing greater emphasis on early consultation and more systematic scoping of the assessment. . . Consideration should be given to the establishment of an independent statutory body to set up and maintain standards relating to scoping, the determination of significant impacts, review of ESs and monitoring/post-auditing. (CEC 1993)

The EC is considering making the requirements of Annex II of the Directive (information required in an EIA) compulsory; this would help to establish a better scope for many statements, although it might still leave them weak on the coverage of socio-economic impacts. A more powerful measure would be a requirement for a formal scoping step in the EIA process, thereby bringing EC procedures much more in line with those seen to be operating more effectively worldwide (see Chs 2 and 11). Formal scoping raises the ancillary issue of who would carry it out. In its recently published *Mock Directive*, the CPRE (1992) suggests a revised Article 1 of Directive 85/337 requiring each Member State to establish an Environmental Assessment Authority to ensure that the requirements of the Directive are carried out. In addition, the CPRE proposes a revision to Article 7 of the Directive that would require the proponent to publish a scoping document on which the competent authority, the public, the Environmental Assessment Authority and other relevant bodies could comment. Such an independent body could also have a review rôle (see below).

An unsuccessful attempt to introduce an independent Council of Environmental Assessment, along the lines of the Dutch Commission, was proposed as an amendment to the 1991 Planning and Compensation Bill in the UK (Coles et al. 1992). The Council would have had a small specialist staff with access to experts as appropriate. It would have assisted developers to scope out the assessments. These would then have been carried out by developers; the Council would have provided a review of the technical quality of the report, within the 16-weeks time frame, for the local authority. This system was seen as having several advantages:

- Independent project specification would both protect environmental interests and enable the developer to obtain competitive quotations for the work. This should substantially offset the costs of commissioning the project specification.
- A more consistent approach to EA across the UK would be applied if a single body were involved, in consultation with the local authority, other consultees and the public, in project specifications. This consistency of approach would also apply to the review procedure and would result in more consistent quality standards.
- The time for determination of applications should be reduced, since the technical assessment would have been completed within the 16 weeks and the scope of the EA would have been agreed in advance, removing many of the arguments over whether the scope of the environmental assessment was adequate (Coles et al. 1992).

Whatever the institutional arrangements for scoping, more guidance will also be needed on the nature of impacts and when they are appropriate to a particular project. The guidelines emanating from the IEA should be very useful in this respect; for instance the new road traffic impacts guidelines suggest that "two broad rules-of-thumb could be used to delimit the scale and extent of the assessment . . . Rule 1. include highway links where traffic flows will increase by more than 30% (or the number of Heavy Goods Vehicles will increase by more than 30%). Rule 2. include any other specifically sensitive area where traffic flows will increase by 10% or more" (IEA 1993a).

Alternatives

Alternatives are often omitted in EIA; for instance Jones et al. (1991) note that only about one-third of UK EISs discuss alternatives (see Table 8.1). When alternatives are discussed, they are not well handled. The EC's five-year review notes that "evaluation of the environmental impacts of certain projects is taking place too late in the development planning and decision-making process. In effect this has the result of removing from consideration the possible adoption of alternatives both to the individual project under consideration as well as to its particular location or route" (CEC 1993). If the CEC makes Annex III compulsory, issues such as alternatives will have to be addressed as standard practice. The CEC may also include the "no-action" alternative as a requirement.

Methods of assessment

This catch-all category relates to several key steps in the EIA process which are not being well caught by current practice and procedures. Uncertainty of the unknown often results in the EIA process starting too late in the development process; it needs to be integrated by developers and their consultants much more into planning and decision-making. The EIA process and the resultant EISs often lack balance. EIAs tend to focus on the more straightforward process of describing the project and the baseline environment, with much less consideration of *impact identification, prediction and evaluation*. The forecasting methods used in EIA are not explained in the majority of cases (see Table 8.1). It is to be hoped that there will be advances in the application of concepts and techniques from theory to operational practice, in the areas of predicting impact magnitude and determining importance (including the rapidly evolving array of multi-criteria and monetary evaluation techniques). A good Method Statement, explaining how the study has been conducted – in terms of techniques, consultation, relative rôles of experts and others – should be a basic element in all EISs.

There is also a need to develop the scope of, and commitment to, *mitigation measures*: "mitigation measures of a wider nature are infrequently and inadequately integrated into the planning and design of projects" (CEC 1993). Much can be gained from *monitoring* the outcomes of developments after implementation. Such monitoring would provide a major contribution to the development of effective predictive methods. The introduction of a compulsory requirement to include a formal monitoring programme in an EIS, as required in California, would be a major step forward in UK and EC EIA procedures.

Participation

One of the fundamental issues in EIA is a concern that the process is not as "honest and unbiased" as it should be, and that it is too developer-driven. The calls for independent environmental assessment authorities are partly motivated by this concern. There is certainly a need to improve public participation in the UK, and at a much earlier stage than is usually the case. There is also an urgent need, highlighted by the report to the DoE (1991), for a better flow of information in the process. The report to the DoE suggests that government guidance should be modified to advise local planning authorities to involve the public and voluntary groups more in the EIA process by ensuring that copies of EISs are readily available, and by providing information and advice. This issue is echoed in the EC's five-year review.

To facilitate greater access to EISs, the report to the DoE also suggests that the developers should be required to submit three copies of an EIS, "one of which is to be held in a national (UK) repository, and another in the appropriate region, both available for public inspection from within seven days of receipt" (DoE 1991). In response, the DoE's 1992 consultation paper does propose that three copies of the EIS should be submitted by the developer to the planning authority, thus allowing a copy to be forwarded to the central DoE Library in London at an early stage. There may also be some merit in proposing a cost limit for EISs; the exorbitant price of some EISs effectively rules out even the most committed participant. Although there is much more to positive public participation than information, these small changes may represent a useful start. Other proposals in the field of participation include the introduction into the process of formal third-party appeals, and some streamlining of EC/European Court of Justice procedures to provide much quicker decisions when third parties resort to this form of challenge to a decision.

Review

The review of the quality of EISs poses problems, especially for local authorities that may only receive one or two statements per year. Some of the proposals noted earlier, such as an independent review body, could help. The EC notes that "adequate control of the ES and of the EIA process as a whole is not always present" and that, in the UK "[m]easures should be taken to improve the quality and objectivity of ESs, including the provision of guidance for the preparation and evaluation of ESs" (CEC 1993). The DoE has already commissioned research on guidance for authorities on the review of environmental information. Some local authorities may take their own initiatives in linking with other authorities, voluntary groups, consultancies and academic institutions, to develop their own "review facility". The IEA also offers an EIS review service to its members.

Decision-making

Of course, a good EIS does not necessarily mean a good decision: "the contribution of the EIA process to the eventual decision-making and the rôle of monitoring project implementation are not as clear or as effective as they could be" (CEC 1993). It is not clear what use local planning authorities do make of EISs, or of statutory consultations on EISs, in their decision-making. But whatever their procedures, they are not helped by the time constraints operating in systems such as that in the UK.

A number of suggestions have been made to improve and to open up the decision-making "black box". A particularly strong proposal in the 1991 report to the DoE was that the 16-week period for the determination of applications should be automatically extended where the local authority demonstrates an EIS to be inadequate. Further information requested from the developer would be circulated to statutory consultees and to the DoE. If implemented, a change such as this would apply strong pressure on project proponents to provide a good-quality EIS in the first place. The EC also recommends that "[m]ore research should be undertaken of: the actual use made of the ES and consultation findings in the authorization of projects and of means of increasing their effectiveness; the costs, time and other resources associated with EA implementation in order to provide guidance on its cost-effective development" (CEC 1993).

12.4 Environmental audits – and baseline data

One new area of environmental activity that will undoubtedly have a major impact on EIA is environmental auditing. In contrast with the orientation of EIA to future development actions, environmental auditing involves the review and assessment of an *existing* organization's impacts on the environment. The EC defines environmental audit as:

> a management tool comprising a systematic, documented, periodic, and objective evaluation of how well organization, management and equipment are performing with the aim of contributing to safeguard the environment by facilitating management control of environmental practices, and assessing compliance with company policies, which would include meeting regulatory requirements and standards applicable. (CEC 1991)

Environmental audits were first carried out by private firms in the USA for financial and legal reasons, as an extension of financial audits. In the past decade, they have been increasingly carried out for private firms in Europe as well and, since the late 1980s, for local authorities in response to public pressure to be "green". As more and more environmental audits are carried out, they will provide increasing levels of environmental information that can be used in EIA. This section briefly discusses the two main types of environmental audit, those carried out for private companies and those for local authorities. It reviews existing and forthcoming standards and regulations on environmental auditing, and concludes by discussing how environmental auditing information can be used in EIA.

Industry environmental auditing

Environmental audits carried out by private sector firms vary widely depending on the purpose of the audit. They include acquisition/divestiture audits, which test environmental liabilities that could arise from the purchase or sale of a company; risk audits, which consider safety and occupational health; compliance audits, which test compliance with relevant environmental and safety standards; corporate audits, which consider the workings of the entire organization; and associate audits, which assess subsidiary or supplier companies.

Generally, environmental audits are seen as part of an environmental management system that includes the formulation of an environmental policy, establishment of an environmental management programme to achieve the policy, and testing of the adequacy of their implementation through an environmental audit. In its fullest form, an environmental management system would entail the following steps:

- full management commitment
- a decision on the scope of the audit
- a baseline review of activities, impacts and regulations
- formulation/refinement of a corporate environmental policy, targets and objectives, and a management system to implement the policy
- identification and assessment of options for implementing the environmental policy
- preparation of a management plan
- a report on the findings, including a description of company activities, relevant environmental issues, data on company emissions, etc., company policies and targets, and evaluation of environmental performance
- publication of the report, and verification by an external authority.

Of these, the environmental audit would represent the penultimate and ultimate steps. To date, few companies have set up a full environmental management system, and even fewer have made the audit findings publicly available, claiming that they need to protect commercially sensitive information. It is likely that the proposed EC "eco-management and audit" regulations will change this, and that they will lead to increasing numbers of audits being prepared.

Local authority environmental auditing

Local authority environmental audits have taken a more consistent format than those of the private sector, primarily because they all aim to do the same thing: provide information on baseline conditions in the relevant area, and suggest ways in which the local authority could change its operations to become "greener". Generally, these audits include, in varying levels of depth:

- a state of the environment report which reviews baseline environmental conditions in the area, preferably in conjunction with a regularly updated environmental data base
- a policy impact assessment which evaluates the local authority's policies and practices and suggests actions to improve matters where necessary
- an environmental management system like those of private firms, to implement, monitor and review the audit findings.

Of these, the environmental management system is a recent addition and only minimally used to date.

A survey by the County Planning Officers' Society in late 1990 showed that 87% of county councils had carried out or were planning to carry out some form of environmental audit, but by 1992 only about a dozen environmental audits had actually been completed, at costs ranging from about £20,000 to more than £300,000 (Grayson 1992). The use of environmental audits by local authorities is limited by cutbacks in central government funding, growing public concerns over economic rather than environmental issues, and the realization that environmental data become rapidly dated unless constantly (and expensively) updated. All of these factors mean that local authorities are re-evaluating whether an environmental audit is the best way of fulfilling their environmental responsibilities. In particular, those authorities that carried out audits in the past have found it difficult to implement the audits' recommendations because of lack of funding and motivation. Again, the establishment of environmental auditing standards may help to reverse this rather discouraging trend. Auditing advice to local authorities includes that by the Association of County Councils et al. (1990), FoE (1990) and the Local Government Management Board (1991).

Standards and regulations on environment auditing

Two recent developments in environmental auditing standards and regulations are likely to have important repercussions in terms of the number of audits carried out, their quality, and the public availability of audit results: British Standard 7750 on Environmental Management Systems of early 1992, and the proposed EC Eco-Management and Audit regulation, which was agreed in early 1993. The two systems are broadly intended to be co-ordinated and linked activities, but at present they still have inconsistencies.

BS7750 is a direct evolution from the well known British Standard 5750 on Quality Systems. Whereas BS5750 addresses an organization's products and services, BS7750 focuses more on by-products such as wastes and emissions. Both standards establish criteria for improving the organization's management system. Fulfilment of BS7750 entails the following steps:

- commitment by the organization to undertake the audit
- a preparatory review and assessment of relevant regulatory requirements, environmental effects, environmental management practices and procedures, and feedback from investigations of previous incidents and non-compliance
- formulation of an environmental policy
- a full inventory and assessment of the organization's activities and environmental effects
- assessment of whether the activities conform to relevant regulations and requirements
- formulation of targets and objectives
- development of an environmental management programme and supporting manual
- application of the management plan in the organization's operations and record-keeping
- an audit to test whether the organization achieves its targets and objectives, which feeds back into environmental policy formulation to form a cycle.

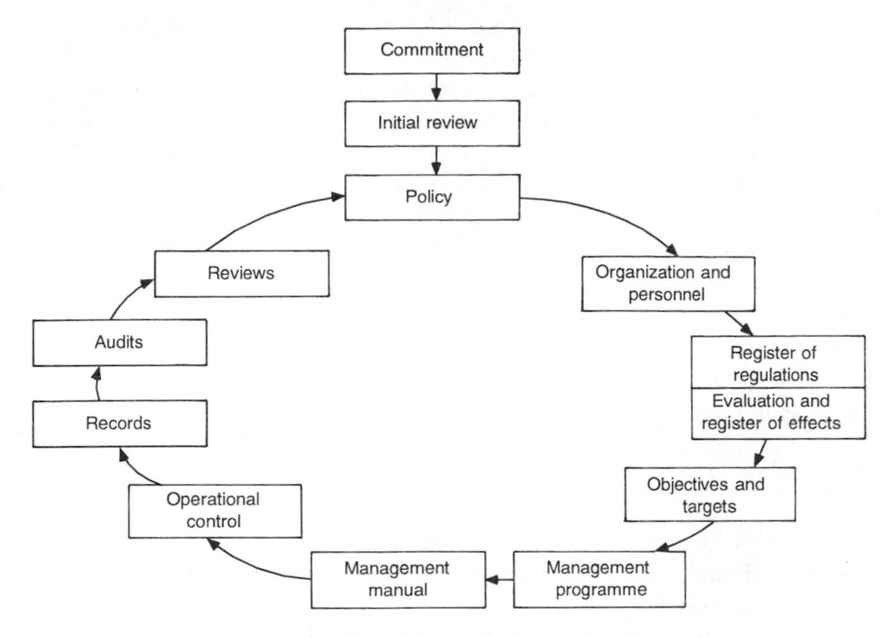

Figure 12.1 Outline of the environmental audit process.

Figure 12.1 summarizes this process. Although BS7750 requires these steps to be fulfilled, it allows organizations to determine what environmental standards they plan to achieve and how they are to be achieved. Individual sites can be taken as separate "organizations", so part of a firm can achieve BS7750 even if other parts do not. The standard requires outside verification only to the extent that relevant environmental management records must be established and maintained, and could thus be checked. The records do not need to be made public. Achievement of BS7750 is voluntary, but it is likely to be increasingly seen as a sign that an organization is carrying out good environmental management. In particular, BS7750 organizations are likely to require their subsidiary or supplying firms to achieve the standards as well. BS7750 will probably provide a basis for a similar International Standards Organization (ISO) standard. By early 1993, a number of organizations were testing the requirements of BS7750. Reportedly some of these organizations were having considerable difficulties implementing the audits, because of the wide range of activities to be audited, the unfamiliarity of some sectors (especially service industries) with auditing, and the lack of direct financial benefits accruing from the audits.

The EC's draft *Eco-management and audit regulation*, which would become directly operational in the Member States (i.e. Member States would not need to pass enabling legislation), also establishes a voluntary auditing system but applies primarily to manufacturing and processing industries. The regulation applies on a site-by-site basis, not to the entire organization. The regulation proposed in early 1993 would entail the following steps:

- a comprehensive review of all activities, their environmental impacts, and compliance with relevant regulations
- preparation/refinement of an environmental policy, action plan and objectives
- preparation of an annual audit statement following the initial review and subsequent audits
- verification of the statement by an accredited and independent verifier
- validation of the statement if the company has fully complied with the regulations and prepared an adequate statement (Simmons & Simmons 1993).

All of this is to be done in every "audit cycle" of no more than three years. Again, the standards to be achieved and the methods to be used are left to the discretion of the organization. However, in contrast to BS7750, the proposed EC regulation requires each audit to be independently verified, and auditing information to be made publicly available.

Use of environmental auditing information in EIA

It is clear that, in the process of environmental auditing, both private and public sector organizations will increasingly generate environmental information that will also be useful in carrying out EIAs. Local authority state of the environment reports provide data on environmental conditions in the area that can be used in EIA baseline studies. Generally, a state of the environment report will contain information on such topics as local air and water quality, noise, land-use, landscape, wildlife habitats and transport. Unfortunately, unless this information is regularly revised it will quickly become outdated. It is also often collected only on a large-scale (e.g. county-wide) basis, and so may not be suitable for the specific site(s) in question. However, state of the environment reports do generally identify sources for environmental data that can be contacted for the most up-to-date information. Similarly, the reports may be useful in determining suitable locations for new developments, by identifying sites that are particularly environmentally sensitive and should clearly be avoided, or that are environmentally robust and more suitable for development. Local authority policy impact assessments are useful in clarifying the authority's views on environmental matters and highlighting future policy directions that may influence the planning decision and the future operation of a project. Policy impact assessments of local authorities or government departments are also likely to generate a need for EIAs of policies, plans and programmes; this is discussed in Chapter 13.

Private firms' environmental audit findings have mostly been kept confidential to date. These audits are only likely to be useful for EIA if a firm with an environmental audit intends to open a similar facility elsewhere. However, the new auditing standards are likely to have a major effect in making auditing information more accessible both to the public and to other organizations. This information could include the levels of wastes and emissions produced by different types of industrial processes, the types of pollution abatement equipment and operating procedures used to minimize these by-products, and the effectiveness of the equipment and operating procedures. This type of information will be useful for determining the impact of similar future developments and mitigation measures. Some of these audits are also likely to provide models of "best practice" that other firms can aspire to in their existing and future facilities.

Overall, environmental auditing is likely to increase the level of environmental monitoring, environmental awareness, and availability of environmental data. All of this can only be of help in EIA.

12.5 Summary

As in a number of other countries discussed in Chapter 11, the introduction of EIA for projects in the UK, set in the wider context of the EC, has progressed rapidly up the learning curve. Understandably however, practice has highlighted problems as well as successes. The resolution of problems and future prospects are determined by the interaction between the various parties involved. There are already several innovative proposals with regard to screening, scoping, review and other steps in the EIA process. Forthcoming guidance by the EC and UK should help to counter some of the problems and improve the theory and practice of EIA for new developments. In particular, the EC is drafting amendments to Directive 85/337 to secure a better consideration of alternatives, provide for a system of scoping, provide for the monitoring of the impacts of the project and auditing against those predicted in the EIA, and address difficulties in the Member States' discretion concerning Annex II projects (IEA 1993b). Assessment can also be aided by the recent development of environmental auditing for existing organizations, be they private-sector firms or local authorities. The information from environmental auditing could provide a significant change in the quality and quantity of baseline data for EIA.

As EIA activity spreads, so more groups will become involved. Training is vital in both the EIA process, which may have some commonality across countries, and in procedures that may be more closely tailored to individual national contexts. The recently instituted EC funded LIFE programme A4 on education, training and information "to promote environmental training in administrative and professional circles" is a welcome development.

References

Association of County Councils, District Councils and Metropolitan Authorities 1990. *Environmental practice in local government*. London: Association of County Councils.

British Standards Institution 1992. *British Standard 7750: specification for environmental management systems*: Linford Wood, Milton Keynes: BSI.

CEC (Commission of the European Communities) 1985. On the assessment of the effects of certain public and private projects on the environment. *Official Journal* L175, 5 July 1985.

CEC 1991. Draft proposal for a Council regulation establishing a Community scheme for the evaluation and improvement of environmental performance in certain activities and the provision of relevant information to the public (*Eco-Audit*). XI/83/93, Rev. 4, Brussels: CEC.

CEC 1993. Report from the Commission of the Implementation of Directive 85/337/EEC on the assessment of the effects of certain public and private projects on the environment, and annex for the United Kingdom. COM(93)28 final, Vol. 12. Brussels: CEC

Coles, T. F., K. G. Fuller, M. Slater 1992. *Practical experience of environmental assessment in the UK*. East Kirkby, Lincolnshire, England: Institute of Environmental Assessment.

CPRE 1991. *The environmental assessment Directive – five years on*. London: Council for the Protection of Rural England.

CPRE 1992. *Mock Directive*. London: Council for the Protection of Rural England.

DoE (Department of the Environment) 1991. *Monitoring environmental assessment and planning*. London: HMSO.

DoE 1992. *Environmental assessment and planning:extension of application*. Consultation paper, June 1992. London: Department of the Environment.

FoE 1990. *Environmental audits of local authorities: terms of reference*. London: Friends of the Earth.

Grayson, L. 1992. Environmental auditing: a guide to best practice in the UK and Europe. London/Letchworth: The British Library/Technical Communications.

IEA (Institute of Environmental Assessment) 1993a. *Guidelines for the environmental assessment of road traffic* (Guidance Note 1). East Kirkby, Lincolnshire, England: Institute of Environmental Assessment.

IEA 1993b. Personal communication, March 1993.

Local Government Management Board 1991. *Environmental audit for local authorities: a guide*. Luton: Local Government Management Board.

Simmons & Simmons 1993. *Environmental Newsletter* **16**.

Wood, C. & C. Jones 1992. The impact of environmental assessment on local planning authorities. *Journal of Environmental Planning and Management* **35**(2), 115–27.

CHAPTER 13
Widening the scope: strategic environmental assessment

13.1 Introduction

One of the most recent trends in EIA is its application at earlier, more strategic stages of development – at the level of policies, plans and programmes. This so-called strategic environmental assessment (SEA) has been carried out as an extension of project EIA in a relatively low-key manner in the USA since the enactment of the NEPA. However, in the EC it has recently come to be viewed as a valuable technique for achieving sustainable development, and a directive on SEA is being discussed. The UK government is also taking steps to appraise the environmental impacts of its policies. SEA is likely to be an area of strong growth in the years ahead, and this in turn will influence – and improve – the process of project EIA.

This chapter discusses the need for SEA, and some of its limitations. It reviews the "best practice" status of SEA in other countries, and discusses the most recent legislative advances in SEA, the proposed EC Directive on SEA of November 1992, and the UK DoE's guidebook *Policy appraisal and the environment* of September 1991. It considers the application of SEA in the UK, and in particular the environmental appraisal of development plans required by Planning Policy Guidance Note 12 on development plans. The chapter concludes with proposals on links between SEA and sustainable development. By necessity this chapter must radically simplify many of the issues surrounding SEA. The reader is referred to Therivel et al. (1992) and various articles in the September 1992 issue of *Project Appraisal* for a discussion in greater depth.

13.2 Strategic environmental assessment (SEA)

Definitions

Strategic environmental assessment can be defined as "the formalized, systematic and comprehensive process of evaluating the environmental impacts of a policy, plan or programme and its alternatives, including the preparation of a written report on the findings of that evaluation, and using the findings in publicly accountable decision-making" (Therivel et al. 1992). It is, in other words, the EIA of policies, plans and programmes, keeping in mind that the process of evaluating environmental impacts at a strategic level is not necessarily the same as that at a project level. Although policies, plans and programmes are generally all described as "strategic" in this and other texts, they are not the same thing, and may themselves require different forms of environmental appraisal. A policy is generally defined as an inspiration and guidance for action, a plan as a set of co-ordinated and timed objectives for the implementation of the policy, and a programme as a set of projects in a particular area (Wood 1991). Here, policies, plans and programmes will be addressed as PPPs unless otherwise noted. PPPs may be sectoral (e.g. transport, mineral extraction), spatial (e.g. national, local), or indirect (e.g. education, research and development, privatization).

In theory PPPs are tiered; a policy provides a framework for the establishment of plans, plans provide frameworks for programmes, and programmes lead to projects. For instance, the UK government's road policies, set out in its White Papers on roads, give rise to suggested road schemes which are incorporated in a national roads programme. This in turn forms a basis for the proposal of specific routes, for which project EIAs are prepared. In practice, as will be discussed later, these tiers are amorphous and fluid, without clear cut-off points. The EIAs for these different PPP tiers can themselves be tiered, as shown in Figure 13.1, so that issues considered at higher tiers need not be reconsidered at the lower tiers.

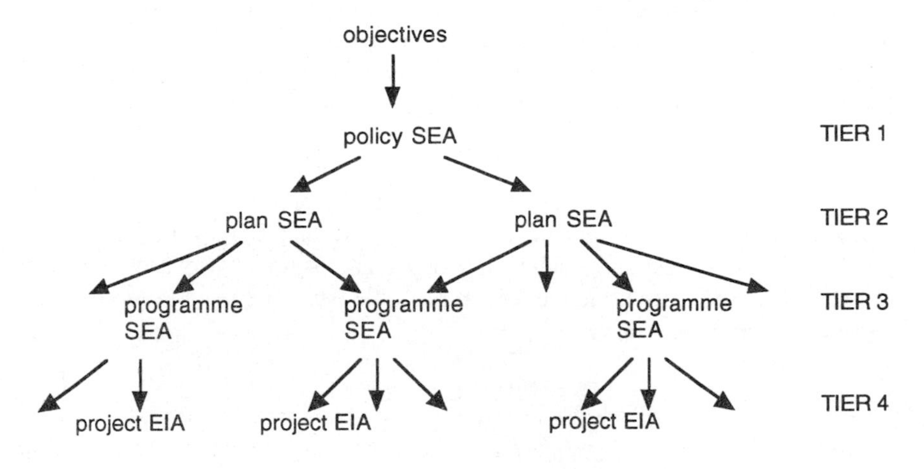

Figure 13.1 Tiers of SEA and EIA.

Need for SEA

Various arguments have been put forward for a more strategic level of EIA, most of which *relate to problems with the existing system of project EIA*. Project EIAs react to development proposals rather than anticipating them, so they cannot steer development towards environmentally "robust" areas or away from environmentally sensitive sites.

Project EIAs do not adequately consider the cumulative impacts[1] caused by several projects, or even by one project's subcomponents or ancillary developments. As noted in Chapter 10, under present regulations different EIAs may be prepared for a power station, the gas pipeline providing the power station's fuel, the facilities for receiving and processing the gas, and the transmission lines carrying electricity away from the power station. Separate EIAs can also be prepared for different sections of one road. Small individual mineral extraction operations may not need an EIA, but the total impact of several of these projects may well be "significant". At present in the UK there is no legal requirement to prepare comprehensive cumulative impact statements for these types of projects.

Project EIAs cannot fully address alternative developments or mitigation measures, because in many cases these alternatives will be limited by choices made at an earlier, more strategic level. In many cases a project will already be planned quite specifically, with irreversible decisions taken, by the time an EIA is prepared.

Project EIAs cannot address the impacts of potentially damaging actions that are not regulated through the approval of specific projects. Examples of such actions include farm management practices, privatization, and new technology. Project EIAs often have to be carried out in a very short period of time because of financial constraints and the timing of planning applications. This limits the amount of baseline data that can be collected, and the quality of analysis that can be undertaken. For instance, the planning timeframes of many projects have required their ecological impact assessments to be carried out in the winter months, when it is difficult to identify plants and when many animals are either dormant or have migrated. The amount and type of public consultation undertaken in project EIA may be similarly limited.

By being carried out earlier in the decision-making process and encompassing all of the projects of a certain type or in a certain area, SEA can ensure that alternatives are adequately assessed, cumulative impacts are considered, the public is fully consulted, and decisions concerning individual projects are made in a proactive rather than reactive manner. As will be discussed later, SEA is also seen as a central step in the *achievement of sustainable development*.

Problems with SEA

On the other hand, the implementation of SEA is fraught with both technical and procedural problems. On the *technical* side, the many potential future developments spread over a large area can lead to great analytical complexity. Information about existing and projected future environmental conditions, and about the nature, scale and location of future development proposals is usually very limited, so the impacts of these developments cannot be predicted precisely. The large number and variety of alternatives to be considered further complicates the process, as do requirements for public participation. There is a general lack of information about SEA, and a dearth of case studies where SEA has been successfully applied, particularly to policies, so there are few models of how to carry out SEA.

More intractable than these technical/information problems are those inherent in the *policy-making process*. Many PPPs are nebulous and they evolve in an incremental and unclear fashion, so there is no clear time when their environmental impacts can best be assessed: "the dynamic nature of the policy process means issues are likely to be redefined throughout the process, and it may be that a series of actions, even if not formally sanctioned by a decision, constitute policy" (Therivel et al. 1992).

PPPs do not have clear boundaries at which they stop and other policies begin. For instance, it is impossible to distinguish fully between policies for transport, energy, and land-use, as they all affect each other. Furthermore, the actual effects of policies are strongly influenced by how the policies are interpreted when they are implemented. The government's emphasis on deregulating many government activities also means that increasing numbers of interest groups with increasingly diverse aims are involved in policy formulation. As a result of these factors, policies may be more fragmented, less well understood and can often have unintended and unpredictable outcomes.

Finally, and most importantly, policy-making is a political process. Decision-makers will weigh up the implications of a PPP's environmental impacts in the wider context of their own interests and those of their "constituents". The UK government's negative response to the EC proposals for SEA (see below) certainly suggests that SEA is currently not perceived as politically beneficial to the existing administration.

13.3 Evolving systems of SEA

Despite these problems, SEA is carried out in certain countries, and looks likely to be increasingly used worldwide. This section looks at existing systems of SEA and recent proposals for SEA systems.

Existing systems of SEA

In the *USA*, hundreds of "programmatic environmental impact statements" have been prepared by government agencies under the National Environmental Policy Act since 1970, primarily as an extension of project EIA to the programme and plan level. Individual government agencies have regulations and guidelines to implement the NEPA. For instance the Department of Housing and Urban Development has prepared a particularly useful *Areawide environmental impact assessment guidebook* (US HUD 1981) that proposes the following steps:

- determine the need and feasibility of the study
- establish area boundaries, analysis units, and an environmental database
- identify areawide alternatives and allocate areawide "totals" to analysis units
- scope and establish a work plan
- undertake environmental analysis (document baseline conditions, establish units of demand/consumption, estimate impacts)
- carry out impact synthesis and evaluation, and compare alternatives, and
- make recommendations on mitigation measures and preferred alternatives.[2]

The Forest Service carries out SEAs for its national, regional and local plans, as well as for individual projects. The federal government has also recently published reports on SEA-related subjects such as cumulative impact assessment (USEPA 1992) and the assessment of biodiversity (CEQ 1993).

About one-third of the USA's 50 states have their own EIA regulations, but only a few of these also cover PPPs. California's has been the most effective SEA system: approximately 120 "programmatic environmental impact reports" are prepared annually under the California Environmental Quality Act 1986 (State of California 1986). Like project EIAs, these include a description of the action, a section on the baseline environment, an evaluation of the action's impacts, alternatives, an indication of why some impacts were not evaluated, organizations consulted, responses of these organizations to the EIS, and the agency's response to the responses.

The Netherlands has required the preparation of a "section on the environment" for plans for waste management, drinking-water supply, energy and electricity supply, and some land-use plans since 1987. In 1989, the Dutch National Environmental Policy Plan (Ministry of Housing, Physical Planning and Environment 1989) expanded this requirement to include all "policy proposals which might have important consequences for the environment". More importantly, the NEPP includes a clause which requires existing policy areas to be assessed to determine how well they fulfil the objectives of sustainable development:

> The government will give an account of how the recommendations contained in the Brundtland Report are to be given substance in each ministry and area of policy. At the same time there will also be an assessment of the extent to which the instruments of the various policy areas contribute to effecting sustainable development. Exploratory work will . . . be carried out in the following areas of policy: physical planning policy, housing, technology, markets and prices, energy, science, traffic, fiscal policy, agriculture, justice/enforcement, education, industry.
>
> (Ministry of Housing, Physical Planning and Environment 1989)

These requirements have not been fully implemented as yet. However, their final form and their effectiveness are likely to be of great interest to other countries considering implementing SEA.[3]

Other countries carrying out SEA on a limited or preliminary basis include France, Germany, Japan and New Zealand. Various international bodies and development agencies are also setting up SEA procedures. However, little practical experience has resulted from these systems to date (Therivel et al. 1992). It is against this background of relatively limited application of SEA and development of SEA methodologies that the proposed EC Directive on SEA, and other guidance in the UK, must be seen.

Proposed EC Directive on SEA

Although the original version of EC Directive 85/337 was intended to also apply to PPPs, the final version applies only to projects. However, the EC's Fourth Action Programme on the Environment of 1987 (CEC 1987) stated that EIA "will also be extended, as rapidly as possible, to cover policies and policy statements, plans and their implementation, procedures, programmes . . . as well as individual projects", and the Fifth Action Programme of 1992 (CEC 1992a) reiterated this aim within its overall framework for achieving sustainable development.

As a first step to implementing this aim, the EC's Directorate General XI released a proposal for a Directive on SEA to experts in the Member States in March 1991 (CEC 1991). A later, quite different, version was released in November 1992 (CEC 1992b). The newest proposal would apply to sectoral "planning actions" on agriculture, industry, energy, transport, tourism, water resources, and waste disposal; and "planning actions which plan, promote, regulate or otherwise influence the future use and development of land in urban or rural areas" (CEC 1992b). At the discretion of Member States, it could also apply to major revision of these actions or to other actions. The proposed Directive identifies four bodies responsible for each SEA. A lead authority, namely the authority responsible for preparing the PPP, would undertake the SEA. A designated environmental authority would advise on the adequacy of the SEA, the likely environmental impacts of the action, and suggested mitigation measures. Finally, the public, and other Member States if they are likely to be affected by the action, are informed about the contents of the EIS, the decision, and any mitigation, monitoring or other conditions forming part of the decision.

The proposed Directive would require impacts on human beings, flora, fauna, landscape, natural resources, cultural heritage and material assets to be assessed for the proposed option and its main alternatives. The SEA would include discussion of:

- the PPP and its main objectives
- environmental protection objectives and related measures established at the EC or Member State level, relevant to the proposed action
- existing environmental problems, especially those related to protected and/or sensitive areas

- likely significant environmental impacts of the proposed action, main alternatives, and mitigation measures
- monitoring arrangements, and an outline of projects or other measures expected as a result of implementing the action
- an outline of difficulties encountered in compiling information, and
- a summary of the above (CEC 1992b).

The type and extent of public consultation would be decided by the Member States. The competent authority would reach a decision on the proposed PPP, taking the findings of the SEA and various parties' opinions into account.

Despite the proposed Directive's similarities to Directive 85/337 on EIA for projects, the Member States' response to the proposed Directive has reportedly been chilly, and the November 1992 proposal was rejected just one month later. In the UK, central government is concerned that, for most PPPs, there is no clear moment when a decision is made, and that any SEA system would have to be extremely flexible to accommodate the large variety of types of decisions, and the inherent uncertainty of some decisions. There is also concern that SEA would require expertise that agencies currently do not have, and that it would require them to interact more with other bodies to gain information and agreement, which would be expensive and politically contentious. Reportedly the response in other countries has been no more encouraging. It is unlikely that an SEA Directive will be adopted soon, although interest in the subject remains high.

UK government guide: policy appraisal and the environment

Despite its lack of enthusiasm for the proposed EC Directive, the UK government has promoted the environmental appraisal of its own policies. The government's White Paper on the Environment of September 1990 (DoE 1990) promised that it would carry out "a review of the way in which the costs and benefits of environmental issues are assessed within the Government". A year later the DoE published *Policy appraisal and the environment* and distributed copies to central government mid-level managers (DoE 1991). The guidebook's procedures are not mandatory, but they aim to assist civil servants in considering the environmental repercussions of their decisions and to promote a "cultural change" in how civil servants formulate policies.

Policy appraisal and the environment suggests that the department or agency from which the PPP originates should carry out the policy appraisal. Policy appraisal should apply to a wide range of PPPs, since "in general, any policy or programme which concerns changes in the use of land or resources, or which involves the production or use of materials or energy, will have some environmental impact". The guidebook does not clearly state which aspects of the environment should be considered in a policy appraisal, but does list "environmental receptors", namely air and atmosphere, water resources, water bodies (size and situation), soil, geology, landscape, climate, energy (light and other electromagnetic radiation, noise and vibration), human beings (physical and mental health and wellbeing), cultural heritage, and other living organisms (flora and fauna). According to the guidebook, policy appraisal should involve the following steps:

- summarizing the policy issue
- listing the objectives
- identifying the constraints
- specifying the options
- identifying the costs and benefits
- weighing up the costs and benefits
- testing the sensitivity of the options
- suggesting the preferred option
- setting up any monitoring necessary, and
- evaluating the policy at a later stage.

The guide gives no clear procedures for public consultation. Although the guide recognizes the limitations of monetary valuation, it does seem to advocate its use: more than half of its text concerns cost–benefit analysis, and the general impression is that it is primarily a compendium for valuing environmental resources (Therivel et al. 1992). At the time of writing, the guide was not being widely used, or even being used formally on case studies. However, this is likely to change with the publication in February 1992 of Planning Policy Guidance Note 12 (PPG12) on development plans and regional planning guidance (DoE 1992).

13.4 SEA and development plans

Planning Policy Guidance Note 12

PPG12's ostensible purpose is to explain the provisions of the Planning and Compensation Act 1991 as they relate to development plans. However, it also amends the Town and Country Planning Act 1990 to require local authorities, in drawing up their development plans, to have regard to environmental considerations, and to include policies for the conservation of the natural beauty and amenity of the land. PPG12 notes that the preparation of development plans "can contribute to the objective of ensuring that development and growth are sustainable" and explains that development plans should take environmental considerations into account:

> Most policies and proposals in all types of plan will have environmental implications, which should be appraised as part of the plan preparation process. Such an environmental appraisal is the process of identifying, quantifying, weighing up and reporting on the environmental and other costs and benefits of the measures which are proposed. All the implications of the option should be analysed, including financial, social and environmental effects. A systematic appraisal ensures that the objectives of a policy are clearly laid out, and the trade-offs between options identified and assessed. Those who later interpret, implement and build on the policy will then have

a clear record showing how the decision was made; in the case of development plans, this should be set out in the explanatory memorandum or reasoned justification. But the requirement to 'have regard' does not require a full environmental impact statement of the sort needed for projects.

(DoE 1992)

PPG12 refers local authorities to *Policy appraisal and the environment* for guidance on the treatment of environmental issues in developing their planning policies. A survey by Bowen (1993) of over 40 local authorities in the outer London counties suggested that local authorities were still trying to come to grips with the advice in PPG12. Few were familiar with *Policy appraisal and the environment*, and almost all were concerned about resourcing such activities and about the potential delays to development plan work that might result. The majority urgently needed more guidance from central government. However, at the time of writing (early 1993), some local authorities had carried out or were planning to carry out environmental appraisals of their development plans, including Lancashire, Kent and Cambridgeshire County Councils. Of these, Lancashire's was the most widely discussed and is likely to be a starting point for other local authority appraisals.

SEA and the Lancashire Structure Plan

Lancashire's appraisal was based on a matrix, which lists the 164 policy statements of its structure plan on the vertical axis, and eleven environmental components on the horizontal axis. Figure 13.2 shows a section of the matrix. The impact of each policy statement on each environmental component was recorded in the relevant cell, using numerical scores from +2 (sizeable benefit) to -2 (sizeable cost). The scores take account of effects that are indirect as well as direct, longer-term as well as immediate, and wide-scale as well as local. The penultimate column of the matrix sums up each policy's scores to form a "sustainability score", which gives an indication of the policy's overall impact on environmental resources. The final column indicates whether revision of the policy was felt necessary because of high negative scores or poor wording of the policy (Pinfield 1992).

The structure plan's various policy areas can then be ranked in terms of how well they perform on the sustainability score. In Lancashire's case, policies on rural landscapes scored highest, and those on minerals and waste disposal scored lowest. Similarly, the environmental resources can be ranked in terms of the structure plan's impact on them. Lancashire's plan had the most positive effect on man-made features and the most negative effects on geology.

This technique has some strong limitations, as stated by Pinfield (1992). The use of a single matrix is rudimentary and subjective. The methodology assumes that all environmental components and policies should be given the same weighting, and that scores from one cell (e.g. energy) can be added to scores from other cells (e.g. wildlife); obviously, in many cases these assumptions are not correct. There is no clear distinction between a strong impact (±2) and a weak one (±1). The approach is also very much an internal one, largely produced by one officer of the authority.

Structure plan policy	Environmental component												
	Geology	Soil	Air	Water	Energy	Land	Wildlife	Landscape	Man-made features	Open space	Human beings	Sustainability score	Policy revision
1	–	+2	–	–1	+1	+2	+1	+2	+1	+1	+1	+10	
2													
3													
. . .													

Figure 13.2 Section of Lancashire Structure Plan environmental impact matrix. (*Source*: Pinfield 1992)

However, Pinfield argues that the methodology has advantages, particularly in its simplicity and relatively speedy application (it was prepared within two months). It could be expected that future appraisals of development plans will expand on this technique. Further information on Lancashire's appraisal is given in a technical paper (Lancashire County Council 1992) and by Pinfield (1992).

SEA and the Kent Structure Plan

The SEA of the policies in the Kent Structure Plan 3rd Review Consultation Draft provides another example of innovative thinking on the environmental appraisal of development plans (Kent County Council 1993). The Kent approach is set in the context of the recent advice of the Local Government Management Board (1992) – "A concept of sustainable development transforms a local authority's approach from a series of ad hoc steps to a strategy and from the need for controls alone to a need for policies. The local authority has to plan, to co-ordinate and to manage for sustainable development." Figure 13.3 provides an overview of Kent's planned approach to SEA.

The first step, as in the Lancashire approach, was to evaluate the structure-plan policies in a matrix, in terms of their impact on the environment and their contribution to the development of sustainability. In the Kent case, the environmental components were grouped into three scales:

- local (including impact on the quality of life of people, townscape, noise, etc.)
- county-wide (including impact on air and water quality, ecology, etc.)
- global (including impact on renewable and non-renewable resources, etc.).

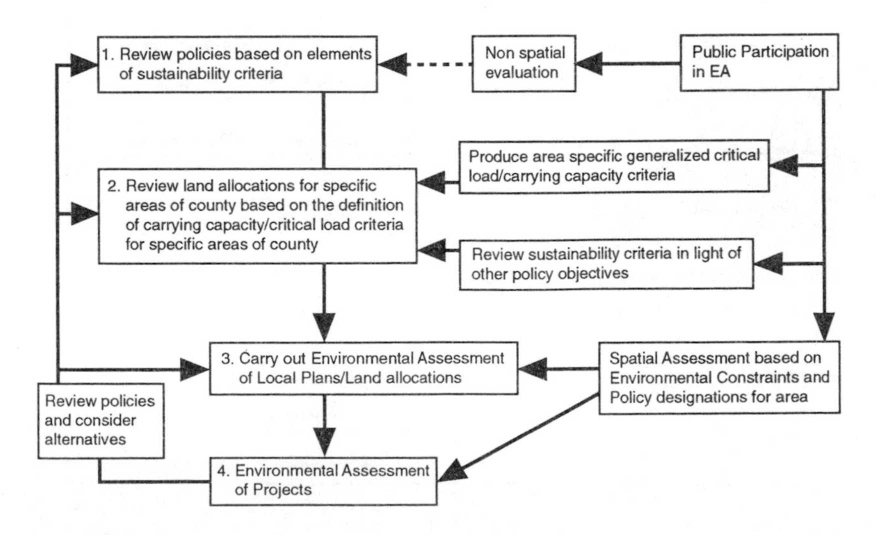

Figure 13.3 SEA of the Kent Structure Plan 3rd Review in relation to assessment of local plans and projects. (*Source:* Kent County Council 1993).

The county's 130 structure plan policies were then scored against the various criteria, also using a five-point scale, but with ticks and crosses rather than the numbers of the Lancashire approach (see Fig. 13.4). The appraisal exercise clearly indicated the emphasis of structure plan policies in relation to environ-

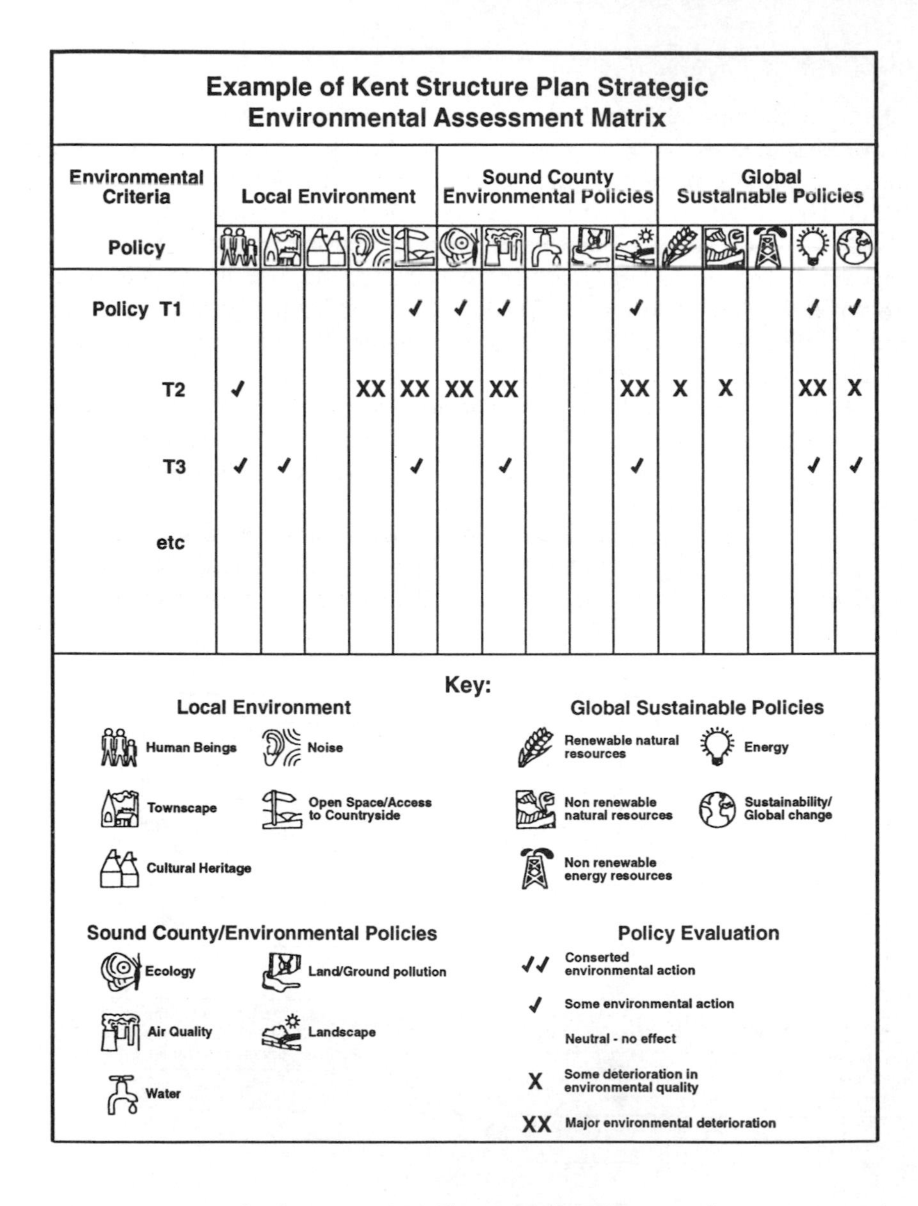

Example of Kent Structure Plan Strategic Environmental Assessment Matrix

Environmental Criteria / Policy	Local Environment					Sound County Environmental Policies					Global Sustainable Policies				
Policy T1					✓	✓	✓			✓				✓	✓
T2	✓			XX	XX	XX	XX			XX	X	X		XX	X
T3	✓	✓			✓		✓			✓				✓	✓
etc															

Key:

Local Environment: Human Beings; Noise; Townscape; Open Space/Access to Countryside; Cultural Heritage

Global Sustainable Policies: Renewable natural resources; Energy; Non renewable natural resources; Sustainability/ Global change; Non renewable energy resources

Sound County/Environmental Policies: Ecology; Land/Ground pollution; Air Quality; Landscape; Water

Policy Evaluation:
✓✓ Conserted environmental action
✓ Some environmental action
Neutral - no effect
X Some deterioration in environmental quality
XX Major environmental deterioration

Figure 13.4 Example of Kent Structure Plan strategic environmental assessment matrix. (*Source:* Adapted from Kent County Council)

mental impacts. For example, economic policies were generally positive on local criteria but less so on county and global criteria; some transport policies were seen as generating major deterioration in environmental quality at all levels. A summary figure was used to show the overall emphasis of structure plan policies, with any policy having even a single negative impact on the environment being shown as negative, and those with only positive impacts being shown as positive.

The Kent study also has its limitations as noted in the study. Environmental issues must be considered in relation to social and economic needs; the views of the public on sustainability and the structure plan were not assessed; the approach did not pick up the implications of the interaction of policies; and it was only the first step in a process, as noted in Figure 13.3. At future stages, land allocations for specific areas of the county will be reviewed. This will involve the crucial and central issue in SEA of identifying critical load/carrying capacity criteria, in relation to factors such as air quality, water quality and traffic congestion. EIA of local plans and their land allocations, and then of projects, would follow in this tiered approach.

13.5 SEA and sustainable development[3]

In addition to resolving many of the difficulties associated with project EIA, SEA can be a central step in the achievement of sustainable development. SEA systems can be divided into two broad types, based on whether they *expand the existing system of project EIA (incremental system)* or whether they *"trickle down" the objective of sustainable development*. Figure 13.5 summarizes the differences between "incremental" and "trickle-down" systems of SEA. The former aims to head towards sustainable development by extending "Green" policies; the latter aims to implement sustainable development directly by making it a central objective of each PPP.

Of the various existing and proposed systems of SEA discussed in §13.2, most could be described as being incremental, in that they expand existing systems of project EIA to include plans and programmes (rarely policies), in an attempt to lead to "more sustainable" practices. In the USA, for instance, the same laws and procedures apply to projects and PPPs, and the proposed EC Directive on SEA is clearly an expansion of Directive 85/337 on EIA of projects.

The Dutch SEA system, described in §11.2 and §13.2, is the first system developed to assist in the implementation of a central objective – sustainable development – at the level first of individual policies, then plans, then programmes and ultimately projects. The carrying capacity studies of the report *Concern for tomorrow* (National Institute for Health and Environment 1988) determined the changes needed to achieve sustainable development. For instance the report noted that emissions of various air pollutants would need to be reduced by 70–90% to achieve sustainability. The National Environmental Policy Plan of 1989 in turn set out

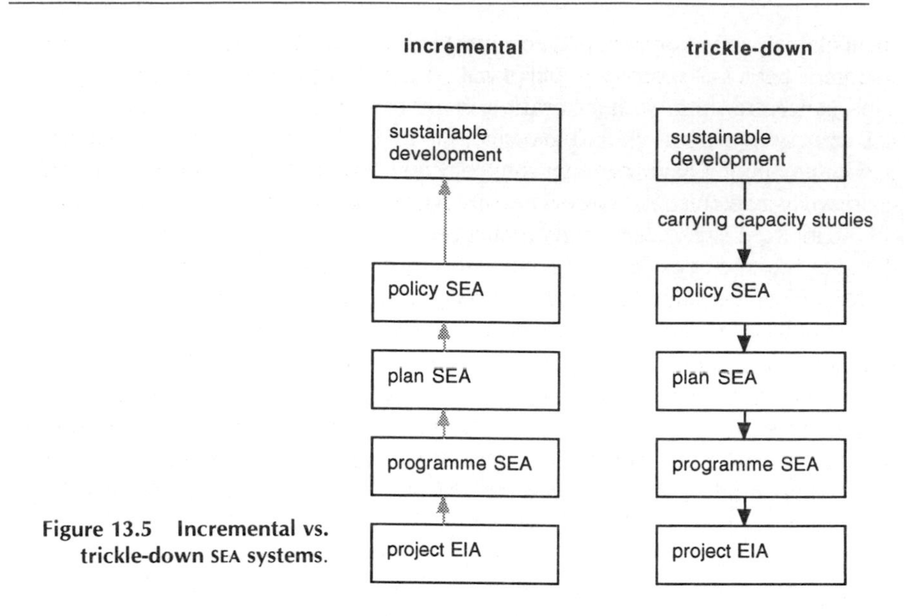

Figure 13.5 Incremental vs. trickle-down SEA systems.

specific actions for achieving these changes, including the requirement that policy instruments would be assessed for their contribution to effecting sustainable development (see §13.2). This form of "trickle-down" SEA system is likely to become a key method for implementing sustainable development.

Broadly, a "trickle-down" SEA system would involve the following steps:

- a commitment to the objective of sustainable development
- a determination of the parameters within which sustainable development is to be achieved (e.g. area, resource, time)
- a determination of carrying capacity within these parameters
- SEA of all relevant PPPs using alternative development scenarios which do not exceed the carrying capacity
- choice of one scenario that optimizes socio-economic factors
- EIA of individual projects within the constraints set by the SEA, and
- a monitoring programme that would give feedback to modify any/all of the above steps (Therivel et al. 1992).

The Kent County Council approach, noted in the previous section (Fig. 13.3), includes several elements of this approach.

Obviously, each of the above steps has strong practical limitations. The political systems of many countries focus on short-term objectives that react to specific events or problems, rather than on long-term preventive goals such as sustainable development. Effective parameters for sustainable development are virtually impossible to set. At a regional or national level, data collection and political agreements may be feasible, but will constantly be influenced by outside factors[4]. At a global level, data may be consistent (e.g. trans-boundary pollution will be accounted for) but the amount of data needed is overwhelming, and political agreement is likely to be impossible. Even if parameters are agreed upon, the determination of carrying capacity within those parameters is very complex: carrying capacity is affected by such factors as personal consumption, technical innovation, and trade in resources within and outside the region. In many cases, the determination of carrying capacity requires much more environmental information than is presently available. The generation of alternative development scenarios and the prediction of the impacts of these scenarios have all the limitations discussed in §13.2. And, as discussed in Chapter 7, monitoring is only in its infancy.

Despite these limitations, SEA is still likely to be one of the most direct and effective ways of ensuring that human activities are carried out at a level that is environmentally sustainable over time. Many countries are already carrying out research on how to overcome these problems, and the Dutch initiative is a particularly valuable model in this sense. The NEPP shows how an overarching objective, sustainable development, can be trickled down using a variety of policies, methodologies and technical requirements. More importantly, it shows that the concepts of sustainable development and SEA can be effectively implemented even if the relevant methodologies are still relatively underdeveloped, as long as they are carried out in the spirit of "best practice". *Concern for tomorrow* pointed to a broad direction and order of magnitude of change, so that implementation could be rapidly begun, even while further more precise studies were being carried out. Other countries could expand their research on EIA and carrying capacity to develop their own "trickle-down" SEA systems, and begin to act on the preliminary findings of these studies, even as they refine the concepts through further studies.

13.6 Summary

Strategic environmental assessment (SEA) is the EIA of policies, plans and programmes. SEA is often considered to be a tiered or nested process, where EIAs of individual projects are carried out within the framework established by the SEA of a programme (group of similar projects), which in turn takes place within the framework of an SEA of a plan (co-ordinated and timed objectives), and before that of a policy (guidance for action).

SEA is still relatively new. SEA systems have been established in the USA (particularly California), the Netherlands, and to a lesser extent in some other countries, but these are generally limited to plans and programmes (not policies) and are not yet well developed. The EC is considering adopting a Directive on SEA. In September 1991, the UK Department of the Environment published a guide on Policy Appraisal and the Environment. A subsequent Planning Policy Guidance Note (PPG12) encouraged local authorities to undertake environmental appraisals of their plans and there is evidence of some interesting activity in such appraisal in some authorities.[5]

However, SEA can have a greater rôle than merely expanding EIA to the more strategic levels of decision-making. As shown in particular by the Dutch National Environmental Policy Plan of 1989, SEA can be a key step towards achieving sustainable development. It can help to "trickle down" the objective of sustainable development from policies to plans, programmes and ultimately projects. SEA is the link between EIA of projects, as it is presently being carried out, and the achievement of a level of human activities that maintains the quality of the environment.

References

Bowen, S. 1993. *Implementation of strategic environmental assessment*. MSc dissertation, School of Planning, Oxford Brookes University.

CEC 1987. *Fourth action programme on the environment*, COM(86)485 final. Brussels: Commission of the European Communities.

CEC 1991. *Draft proposal for Directive on the environmental assessment of policies, plans and programmes*, XI/194/90-EN-REV. 4. Brussels: Commission of the European Communities.

CEC 1992a. *Towards sustainability: a European Community programme of policy and action in relation to the environment and sustainable development*, COM(92)23 final. Brussels: Commission of the European Communities.

CEC 1992b. *Proposal for a directive on the environmental assessment of actions approved during the planning process*. Document XI/745/92. Brussels: Commission of the European Communities.

CEQ 1993. *Incorporating biodiversity considerations into environmental impact assessment under the National Environmental Policy Act*. Washington DC: Council on Environmental Quality.

DoE (Department of the Environment) 1990. *Our common inheritance*. London: HMSO.

DoE 1991. *Policy appraisal and the environment*. London: HMSO.

DoE 1992. *Planning Policy Guidance Note 12: Development plans and regional planning guidance*. London: HMSO.

Kent County Council 1993. *Kent structure plan 3rd review consultation draft: strategic environmental assessment of consultation draft policies*. Maidstone: Kent County Council.

Lancashire County Council 1992. *Lancashire structure plan technical report No. 13 – environmental appraisal of the plan*. Preston: Lancashire County Council.

Ministry of Housing, Physical Planning and Environment 1989. *National environmental policy plan – to choose or to lose*. The Hague: Netherlands Ministry of Housing, Physical Planning and Environment.

National Institute for Health and Environment 1988. *Concern for tomorrow*. Bilthoven, The Netherlands: Government of the Netherlands.

Pinfield, G. 1992. Strategic environmental assessment and land-use planning. *Project Appraisal* 7(3), 157–63.

State of California 1986. *The California Environmental Quality Act*. Sacramento: Office of Planning and Research.

Therivel. R., E. Wilson, S. Thompson, D. Heaney, D. Pritchard 1992. *Strategic environmental assessment*. London: RSPB/Earthscan.

USEPA (Environmental Protection Agency) 1992. *A synoptic approach to cumulative impact assessment*. Washington DC: Office of Research and Development.

USHUD (Department of Housing and Urban Development) 1981. *Areawide environmental assessment guidebook*. Washington, DC: Office of Policy Development and Research.

Wood, C. 1991. EIA of policies, plans and programmes. *EIA Newsletter* **5**, 2–3.

UN World Commission on Environment and Development [Brundtland Commission] 1987. *Our common future*. Oxford: Oxford University Press.

Notes

1. Cumulative impacts can be of several types: additive impacts of projects that individually have an insignificant impact but in total have a significant impact; synergistic impacts where several projects' total impacts exceed the sum of their individual impacts; threshold/saturation impacts where the environment may be resilient up to a certain level and then becomes rapidly degraded; induced impacts where one project may trigger secondary development; time- or space-crowded impacts where the environment does not have the time to recover from one impact before it is subject to the next one; and indirect impacts where the impact is triggered at some time or distance away from where it ultimately occurs.
2. Note how the main distinction between project EIA procedures and this (and other) SEA procedures is in the early steps: defining need, objectives, boundaries, constraints.
3. Therivel et al. (1992) suggest that SEA should aim to achieve sustainability rather than sustainable development, arguing that some of the assumptions behind the concept of sustainable development have not been proven. The space does not exist here to pursue this argument; the term sustainable development is used to mean both sustainability and sustainable development.
4. For instance the Dutch NEPP includes two very different costing estimates for its implementation: one if other countries aim towards sustainable development, and another if they do not.
5. In November 1993, the UK Department of the Environment announced the publication of *Environmental appraisal of development plans: a good practice guide*. The DoE sees the guide as providing "advice on a straightforward range of techniques and procedures. It clearly shows the benefits of environmental appraisal, and how it can be easily integrated into each stage of the plan-making process. Environmental appraisal can be an important tool in helping to meet the wider goal of sustainable development. Local authorities should carry out plan appraisals to ensure that they contribute to this aim."

APPENDIX 1
The text of the EC Directive

Council Directive of 27 June 1985 on the assessment of the effects of certain public and private projects on the environment (85/337/CEC)

THE COUNCIL OF THE EUROPEAN COMMUNITIES,

Having regard to the Treaty establishing the European Economic Community, and in particular Articles 100 and 235 thereof,
Having regard to the proposal from the Commission[1],
Having regard to the opinion of the European Parliament[2],
Having regard to the opinion of the Economic and Social Committee[3],

Whereas the 1973[4] and 1977[5] action programmes of the European communities on the environment, as well as the 1983[6] action programme, the main outlines of which have been approved by the Council of the European Communities and the representatives of the governments of the Member States, stress that the best environmental policy consists in preventing the creation of pollution or nuisances at source, rather than subsequently trying to counteract their effects; whereas they affirm the need to take effects on the environment into account at the earliest possible stage in all the technical planning and decision-making processes; whereas to that end, they provide for the implementation of procedures to evaluate such effects;

Whereas the disparities between the laws in force in the various Member States with regard to the assessment of the environmental effects of public and private projects may create unfavourable competitive conditions and thereby directly affect the functioning of the common market; whereas, therefore, it is necessary to approximate national laws in this field pursuant to Article 100 of the Treaty;

Whereas, in addition, it is necessary to achieve one of the Community's objectives in the sphere of the protection of the environment and the quality of life;

Whereas, since the Treaty has not provided the powers required for this end, recourse should be had to Article 235 of the treaty;

Whereas general principles for the assessment of environmental effects should be introduced with a view to supplementing and coordinating development consent procedures governing public and private projects likely to have a major effect on the environment;

Whereas development consent for public and private projects which are likely to have significant effects on the environment should be granted only after prior assessment of the likely significant environmental effects of these projects has been carried out; whereas this assessment must be conducted on the basis of the appropriate information supplied by the developer, which may be supplemented by the authorities and by the people who may be concerned by the project in question;

Whereas the principles of the assessment of environmental effects should be harmonized, in particular with reference to the projects which should be subject to assessment, the main obligations of the developers and the content of the assessment;

Whereas projects belonging to certain types have significant effects on the environment and these projects must as a rule be subject to systematic assessment;

Whereas projects of other types may not have significant effects on the environment in every case and whereas these projects should be assessed where the Member States consider that their characteristics so require;

Whereas, for projects which are subject to assessment, a certain minimal amount of information must be supplied, concerning the project and its effects;

Whereas the effects of a project on the environment must be assessed in order to take account of concerns to protect human health, to contribute by means of a better environment to the quality of life, to ensure maintenance of the diversity of species and to maintain the reproductive capacity of the ecosystem as a basic resource for life;

Whereas, however, this Directive should not be applied to projects the details of which are adopted by a specific act of national legislation, since the objectives of this Directive, including that of supplying information, are achieved through the legislative process;

Whereas, furthermore, it may be appropriate in exceptional cases to exempt a specific project from the assessment procedures laid down by this Directive, subject to appropriate information being supplied to the Commission.

HAS ADOPTED THIS DIRECTIVE:

Article 1

1. This Directive shall apply to the assessment of the environmental effects of those public and private projects which are likely to have significant effects on the environment.
2. For the purposes of this Directive:

"project" means:

- the execution of construction works or of other installations or schemes,
- other interventions in the natural surroundings and landscape including those involving the extraction of mineral resources;

"developer" means:
the applicant for authorization for a private project or the public authority which initiates a project;

"development consent" means:
the decision of the competent authority or authorities which entitles the developer to proceed with the project.

3. The competent authority or authorities shall be that or those which the Member States designate as responsible for performing the duties arising from this Directive.
4. Projects serving national defence purposes are not covered by this Directive.
5. This Directive shall not apply to projects the details of which are adopted by a specific act of national legislation, since the objectives of this Directive, including that of supplying information, are achieved through the legislative process.

Article 2

1. Member States shall adopt all measures necessary to ensure that, before consent is given, projects likely to have significant effects on the environment by virtue inter alia, of their nature, size or location are made subject to an assessment with regard to their effects.
   These projects are defined in Article 4.
2. The environmental impact assessment may be integrated into the existing procedures for consent to projects in Member States, or, failing this, into other procedures or into procedures to be established to comply with the aims of this Directive.
3. Member States may, in exceptional cases, exempt a specific project in whole or in part from the provisions laid down in this Directive.
   In this event, the Member States shall:
   (a) consider whether another form of assessment would be appropriate and whether the information thus collected should be made available to the public;
   (b) make available to the public concerned the information relating to the exemption and reasons for granting;
   (c) inform the Commission, prior to granting consent, of the reasons justifying the exemption granted, and provide it with the information made available, where appropriate, to their own nationals.

   The Commission shall immediately forward the documents received to the other Member States.
   The Commission shall report annually to the Council on the application of this paragraph.

Article 3

The environmental impact assessment will identify, describe and assess in an appropriate manner, in the light of each individual case and in accordance with the Articles 4 to 11, the direct and indirect effects of a project on the following factors:

- human beings, fauna and flora,
- soil, water, air, climate and the landscape,
- the inter-action between the factors mentioned in the first and second indents,
- material assets and the cultural heritage.

Article 4

1. Subject to Article 2(3), projects of the classes listed in Annex I shall be made subject to an assessment in accordance with Articles 5 to 10.
2. Projects of the classes listed in Annex II shall be made subject to an assessment, in accordance with Articles 5 to 10, where Member States consider that their characteristics so require.
   To this end Member States may inter alia specify certain types of projects as being

subject to an assessment or may establish the criteria and/or thresholds necessary to determine which of the projects of the classes listed in Annex II are to be subject to an assessment in accordance with Articles 5 to 10.

Article 5

1. In the case of projects which, pursuant to Article 4, must be subjected to an environmental impact assessment in accordance with Articles 5 to 10, Member States shall adopt the necessary measures to ensure that the developer supplies in an appropriate form the information specified in Annex III inasmuch as:
   (a) the Member States consider that the information is relevant to a given stage of the consent procedure and to the specific characteristics of a particular project or type of project and of the environmental features likely to be affected;
   (b) the Member States consider that a developer may reasonably be required to compile this information having regard inter alia to current knowledge and methods of assessment.
2. The information to be provided by the developer in accordance with paragraph 1 shall include at least:
   - a description of the project comprising information on the site, design and size of the project,
   - a description of the measures envisaged in order to avoid, reduce and, if possible, remedy significant adverse effects,
   - the data required to identify and assess the main effects which the project is likely to have on the environment,
   - a non-technical summary of the information mentioned in indents 1 to 3.
3. Where they consider it necessary, Member States shall ensure that any authorities with relevant information in their possession make this information available to the developer.

Article 6

1. Member States shall take the measures necessary to ensure that the authorities likely to be concerned by the project by reason of their specific environmental responsibilities are given an opportunity to express their opinion on the request for development consent. Member States shall designate the authorities to be consulted for this purpose in general terms or in each case when the request for consent is made. The information gathered pursuant to Article 5 shall be forwarded to these authorities. Detailed arrangements for consultation shall be laid down by Member States.
2. Member States shall ensure that:
   - any request for development consent and any information gathered pursuant to Article 5 are made available to the public,
   - the public concerned is given the opportunity to express an opinion before the project is initiated.
3. The detailed arrangements for such information and consultation shall be determined by the Member States, which may in particular, depending on the particular characteristics of the projects or sizes concerned:
   - determine the public concerned,
   - specify the places where the information can be consulted,
   - specify the way in which the public may be informed, for example, by bill-posting within a certain radius, publication in local newspapers, organization of exhi-

bitions with plans, drawings, tables, graphs, models,
- determine the manner in which the public is to be consulted, for example, by written submissions, by public enquiry,
- fix appropriate time limits for the various stages of the procedure in order to ensure that a decision is taken within a reasonable period.

Article 7

Where a Member State is aware that a project is likely to have significant effects on the environment in another Member State or where a Member State likely to be significantly affected so requests, the Member State in whose territory the project is intended to be carried out shall forward the information gathered pursuant to Article 5 to the other Member State at the same time as it makes it available to its own nationals. Such information shall serve as a basis for any consultations necessary in the framework of the bilateral relations between two Member States on a reciprocal and equivalent basis.

Article 8

Information gathered pursuant to Articles 5, 6 and 7 must be taken into consideration in the development consent procedure.

Article 9

When a decision has been taken, the competent authority or authorities shall inform the public concerned of:
- the content of the decision and any conditions attached thereto,
- the reasons and consideration on which the decision is based where the Member States' legislation so provides.

The detailed arrangements for such information shall be determined by the Member States.

If another Member State has been informed pursuant to Article 7, it will also be informed of the decision in question.

Article 10

The provisions of this Directive shall not affect the obligation on the competent authorities to respect the limitations imposed by national regulations and administrative provisions and accepted legal practices with regard to industrial and commercial secrecy and the safeguarding of the public interest.

Where Article 7 applies, the transmission of information to another Member State and the reception of information by another Member State shall be subject to the limitations in force in the Member State in which the project is proposed.

Article 11

1. The Member States and the Commission shall exchange information on the experience gained in applying this Directive.
2. In particular, Member States shall inform the Commission of any criteria and/or thresholds adopted for the selection of the projects in question, in accordance with Article 4(2), or of the types of projects concerned which, pursuant to Article 4(2), are subject to assessment in accordance with Articles 5 to 10.
3. Five years after notification of this Directive, the Commission shall send the European Parliament and the Council a report on its application and effectiveness. The report shall

be based on the aforementioned exchange of information.

4. On the basis of this exchange of information, the Commission shall submit to the Council additional proposals, should this be necessary, with a view to this Directive's being applied in a sufficiently coordinated manner.

Article 12

1. Member States shall take the measures necessary to comply with this Directive within three years of its notification[7].
2. Member States shall communicate to the Commission the texts of the provisions of national law which they adopt in the field covered by this Directive.

Article 13

The provisions of this Directive shall not affect the right of Member States to lay down stricter rules regarding scope and procedure when assessing environmental effects.

Article 14

This Directive is addressed to the Member States.

ANNEX A
EC DIRECTIVE 85/337: ANNEXES I, II AND III

(slightly shortened from CEC 1985)

Annex I

1. Crude oil refineries, facilities for the gasification and liquefaction of coal or bituminous shale.
2. Thermal power stations and other combustion installations with a heat output of 300 megawatts or more, and nuclear power stations and other nuclear reactors.
3. Installations for the permanent storage or disposal of radioactive waste.
4. Integrated works for the initial melting of cast-iron and steel.
5. Installations for the extraction, processing, and transformation of asbestos.
6. Integrated chemical installations.
7. Construction of motorways, express roads, lines for long-distance railway traffic, and airports.
8. Ports and inland waterways.
9. Waste disposal installations for the incineration, chemical treatment or landfill of toxic and dangerous waste.

Annex II

1. Agriculture
   (a) Projects for restructuring rural land holdings.
   (b) Projects for the use of uncultivated land or semi-natural areas for intensive agricultural purposes.

(c) Water management projects for agriculture.
(d) Initial afforestation and land reclamation.
(e) Poultry-rearing installations.
(f) Pig-rearing installations.
(g) Salmon breeding.
(h) Reclamation of land from the sea.

2. Extractive industry
(a) Extraction of peat.
(b) Deep drillings.
(c) Extraction of minerals other than metalliferous and energy-producing minerals.
(d) Extraction of coal and lignite by underground mining.
(e) Extraction of coal and lignite by open-cast mining.
(f) Extraction of petroleum.
(g) Extraction of natural gas.
(h) Extraction of ores.
(i) Extraction of bituminous shale.
(j) Extraction of minerals other than metalliferous and energy-producing minerals by open-cast mining.
(k) Surface industrial installations for the extraction of coal, petroleum, natural gas, ores, and bituminous shale.
(l) Coke ovens.
(m) Installations for the manufacture of cement.

3. Energy industry
(a) Industrial installations for the production of electricity, steam and hot water (not in Annex I).
(b) Industrial installations for carrying gas, steam and hot water; transmission of electrical energy by overhead cables.
(c) Surface storage of natural gas.
(d) Underground storage of combustible gases.
(e) Surface storage of fossil fuels.
(f) Industrial briquetting of coal and lignite.
(g) Installations for the production or enrichment of nuclear fuels.
(h) Installations for the reprocessing of nuclear fuels.
(i) Installations for the collection and processing of radioactive waste (not in Annex I).
(j) Installations for hydroelectric energy production.

4. Processing of metals
(a) Iron and steelworks.
(b) Installations for the production of non-ferrous metals.
(c) Pressing, drawing and stamping of large castings.
(d) Surface treatment and coating of metals.
(e) Boilermaking, manufacture of sheet-metal containers.
(f) Manufacture and assembly of motor vehicles.
(g) Shipyards.
(h) Installations for the construction and repair of aircraft.
(i) Manufacture of railway equipment.
(j) Swaging by explosives.
k) Installations for the roasting and sintering of metallic ores.

5. Manufacture of glass

6. Chemical industry
   (a) Treatment of intermediate products and production of chemicals (not in Annex I).
   (b) Production of pesticides and pharmaceutical products, paint etc.
   (c) Storage facilities for petroleum, petrochemical and chemical products.
7. Food industry
   (a) Manufacture of vegetable and animal oils and fats.
   (b) Packing and canning of animal and vegetable products.
   (c) Manufacture of dairy products.
   (d) Brewing and malting.
   (e) Confectionery and syrup manufacture.
   (f) Installations for the slaughter of animals.
   (g) Industrial starch manufacturing installations.
   (h) Fish-meal and fish-oil factories.
   (i) Sugar factories.
8. Textile, leather, wood and paper industries
   (a) Wool scouring, degreasing and bleaching factories.
   (b) Manufacture of fibre board, particle board and plywood.
   (c) Manufacture of pulp, paper and board.
   (d) Fibre-dyeing factories.
   (e) Cellulose-processing and production installations.
   (f) Tannery and leather-dressing factories.
9. Rubber industry
10. Infrastructure projects
    (a) Industrial-estate development projects.
    (b) Urban-development projects.
    (c) Ski-lifts and cable-cars.
    (d) Construction of roads, harbours, and airfields (not in Annex I).
    (e) Canalization and flood-relief works.
    (f) Dams.
    (g) Tramways, railways, suspended lines.
    (h) Oil and gas pipeline installations.
    (i) Installation of long-distance aqueducts.
    (j) Yacht marinas.
11. Other projects
    (a) Holiday villages, hotel complexes.
    (b) Permanent racing and test tracks for cars and motorcycles.
    (c) Installations for the disposal of industrial and domestic waste (not in Annex I).
    (d) Waste water treatment plants.
    (e) Sludge-depositions sites.
    (f) Storage of scrap iron.
    (g) Test benches for engines, turbines or reactors.
    (h) Manufacture of artificial mineral fibres.
    (i) Manufacture etc. of gunpowder and explosives.
    (j) Knackers' yards.
12. Modifications to development projects included in Annex I and projects in Annex I undertaken exclusively or mainly for the development and testing of new methods or products and not used for more than one year.

Annex III

1. Description of the project, including in particular:
   - a description of the physical characteristics of the whole project and the land-use requirements during the construction and operational phases,
   - a description of the main characteristics of the productions processes,
   - an estimate, by type and quantity, of expected residues and emissions resulting from the operation of the proposed project.
2. Where appropriate, an outline of the main alternatives studied by the developer and an indication of the main reasons for his choice, taking into account the environmental effects.
3. A description of the aspects of the environment likely to be significantly affected by the proposed project, including, in particular, population, fauna, flora, soil, water, air, climatic factors, material assets, including the architectural and archaeological heritage, landscape and the inter-relationship between the above factors.
4. A description[8] of the likely significant effects of the proposed project on the environment resulting from:
   - the existence of the project,
   - the use of natural resources,
   - the emission of pollutants, the creation of nuisances and the elimination of waste;

   and the description by the developer of the forecasting methods used to assess the effects on the environment.
5. A description of the measures envisaged to prevent, reduce and where possible offset any significant adverse effects on the environment.
6. A non-technical summary of the information provided under the above headings.
7. An indication of any difficulties encountered by the developer in compiling the required information.

Notes

1. *OJ* No. C 169, 9.7.1980, p. 14.
2. *OJ* No. C 66, 15.3.1982, p. 89.
3. *OJ* No. C 185, 27.7.1981, p. 8.
4. *OJ* No. C 112, 20.12.1973, p. 1.
5. *OJ* No. C 139, 13.6.1977, p. 1.
6. *OJ* No. C 46, 17.2.1983, p. 1.
7. This Directive was notified to the Member States on 3 July 1985.
8. This description should cover the direct effects and any indirect, secondary, cumulative, short-, medium- and long-term, permanent and temporary, positive and negative effects of the project.

APPENDIX 2
EC Member States' EIA systems

This appendix is based on the University of Manchester's annual *EIA Newsletter*, the EC's five-year review of Directive 85/337 (CEC 1993), and the United Nations Economic Commission for Europe (1991) report on EIA.

Belgium

The Laws of Institutional Reform in Belgium have transferred the responsibility for EIA from the federal government to the regional governments in Brussels, Flanders and Wallonia.

In Brussels, a Draft Ordinance on the Evaluation of Urban Environmental Effects has been submitted to the Council of State for legal review. The integration of urban development and environmental concerns is a central issue, as is the need to conduct EIAs of intermediate scope and depth for a wider range of projects.

In Flanders, the regional government enacted EIA through six administrative orders of March 1989. The Environmental Licence Decree of June 1985 provides the legal basis for EIA of industrial installations. Two administrative orders (VLAREM) of March 1989 deal with EIA for industrial projects and certain infrastructure-related projects. Four further administrative orders amend the existing building permit procedures. In 1992, a draft framework regulation on EIA was proposed which would substantially improve existing EIA procedures, as well as apply to policies, plans and programmes.

In Wallonia, EIA was introduced in September 1985 by the Decree on the Organization of EIA in Wallonia. This requires an Initial Environmental Examination report to be prepared to act as a basis for a screening procedure to determine the need for a full EIA. The public is involved after this procedure, to help in determining the alternatives to be addressed in the EIA. The 1985 Decree was superseded by a new Administrative Order of July 1990, and a proposal for yet another, more comprehensive, Administrative Order was agreed in June 1991.

About 28 EISs a year are prepared in Flanders (about one-third Annex I, two-thirds Annex I), and 15 in Wallonia (about 60% Annex I, 40% Annex II).

Denmark

Legal provisions have been in place since July 1989 to implement Directive 85/337. Executive Order No. 446 (June 1989) requires EIA for listed projects, including all those in Annex I of the Directive and some from Annex II. Executive Order No. 119 (February 1991) specifies part of the EIA procedure. Executive Order No. 379 (July 1988) requires EIA for major projects in coastal waters. The Environmental Protection Act requires EIA for many of the Annex II projects not covered by Order 446. Other relevant legislation includes the National and Regional Planning Act, the Urban and Rural Zone Act, The Nature Protection Act, the Raw Materials Act, the Water Supply Act, the Regional Planning Act for the Metropolitan Region, and Executive Order No. 783. EIA procedures for projects adopted by acts of national legislation are still being discussed.

Fewer than 15 EIAs are prepared annually in Denmark, of which about one-third are Annex I and two-thirds Annex II.

France

EIA has been required since 1976 in France as a result of the Environmental Protection Act (Laws 76-629 and 76-663, July 1976), Decrees 77-1133 and 77-1134 (September 1977) on the implementation of Law 76-663, and Decree 77-1141 (October 1977) on the implementation of Law 76-629. These regulations establish a two-tier EIA procedure – a full EIA or a simplified *notice d'impact* – depending on the level of impact of the project. A law of December 1983 establishes procedures for public consultation, including requirements for the publication of EIA for public inquiries; and for certain types of construction works, development plans and land-use plans. These regulations are being reviewed.

About 1000 EIAs are prepared annually in France, of which almost all (96–7%) are Annex II. About one-third of the EISs relate to industrial developments, 20% to agricultural development, 10% to mineral extraction, and 5% to transport infrastructure.

Germany

The Act on the Implementation of EC Directive 85/337 was enacted in Germany in February 1990; it is the general statute providing for consequential changes in 16 other acts. Amendments to the Federal Mining Act (February 1990) and to the Federal Land-Use Planning Act (July 1989) and related ordinances also require EIA. In March 1992, an ordinance was passed which also requires EIA for industrial projects. Still to be adopted is an ordinance concerning nuclear power stations. Draft administrative provisions, which prescribe criteria and procedures for identifying, scoping, describing, assessing and summarizing environmental impacts were released in June 1991, but were still being discussed in mid-1992. EIA has already been carried out for many infrastructure projects.

Most of Germany's regions (Länder) have established their own EIA acts or administrative provisions in response to the federal EIA Act, including Bavaria in 1990, Hessen in 1990, Saarland in 1991, and Schleswig–Holstein in 1991; in some cases these are more stringent than the federal regulations. The need to streamline the EIA system to allow for the rapid reconstruction of former East Germany's infrastructure and economy has been discussed, and a bill has been presented to Parliament.

About 1000 EIAs a year are prepared in Germany.

Greece

Law No. 1650/86 for the Protection of the Environment was passed in 1986 in Greece and provides a legal framework for EIA. This law was implemented in October 1990 after the promulgation of a number of relevant ministerial decisions and circulars. Ministerial Decision 69269/5387 of October 1990 requires EIA for Category A and B projects (Annex I and II projects), and specifies the contents of EISs and procedures for project approval. Ministerial Decision 75308/5512 gives procedures for the publication of EISs. Circulars and manuals clarifying EIA and licensing procedures are to be published.

Ireland

The preparation of EIAs for projects costing over £5 million was required in Ireland by the 1976 Local Government (Planning and Development) Act. More recently, EC Directive 85/337 was implemented through 12 regulations issued between 1988 and 1990. The most important of these regulations were the European Communities (Environmental Impact Assessment) Regulations 1989 (SI No. 340), and the Local Government (Planning and Development Regulations 1990 (SI No. 25). Separate regulations cover motorways (4), fisheries, gas, air navigation and transport, petroleum and other minerals development, the foreshore, and arterial drainage. The Department of the Environment has issued guidance notes on EIA for LPAs and roads authorities.

The number of EIAs prepared in Ireland rose rapidly, from 13 in the second half of 1988, to 40 in 1989, 70 in 1990, and 83 in 1991; of these almost all (98%) were for Annex II projects. Of these, about a quarter were for infrastructure, 17% for Sched. 2.11 (including 21 holiday villages/hotels), 16% for extraction, and 10% for agriculture.

Italy

Provisions for EIA in Italy were first set up by Article 6 of Law No. 349 (1986) which envisaged future EIAs for certain projects. This was provided for in Decree No. 377 (August 1988), which requires EIA for Annex I projects as well as certain dams. This decree came into force in December 1988 through an additional decree which contained regulations for the preparation of EISs and for deciding on environmental compatibility. Bill No. 5181, which proposes to extend these procedures, was approved by Italy's Council of Ministers and presented to Parliament in October 1990. It aims to extend EIA to Annex II projects, and to sectoral and regional plans/programmes, as well as to improve public participation procedures and specify the responsibilities of various levels of government. Decree No. 363 requires EIA for projects relating to intervention in the Mezzogiorno (Southern Italy).

EIA laws have also been enacted in the regions of Veneto (1985, 1990), Valle D'Aosta (1991), Abruzzo (1990), Friuli Venezia Giulia (1990) and Trento (1988). Others have prepared proposals and are waiting for the national bill to be enacted before proceeding.

In the 1989–91 period about 30 EISs per year were prepared in Italy, with 70% for Annex I projects and 30% for Annex II projects. Of these, 57% related to industrial installations, 27% to dams, 10% to power stations, and 6% to other types of works.

Luxembourg

EC Directive 85/337 is not completely incorporated in Luxembourg's regulations, but draft legislation has been proposed. EIAs are already carried out under a law of May 1990 concerning dangerous, dirty or noxious installations, a law of August 1982 on nature conservation, and a law of August 1967 on the creation of a communication network.

The Netherlands

See §11.2.

Portugal

Portugal entered the EC a year after the publication of Directive 85/337. Some EIAs conforming to the Directive were carried out in 1987 and 1988, and the Portuguese Environmental Act (Law No. 11/87), which provides for EIA, was enacted in 1987. However, until 1990, EIA practice in Portugal was carried out in a climate of uncertainty, administrative discretion and inconsistency. In 1990, the Directive was officially implemented through Decree Law No. 186/90 on the EIA process, and Decree Regulation No. 38/90 on the EIA process. These make EIA mandatory for only Annex I projects. Further detailed regulations are expected to be published, specifying the EIA process, the contents of EISs, and the criteria to be used to select Annex II projects for EIA. Most EISs prepared in Portugal have been for highway schemes.

Spain

EIA in Spain is required by Legislative Decree 1302/1986, whose implementation procedures are set forth in Decree 1131/1988. The need for EIA is also set out in Act 25/1988 on highways, and Act 4/1988 on conservation of natural areas and wildlife. According to these pieces of legislation, all Annex I projects require EIA, but not all those in Annex II. Various legal provisions for EIA have been enacted at the regional level, and several regions require EIA for all, or nearly all, Annex II projects.

Approximately 300 national-level EIAs had been carried out by October 1992, of which about half were for mining projects, a quarter for urban projects, 7% for roads, and 5% for chemical projects. Approximately 100 regional-level EIAs had been carried out, primarily for extraction projects, roads, and waste-disposal installations.

The United Kingdom

See Chapter 3.

References

Commission of the European Communities 1993. Report from the Commission of the Implementation of Directive 85/337/EEC on the assessment of the effects of certain public and private projects on the environment, and annex for the United Kingdom. COM(93)28 final, Vol. 12.

United Nations Economic Commission for Europe 1991. *Policies and systems of environmental impact assessment*. New York: United Nations.

APPENDIX 3
Key references

Principles and procedures

Clark, B. D. & R. G. H. Turnbull 1984. Proposals for environmental impact assessment procedures in the UK. In *Planning and ecology*, R. D. Roberts & T. M. Roberts (eds), 135–44. London: Chapman & Hall.

Commission of the European Communities 1985. On the assessment of the effects of certain public and private projects on the environment. *Official Journal* L175, 5.7.1985.

Commission of the European Communities 1992. *Towards sustainability: a European Community Programme of policy and action in relation to the environment and sustainable development*, vol. II. Brussels: Commission of the European Communities.

Department of the Environment/Welsh Office 1988. DoE Circular 15/88 (Welsh Office Circular 23/88) *Environmental assessment*, 12 July 1988.

Department of the Environment/Welsh Office 1989. *Environmental assessment: a guide to the procedures*. London: HMSO.

Fortlage, C. 1990. *Environmental assessment: a practical guide*, Aldershot, England: Gower Technical.

Lee, N. & C. M. Wood 1984. Environmental impact assessment procedures within the European Economic Community. In *Planning and ecology*, R. D. Roberts & T. M. Roberts (eds), 128–34 London: Chapman & Hall.

Munn, R.E. 1979. *Environmental impact assessment: principles and procedures* New York: John Wiley.

O'Riordan, T. & W. R. D. Sewell (eds) 1981. *Project appraisal and policy review*. Chichester, England: John Wiley.

Pearce, D., A. Markandya, E. Barbier 1989. *Blueprint for a Green economy*. London: Earthscan.

Tomlinson, P. 1986. Environmental assessment in the UK: implementation of the EEC Directive. *Town Planning Review* **57**(4), 458–86.

United Nations Economic Commission for Europe 1991. *Policies and systems of environmental impact assessment*. (Environmental Series). Geneva: United Nations Economic Commission for Europe.

United Nations World Commission on Environment and Development 1987. *Our common future*. Oxford: Oxford University Press.
Wathern, P. (ed.) 1988. *Environmental impact assessment: theory and practice*. London: Unwin Hyman.
Westman, W. E. 1985. *Ecology, impact assessment and environmental planning*. New York: John Wiley.

Process

Atkinson, N. & R. Ainsworth 1992. Environmental assessment and the local authority: facing the European imperative. *Environmental Policy and Practice* **2**(2), 111–28.
Barde, J. P. & D. W. Pearce 1991. *Valuing the environment: six case studies*. London: Earthscan.
Bisset, R. 1983. A critical survey of methods for environmental impact assessment. In *An annotated reader in environmental planning and management*, T. O'Riordan & R. K. Turner (eds), 168–85. Oxford: Pergamon Press.
Bisset, R. 1989. Introduction to EIA Methods. Paper presented at the 10th International Seminar on Environmental Impact Assessment, University of Aberdeen, 9–22 July.
Bregman, J. I. & K. M. Mackenthun 1992. *Environmental impact statements*. Chelsea, Michigan: Lewis.
Buckley, R. 1991. Auditing the precision and accuracy of environmental impact predictions in Australia. *Environmental Impact Assessment Review* **11**, 1–23.
Department of the Environment 1991. *Policy appraisal and the environment*. London: HMSO.
Department of the Environment/Welsh Office 1989. *Environmental assessment: a guide to the procedures*. London: HMSO.
Department of Transport 1983. *Manual of environmental appraisal*. London: HMSO.
Fortlage, C. 1990. *Environmental assessment: a practical guide*. Aldershot: Gower Technical.
Hansen, P. E. & S. E. Jorgensen (eds) 1991. *Introduction to environmental management*. New York: Elsevier.
Jain, R. K., L. V. Urban, G. S. Stacey 1977. *Environmental impact analysis*. New York: Van Nostrand Reinhold.
Lee, N. 1987. *Environmental impact assessment: a training guide*. Occasional Paper 18, Department of Planning and Landscape, University of Manchester.
Lee, N. & R. Colley 1990. *Reviewing the quality of environmental statements*. Occasional Paper 4, Department of Planning and Landscape, University of Manchester.
McCormick, J. 1991. *British politics and the environment*. London: Earthscan.
Morris, P. & R. Therivel (eds) 1995. *Methods of environmental impact assessment*. London: UCL Press.
Munn, R. E. 1979. *Environmental impact assessment: principles and procedures*. New York: John Wiley.
O'Riordan, T. & R. K. Turner (eds) 1983. *An annotated reader in environmental planning and management*. Oxford: Pergamon Press.
Parkin, J. 1992. *Judging plans and projects*. Aldershot, England: Avebury.
Rau, J. G. & D. C. Wooten 1980. *Environmental impact analysis handbook*. New York: McGraw Hill.
Salter, J. R. 1992a. Environmental assessment: the challenge from Brussels. *Journal of*

Planning and Environment Law (January), 14–20.
Salter, J. R. 1992b. Environmental assessment: the need for transparency. *Journal of Planning and Environment Law* (March), 214–21.
Salter, J. R. 1992c. Environmental assessment: the question of implementation. *Journal of Planning and Environment Law* (April), 313–18.
Suter II, G. W. 1993. *Ecological risk assessment*. Chelsea, Michigan: Lewis.
Tomlinson, P. 1989. Environmental statements: guidance for review and audit. *The Planner* **75**(28), 12–15.
Tomlinson, P. & S. F. Atkinson 1987a. Environmental audits: proposed terminology. *Environmental Monitoring and Assessment* **8**, 187–98.
United Nations Environment Programme 1981. *Guidelines for assessing industrial environmental impact and environmental criteria for the siting of industry*. Paris: UNEP.
Wathern, P. (ed.) 1988. *Environmental impact assessment: theory and practice*. London: Unwin Hyman.
Westman, W. E. 1985. *Ecology, impact assessment and environmental planning*. New York: John Wiley.
Winpenny, J. T. 1991. *Values for the environment: a guide to economic appraisal*. London: Overseas Development Institute/HMSO.
Wood, C. M. & N. Lee (eds). 1987 *Environmental impact assessment: five training case studies*. Occasional Paper No. 19, Department of Planning and Landscape, University of Manchester.

Practice

Coles, T. F., K. G. Fuller, M. Slater 1992. Practical experience of environmental assessment in the UK. *Proceedings of Conference on Advances in Environmental Assessment*. London: IBVC Technical Services.
Council for the Protection of Rural England 1991. *The environmental assessment Directive: five years on*. London: Council for the Protection of Rural England.
Department of the Environment 1991. *Monitoring environmental assessment and planning*. Report by the EIA Centre, Department of Planning and Landscape, University of Manchester. London: HMSO.
Department of Transport 1993. *Manual of environmental appraisal* (revised version). London: HMSO.
Heaney D. & R. Therivel 1993. *Directory of environmental statements 1988–1992*. Oxford: School of Planning, Oxford Brookes University.
Jones, C. E., N. Lee, C. M. Wood 1991. *UK environmental statements 1988–1990: an analysis*. Occasional Paper No. 29, Department of Planning and Landscape, University of Manchester.
Lee, N. & D. Brown 1992. Quality control in environmental assessment. *Project Appraisal* **7**(1), 41–5.
Sheial, J. 1991. *Power in trust: the environmental history of the Central Electricity Generating Board*. Oxford: Oxford University Press.
Standing Advisory Committee on Trunk Road Assessment 1992. *Assessing the environmental impact of road schemes*. London: HMSO.
United Nations Economic Commission for Europe 1991. *Policies and systems of environmental impact assessment*. Environmental Series 4. Geneva: United Nations Economic Commission for Europe.

Prospects

Council for the Protection of Rural England 1991. *The environmental assessment Directive: five years on*. London: Council for the Protection of Rural England.

Department of the Environment 1991a. *Policy appraisal and the environment*. London: HMSO.

Lee, N. & E. Walsh 1992. Strategic environmental assessment: special collection of articles. *Project Appraisal* 7(3).

Local Government Management Board 1991. *Environmental audit for local authorities: a guide*. Luton: Local Government Management Board.

Pinfield, G. 1992. Strategic environmental assessment and land-use planning. *Project Appraisal* 7(3), 157–63.

Therivel, R., E. Wilson, S. Thompson, D. Heaney, D. Pritchard 1992. *Strategic environmental assessment*. London: RSPB Earthscan.

Useful journals

EIA *Newsletter*, EIA Centre, Department of Planning and Landscape, University of Manchester.

Environmental Impact Assessment Review, published quarterly by Elsevier Science Publishing Company, New York.

Environmental Monitoring and Assessment, published monthly by Kluwer Academic Publishers, USA.

Environmental Policy and Practice, published quarterly by EPP Publications, UK.

The Environmental Professional, journal of the National Association of Environmental Professionals, USA.

Institute of Environmental Assessment Newsletter, IEA, East Kirkby, Lincolnshire, UK.

Journal of Environmental Management, published monthly by Academic Press.

Journal of Environmental Planning and Management (formerly *Planning Outlook*), published three times a year by Carfax Publishing Company for the University of Newcastle upon Tyne.

Journal of Planning and Environmental Law, published monthly by Sweet & Maxwell, London, UK.

APPENDIX 4
Addresses of UK organizations with EIS collections

Crown Estate Office
10 Charlotte Street, Edinburgh EH2 4DR

Department of the Environment
Room P3-006, 2 Marsham Street, London SW1P 3EB

Department of the Environment (Northern Ireland), Town and Country Planning Service, Commonwealth House, 35 Castle Street, Belfast BT1 1GU

Department of Transport, Ports Division
Sunley House, 90–93 High Holborn, London SC1V 6LP

Forestry Authority
231 Corstorphine Road, Edinburgh EH12 7AT

Ministry of Agriculture, Fisheries & Food (MAFF), Flood Defence Division
Eastbury House, 30–34 Albert Embankment, London SE1 7TL

Scottish Office, Environment Department
New St Andrews House, Edinburgh EH1 3SZ

Welsh Office, Planning Division
Cathays Park, Cardiff CF1 3NQ

Institute of Environmental Assessment
The Old School, Fen Road, East Kirkby, Lincs

EIA *Centre,* Department of Planning and Landscape, University of Manchester, Manchester M13 9PL

Impacts Assessment Unit, School of Planning, Oxford Brookes University, Oxford OX3 0BP

Author index

Index